Various Electronic Displays *(for details, see the corresponding pages)*

(a) *LCD terminal display module* (p. 76)

(b) *LCD colour TV set* (p. 79)

(c) *LCD car dashboard* (p. 80)

(d) *ECD wrist watch* (p. 126)

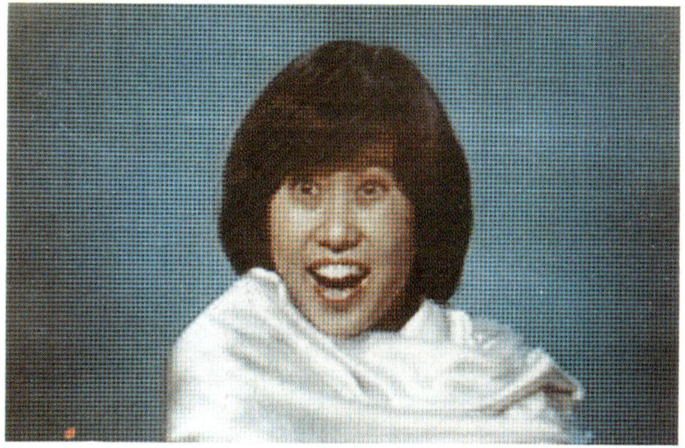

(e) *PDP colour TV display* (p. 174)

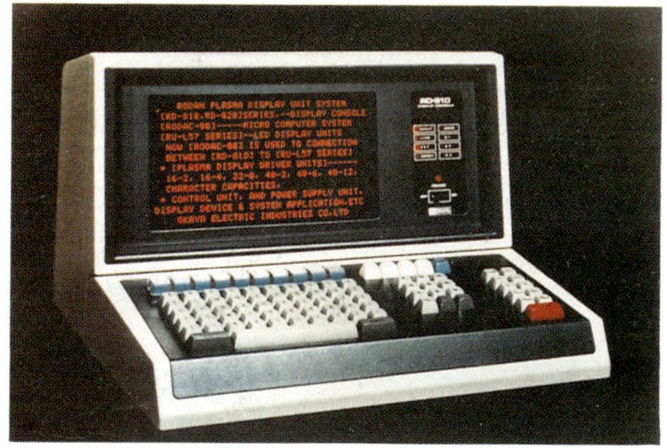

(f) *PDP terminal display device* (p. 172)

(g) *ELD airport arrival and departure guide display* (p. 210)

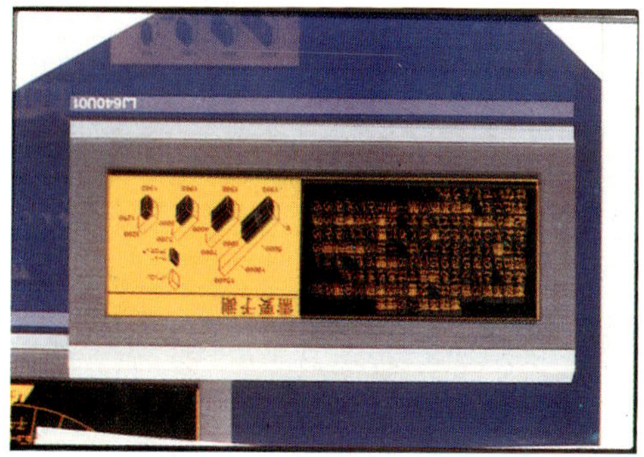

(h) *ELD terminal display module* (p. 212)

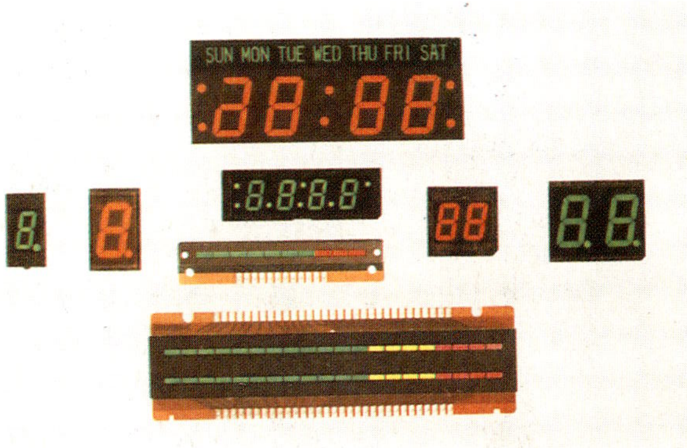

(i) *VFD colour image display* (p. 252)

(k) *LED display device* (p. 274)

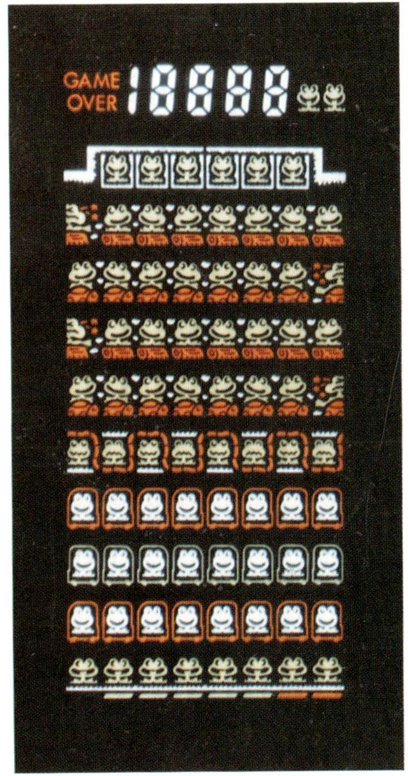

(j) *VFD colour electronic game* (p. 265)

(l) *LED TV display* (p. 317)

(m) *magnetic inversion-type large-screen colour display board* (p. 368)

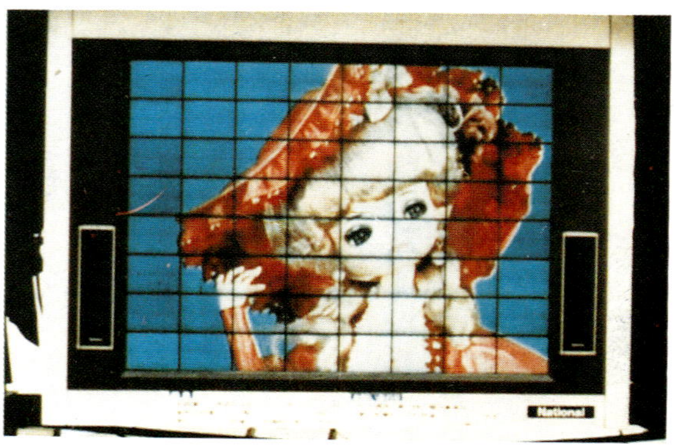

(n) *LCD large screen colour TV* (p. 370)

Electronic Display Devices

Electronic Display Devices

Edited by
Shoichi Matsumoto
Toshiba Corporation

Coauthors
Isao Ohta
Takeo Kamegaya
Tadaji Suzuki
Mitsuru Tamura
Tatsuro Beppu
Hiroshi Morita

Translated by
F. R. D. Apps

JOHN WILEY & SONS
Chichester · New York · Brisbane · Toronto · Singapore

Other Wiley Editorial Offices

John Wiley & Sons, Inc., 605 Third Avenue,
New York, NY 10158-0012, USA

Jacaranda Wiley Ltd, G.P.O. Box 859, Brisbane,
Queensland 4001, Australia

John Wiley & Sons (Canada) Ltd, 22 Worcester Road,
Rexdale, Ontario M9W 1L1, Canada

John Wiley & Sons (SEA) Pte Ltd, 37 Jalan Pemimpin 05-04,
Block B, Union Industrial Building, Singapore 2057

Library of Congress Cataloging-in-Publication Data:

Denshi disupurei gijutsu. English.
Electronic display devices / edited by Shoichi Matsumoto ;
translated by F. R. D. Apps.
p. cm.
Translation of: Denshi disupurei gijutsu.
Includes bibliographical references.
ISBN 0 471 92218 8
1. Information display systems. I. Matsumoto, Shoichi.
II. Title.
TK7882.I6D3813 1990
621.38—dc20 89-27756

British Library Cataloguing in Publication Data:

Electronic display devices.
1. Electronic display systems
I. Matsumoto, Shoichi
621.38′0414

ISBN 0 471 92218 8

Typeset by Thomson Press (India) Ltd, New Delhi
Printed in Great Britain by Biddles Ltd, Guildford

Contents

Chapter 4 PLASMA DISPLAY PANELS (PDPs)

Chapter 5 ELECTROLUMINESCENT DISPLAYS (ELDs)

Chapter 6 VACUUM FLUORESCENT DISPLAYS (VFDs)

Chapter 7 *LIGHT-EMITTING DIODE DISPLAYS (LEDs)*

Chapter 8 *SPECIAL DISPLAYS*

Preface

Electronic display devices play a bridging role (man–machine interface) and have become increasingly important with progress towards the 'information society', development of new media, and the spread of electronic systems. Demand for such devices continues to expand and diversify. Various types of flat panel type electronic display devices of the active and passive types, characterized by small depth and light weight, with low voltage and low power, have attracted attention as novel electronic devices adapted to this new situation.

Such novel flat panel electronic displays are implemented by means of LCDs (liquid crystal displays), ECDs (electrochemical displays), PDPs (plasma displays), ELDs (electroluminescent displays), VFDs (vacuum fluorescent displays), LEDs (light-emitting diode displays), EPIDs (electrophoretic displays), and special displays such as particle rotation type or electrostatic attraction type displays, large screen displays, or flat plate type CRTs. This book explains the principles and application of these displays simply and clearly from a practical point of view and using the latest data. Each chapter of this book has been independently written by an acknowledged leader in research and development of the display device in question.

The first chapter (Overview) sets out basic knowledge regarding the history of development, principles, and construction, and compares the performance and characteristics of each type of electronic display device. Chapters 2 to 8 discuss particular display devices, in the areas of: Principles of Operation and Systems; Materials and Physical Properties; Structure and Manufacturing Processes; Drive Methods and Circuitry; Display Characteristics and Performance; Applications and Uses; and Problems and Outlook. To facilitate understanding, in all chapters many drawings, tables, and photographs are provided, and the illustration pages show colour photographs of various recent applications of electronic display devices. For readers who wish to pursue their study of electronic display devices further, References and Bibliography are given at the end of each chapter.

This book is intended not only as a textbook for those interested in electronic display devices but also as a specialist text for manufacturers and users.

On behalf of all the authors, I would like to take this opportunity of expressing our thanks to the various persons both in Japan and overseas who contributed cooperation and advice in the writing of this book. In particular, I would like to express our heart-felt gratitude to those who made available vital technical data and photographs, etc.

We also record our thanks to the staff of the publishers, Ohmsha, for providing the opportunity to produce this work.

In conclusion, the authors will be delighted if this book should be of some assistance to its readers. We shall also be grateful for readers' corrections and suggestions.

Shoichi Matsumoto
December 1984

Chapter 1

Overview

1.1 WHAT ARE ELECTRONIC DISPLAYS?

1.1.1 Definition and classification

Electronic display devices can be defined in terms of their role as electronic devices that form a so-called 'man–machine interface', transmitting data from a machine so that a human user can visually recognize it. Electronic display devices will therefore play an extremely large role in the information age although existing types of electronic display device are in use in a wide variety of applications (both industrial and domestic), at the same time, in order to meet the ever-diversifying requirements of the information society, the development of new high-powered electronic display devices is continuing all the time.

Next, if we consider the definition of electronic display devices in terms of their operation, as shown in Fig. 1.1 they are electronic devices that convert various electrical signals transmitted from electronic devices into optical signals that can be recognized visually by a human. These electronic devices are generally provided with the ability to map the converted optical signal into a pattern in two-dimensional space; in other words, displaying the optical signal as an image in a form such as digits, characters, or graphics. When the optical signal is displayed by the emission of light it is termed an active or emissive display, whereas when the display works by modulating incident light through reflection, scattering, interference, or similar phenomena, it is termed a passive or non-emissive display.

Table 1.1 shows typical electronic display devices with these functions, classified into active and passive types. Of these, the CRT display, which has the longest history, is still the leading type both economically and for display quality, and the market for the next biggest products, which are the LCD and LED displays, is only about 10 per cent of the CRT market. Thus at present the CRT can still be said by any measure to be the foremost electronic display device.

On the other hand, since the 1960s and 1970s, with the rapid developments in integrated circuit and LSI semiconductor technology, electronic devices have become solid state, working with low supply voltages and low power consumption, as a result all electronic devices have become smaller and lighter, while

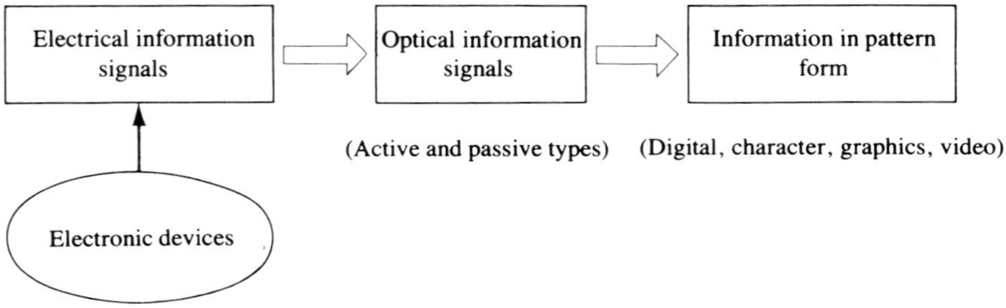

Figure 1.1 Electronic display device operation

Table 1.1 Typical electronic display devices

- Electronic display devices
  - Active displays (emissive)
    - CRT (cathode ray tube)
    - PDP (plasma display panel)
    - ELD (electroluminescent display)
    - VFD (vacum fluorescent display)
    - LED (light-emitting diode)
  - Passive displays (non-emissive)
    - LCD (liquid crystal display)
    - ECD (electrochemical display)
    - EPID (electrophoretic image display)
    - SPD (suspended particle display)
    - TBD (twisting ball display)
    - PLZT (transparent ceramics display)

various multipurpose data-processing devices based on the computer have been developed; there has also been a strong increase in the demand for new electronic display devices appropriate to these applications. A flat panel electronic display device is required that is small and light and will operate with a low drive voltage and low power consumption. In order to meet these new requirements research, development and improvement are continuing into a variety of flat panel electronic display devices, and efforts are being made for their practical implementation.

All of the electronic display devices shown in Table 1.1 apart from the CRT can be classified as flat panel type. In this book we are principally concerned with discussing these new flat panel types of electronic display, but will also touch on the flat type of CRT.

1.1.2 History and development

Table 1.2 illustrates the historical development of various flat panel electronic display devices from their inception to the present, along with the history of some

***Table 1.2** History of flat-panel electronic display device technologies*

Related technologies	
1879	Invention of carbon filament incandescent lamp (Edison, USA)
1897	Invention of cathode ray tube (Braun, Germany)
1904	Invention of vacuum tube (Fleming, Britain)
1910	Invention of tungsten filament lamp (Coolidge, USA)
1935	World's first television broadcast (Germany)
1938	Invention of fluorescent light (General Electric, USA)
1947	Development of transparent electrode glass (Corning, USA)
1948	Invention of transistor (Barden and others, USA)
1952	Beginning of television broadcasts in Japan
1960	Invention of ruby laser (Maiman, USA)
LCD (liquid crystal display)	
1888	Discovery of liquid crystals (Reinitzer, Austria)
1963	Publication of liquid crystal electrooptic effect (Williams, USA)
1968	DS and GH type LCD techniques published (Heilmeier and others, USA)
1971	TN type LCD technique published (Schadt and others, Switzerland)
1972	Implementation of liquid crystal watch and calculator (Muroma and Rockwell, USA)
1982–3	Implementation of black and white liquid crystal TV (Seiko and Casio, Japan)
1983	Implementation of large-area colour liquid crystal video device (Mitsubishi Denki and Matsushita Tsushin, Japan)
1984	Implementation of colour liquid crystal TV (Seiko, Japan)
ECD (electrochemical display)	
1953	Patent for pH indicator electro-coloration effect (Sziklai, USA)
1963	Discovery of tungsten oxide electrochromic effect (Jones, USA)
1969	Announcement of tungsten oxide ECD (Deb, USA)
1973	Announcement of viologen ECD (Schoot and others, Netherlands)
1978	Announcement of iridium oxide ECD (Gottesfeld and others, USA)
1978	Announcement of diphthalocyanin multicolour ECD (Nicholson, USA)
1979	Development of silver electroplating ECD (LETI, France)
1982	Implementation of ECD clock (Seiko, Japan)
PDP (plasma display panel)	
1954	Publication of d.c. drive PDP (Skellet, USA)
1958	Development of the NIXIE discharge display indicator (Burroughs, USA)
1966	Announcement of memory type a.c. drive PDP (Illinois University, USA)
1968	TV image display experiments using d.c. drive PDP (Philips, Netherlands)
1970	Development of self-scan d.c. drive PDP (Burroughs, USA)
1978	Trial d.c. drive colour PDP TV (NHK, Japan)
ELD (electroluminescent display)	
1936	Discovery of electroluminescent phenomenon in ZnS:Cu fluorescent substance (Destriau, France)
1950	Development of dispersion type a.c. drive EL illumination panel (Sylvania, USA)

Table 1.2 *(contd.)*

1968	Development of evaporated thin-film a.c. drive ELD Lumocen (Bell Labs, USA)
1978	Implementation of double-insulated thin-film type a.c. drive ELD (Sharp, Japan)
1980	Implementation of atomic layer epitaxy (ALE) thin-film a.c. drive ELD (Lohja, Finland)
1982	Development of low-voltage operating and multicolour thin-film ELDs (Osaka University, Japan)
VFD (vacuum fluorescent display)	
1967	Development of single-digit VFD (Ise Denshi, Japan)
1972	Development of round multidigit VFD (Futaba Denshi, Japan)
1972	Development of flat multidigit VFD (Futaba Denshi, Japan)
1979	Trial VFD television (Ise Denshi, Japan)
LED (light-emitting diode)	
1923	Discovery of the injection electroluminescence effect in SiC crystals (Lossew, Soviet Union)
1952	Publication of GaP and GaAs and other group III–V compound semiconductors (Welker, Germany)
1954	Announcement of GaP light-emission phenomenon (Wolf and others, USA)
1962	Announcement of GaAs light-emission phenomenon (Pankove, USA)
1962	Invention of semiconductor laser (Nathan and others, USA)
1968	Implementation of GaAsP red LED (Monsanto, USA)
1979	Trial LED television (Sanyo, Japan)
Special displays	
1969	Publication of SPD (suspended particle display) (Marks, USA)
1969	Development of transparent PLZT for displays (Sandia Lab, USA)
1970	Trial DMD (foil deformation type) (RCA, USA)
1973	Publication of EPID (electrophoretic image display) (Matsushita Denki, Japan)
1975	MPD (magnetic particle rotation display) (Magnavox, USA)
1977	Development of Gyriccon TBD (twisting ball display) (Xerox, USA)
1980	DFD (dye foil display) (Philips, Netherlands, and Bendix, USA)
1981	EOD/EWD (electroosmotic/electrowetting displays) (Kohashi, Japan, and Beni, USA)
1982	Trial beam-guide flat CRT (RCA, USA)
1982	Development of plasma-excitation flat CRT (Siemens, West Germany)
1982	Implementation of flat CRT television (Sony, Japan)

related technologies. The technologies for some of the flat panel displays were in fact the result of discoveries made a relatively long time ago: for example the discovery of liquid crystals by Reinitzer in 1888, the discovery of the injection electroluminescence effect in SiC crystals by Lossew in 1923, or the discovery of the electroluminescence effect in a ZnS phosphor by Destriau in 1936.

However, compared with the invention of the cathode ray tube by Braun in 1897, which has enabled the CRT display to have its dominant position today, the origins of many of the flat panel electronic display devices began some 60 or 70 years later in the 1960s, and are thus extremely recent. This timing coincides well with the increased demand for display devices with new forms and functions appropriate to applications in the rapidly developing integrated circuit and LSI semiconductor technologies. This means that the development of small, flat display devices of various types requiring low supply voltages and power was driven forcibly by the requirements of society.

The passive display technologies for LCD and ECD devices arose at about the same time—the former with the announcement of DS (dynamic scattering) and GH (guest–host) types of liquid crystal displays by Heilmeier in 1968 and the announcement of the TN type LCD by Schadt in 1971, and the latter with the announcement of the tungsten oxide solid ECD by Deb in 1969 and the announcement of the viologen liquid ECD by Schoot in 1973. They have reached quite different stages in their application, however. Liquid crystal displays began to be available commercially around 1975, covering a range of applications such as numeric, character, graphic, and video displays, and the technology is now second only to the CRT in its share of the market; on the other hand, electrochemical displays have only just reached the stage of trial commercialization. The principal reasons for this are the inadequate operating life of an ECD and its difficulty of effecting a multiplex drive.

Plasma display panels and electroluminescent displays are both active flat panel displays, and both have the capability of competing with a CRT on the amount of data displayed. The origin of the plasma display panel was in the counter tube (NIXIE display indicator) developed in 1958, but its real development as a flat panel display device did not begin until the announcement from Illinois University in 1966 of a memory type a.c. driven plasma display panel and from the Burroughs Corporation in 1970 of a self-scan d.c. drive plasma display panel. Both a.c. drive and d.c. drive character and graphic display panels began to appear widely from the beginning of 1975.

As for the electroluminescent display, an illumination panel using the dispersion a.c. drive electroluminescent effect discovered by Destriau in 1936 was developed by the Sylvania company in 1950, but thereafter the application as a display device came almost to a halt because of the problems of low luminance and short life. A sharp impulse was provided by the development in 1968 at Bell Labs of the thin-film a.c. driven electroluminescent panel (Lumocen), and activity on application research was resumed; in 1978 a double-insulator thin-film a.c. drive electroluminescent character display terminal was commercialized by the Sharp Corporation, and this had high luminance and long life. Then in 1980, the even higher luminance ALE (atomic layer epitaxy) thin-film a.c. drive electroluminescent panel was developed by the Lohja Company. Compared with the plasma display panel which uses a discharge effect, an electroluminescent display is characterized by being completely solid state and by having the capability of a much reduced drive voltage, but at present it still lags a long way behind the plasma display panel in the market.

Both the VFD and LED active displays were initially implemented for small capacity display devices such as numeric displays and indicators. VFDs were first developed as a single-digit display tube by Ise Denshi in 1967, which was followed by a round multidigit display tube in 1972 and then by a flat multidigit display tube developed in 1974. With the advent of the flat VFD the path to a flat panel display device of large capacity for characters and graphics was opened.

Although the history of the LED began in 1923, practical research towards implementation began with the observation of the GaP light-emitting phenomenon by Wolf *et al.* in 1954 and the publication of the injection electroluminescence effect of GaAs by Pankove in 1962. Then the first practical product was the GaAsP red LED in 1968 followed by the GaP type red LED and also orange, yellow and green types developed up to the present time. Currently the principal applications for LEDs are for pilot lamps and digital displays, but as with the vacuum fluorescent display the range of applications is beginning to spread into character and graphic displays.

Next we shall choose some of the principal flat panel type electronic displays treated as special displays in this book in their order of development. We can list the trials on display devices using transparent PLZT ceramic announced by the Sandia Laboratory in 1969, the electrophoretic image display (EPID) developed by Matsushita Denki in 1973, the twisting ball display (TBD) developed by Xerox in 1977, the dye foil display (DFD) proposed in 1980 by Philips and Bendix, and then the beam guide CRT, a trial product from RCA in 1982, the plasma energizing CRT developed by Siemens, and the flat CRT implemented by Sony, among other flat CRT displays.

This is the historical perspective to the various flat panel electronic displays; more detailed discussions appear in the individual chapters.

1.2 ELECTRONIC DISPLAY DEVICES

1.2.1 Basic principles, construction, and materials

The basic construction and principal materials used and the display principles of the various flat panel displays are shown in Tables 1.3 and 1.4, which include passive and active displays respectively. All of these display devices have in common the fact that a layer of either solid, liquid, or gaseous display material is laminated with electrode layers formed on a substrate.*

The individual display devices are now summarized, referring to Tables 1.3 and 1.4.

(1) Liquid crystal display (LCD)

In a liquid crystal display, the application of a voltage changes the molecular orientation of the liquid crystal, and the resulting change in optical character-

*In the case of PLZT and LED displays the display material itself forms the substrate.

istics such as double refraction, optical rotation, dichroism, or optical scattering, caused by the reorientation, is converted to a visible change. In other words, the LCD is a passive display device that utilizes the modulation of light within a liquid crystal cell. The liquid crystal cell is constructed with a layer of liquid crystal about 10 μm thick sandwiched between two glass substrates on which are formed transparent electrodes, and a molecular orientation film is formed on the surface of the electrode to impart a fixed orientation to the liquid crystal molecules.

The most widely used type of LCD is the twisted nematic (TN), in which the liquid crystal molecular orientation is twisted through 90° between the electrodes. This twisted molecular orientation disappears under the application of a voltage, and by putting the liquid crystal cell between crossed polarizing elements bright ($V = 0$) and dark ($V =$ V) appearances can be obtained. The liquid crystals used in the twisted nematic type are principally biphenyl, phenylcyclohexane and cyclohexylcarboxylic esters nematic liquid crystals. Other types of LCDs use smectic and cholesteric liquid crystals; in the guest–host (GH) type, which uses the change in dichroism with the reorientation of the liquid crystal molecules, dichroic dyes are also used.

(2) Electrochemical displays (ECD)

Electrochemical displays are passive display devices that utilize a reversible colour change brought about by an electrochemical oxidation–reduction reaction which occurs in some liquid and solid materials, whether organic or inorganic. A variety of materials have been proposed for use in electrochemical displays, including oxides of transition metals such as amorphous tungsten oxide (a-WO_3), molybdenum oxide (MoO_3), and iridium oxide (IrO_2), viologen derivatives, organic complexes such as cobalt bipyridine, diphthalocyanins of rare earth metals, silver iodide, and various redox compounds and pigments.

The basic ECD construction shown in Table 1.3 is an example using a transition metal oxide (a-WO_3). On the front glass substrate is formed a transparent electrode and a WO_3 film of thickness 0.3 to 1.0 μm; on the back glass substrate is provided the counterelectrode and a porous white background. The space between these substrates is filled with an electrolyte solution such as sulphuric acid or lithium perchlorate. When a negative potential is applied to the front electrode of this ECD cell, the WO_3 film is turned blue. This coloration is said to be due to the formation of tungsten bronze (H_XWO_3 for $0 < X < 1$) caused by the simultaneous injection of protons and electrons into the WO_3. The coloration remains even if the voltage is turned off, and is removed by applying a reverse potential voltage.

(3) Electrophoretic image display (EPID)

In a colloidal suspension in which fine particles are dispersed in a liquid, an electric double layer is formed so that the dispersed particles are either positively

Table 1.3 *Operating principles, basic construction, and principal materials of passive display devices*

Type	Operating Principle	Basic Construction	Principal Materials
LCD	Optical rotatory power, double refraction, dichroism, or light scattering caused by aligned liquid crystals	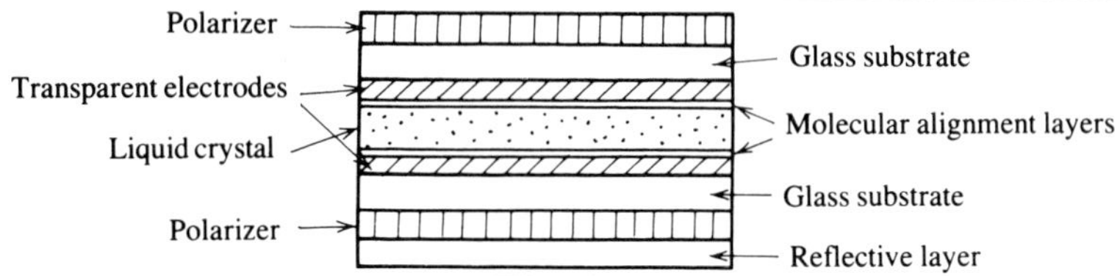 (TN type)	Nematic liquid crystals (e.g. cyanophenyl-cyclohexane), smectic liquid crystals (e.g. cyanobiphenyl), cholesteric liquid crystals (e.g. fatty acid esters of cholesterol)
ECD	Reversible coloration–decoloration phenomena due to an electrochemical oxidation–reduction reaction	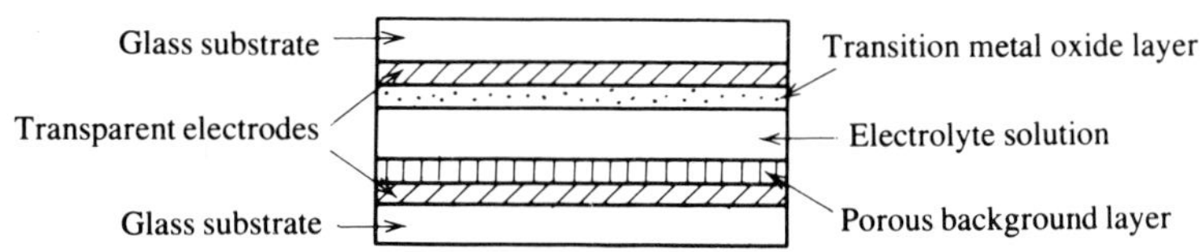	Transition metal oxides (e.g. WO_3, IrO_2), viologen derivatives, metal diphthalocyanin, silver iodide

EPID	Electrophoretic phenomenon of dispersed colloidal particles	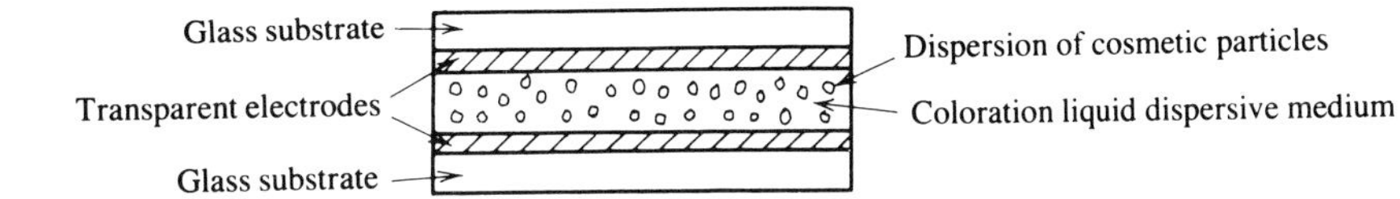	Colloidal solution with pigment particles (e.g. TiO_2) in coloration liquid dispersive medium
SPD	Light absorption or polarization effect by oriented particles	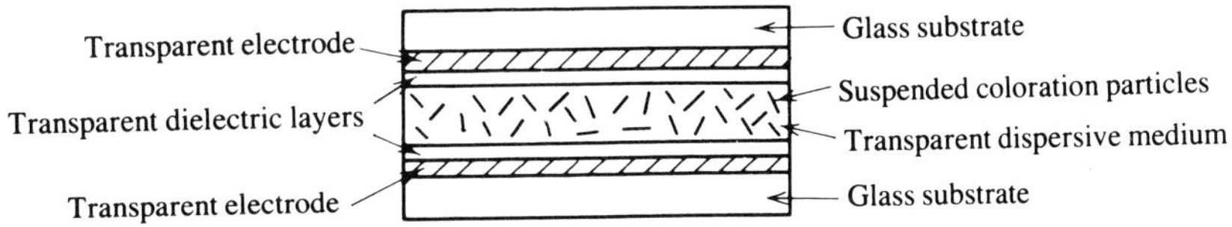	Colloidal solution with needle or plate form coloration particles (e.g. polyiodide organic crystals) in a transparent liquid dispersive medium
PLZT	Double-refraction phenomenon due to domain orientation (Pockels effect)	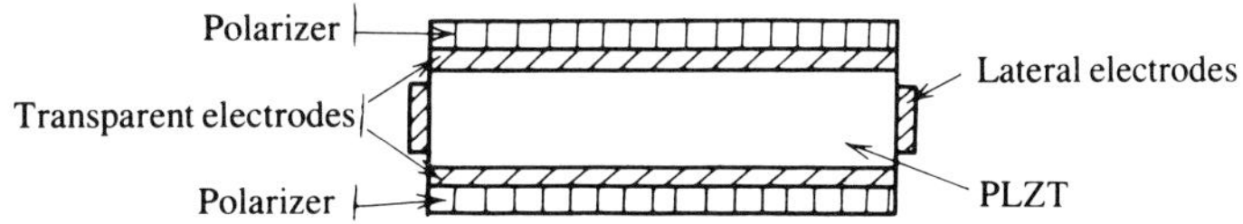	Transparent ferroelectric ceramic PLZT [(Pb,La)(Zr,Ti)O_3]

Table 1.4 *Operating principles, basic construction, and principal materials of active electronic display devices*

Type	Operating Principle	Basic Construction	Principal Materials
PDP	Luminescence by inert gas glow discharge; inert gas ultraviolet excited luminescence in phosphor (photoluminescence)	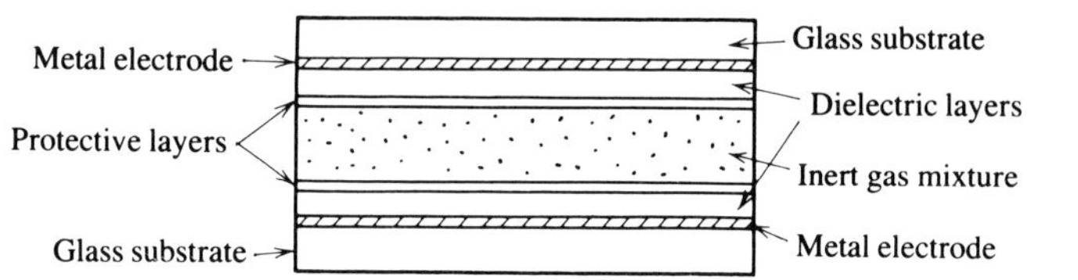 (AC type)	Inert gas mixture, principally neon (discharge luminescence type); inert gas mixture, principally xenon and phosphors (phosphor excited luminescence type)
ELD	Luminescence by direct electric field excitation of phosphor (intrinsic electroluminescence phenomenon)	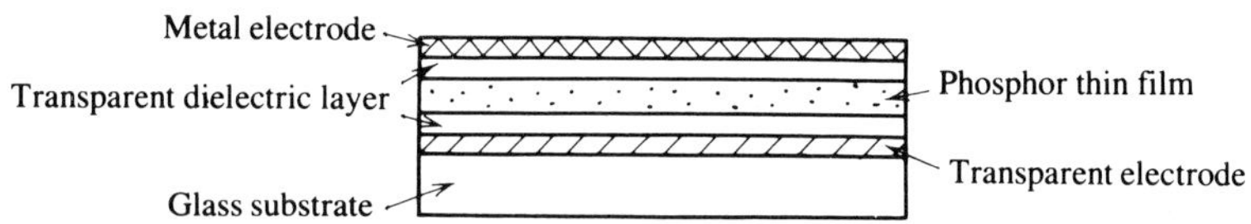 (Three-layer construction type)	Phosphor thin film (e.g. ZnS:Mn); dielectric layer dispersing fluorescent substance

VFD	Luminescence based on excitation of phosphor by slow electrons (cathodoluminescence)	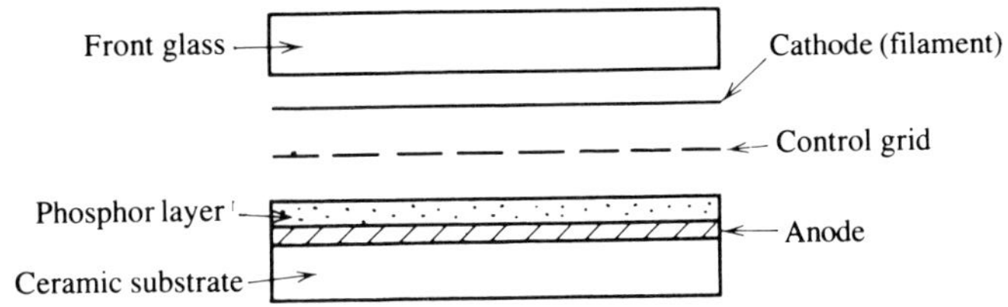	Phosphor for slow electron excitation (e.g. ZnO:Zn)
LED	Recombination radiation by the injection of minority charge carriers (injection type electroluminescence)	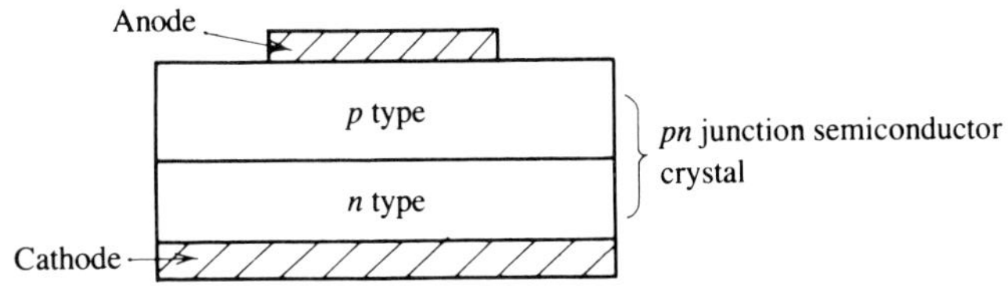	*p–n* junction semiconductor, principally group III–V compound semiconductors (e.g. GaP, GaAsP, GaAlAs)

or negatively charged. In such a colloidal suspension, if a d.c. voltage is applied, the suspended particles will migrate within the colloid because of the Coulomb force. An EPID utilizes this electrical migration phenomenon.

In the basic construction of an EPID cell, the space between two electrode substrates, at least one of which is transparent, is filled with a colloidal suspension of thickness approximately 50 μm, which is, for example a suspension of a white pigment of submicrometre particle size dispersed in an organic solvent that has been dyed blue. When a d.c. voltage is applied to this cell the charged white particles will migrate towards one of the electrode substrates according to the polarity of the charge, and will adhere to the substrate. Therefore those display portions to which the voltage is applied will be white and the background portions where the voltage is not applied will be blue, and the reverse colour contrast display is also possible. This display is retained for a long time after the voltage is cut off.

(4) Suspended particle display (SPD)

This type of display uses two transparent electrode substrates on which are formed dielectric thin films, and the space between these is filled with a colloidal suspension of thickness approximately 50 μm, consisting of needle- or plate-shaped coloured particles suspended in a colourless liquid.

When no voltage is applied, the suspended particles adopt random orientations through Brownian motion, and incident light is either absorbed or scattered, so that the cell is opaque. When, on the other hand, a voltage is applied, the suspended particles become dielectrically polarized along their long axis and orient themselves in the direction of the electric field, thus making the cell transparent to incident light. The voltage applied can be ordinary a.c., and there is no memory function. Originally needle-shaped metal particles were used as the suspended particles, but more recently blue organic iodide crystals have been used.

(5) Transparent ceramics display (PLZT)

The transparent ferroelectric ceramic PLZT [$(Pb,La)(Zr,Ti)O_3$] normally has a random domain orientation, and therefore is optically isotropic overall. If, however, a voltage is applied, individual domains align themselves with the direction of the electric field, and their optical axes are aligned, so that a double refraction similar to that of a uniaxial crystal is produced (the Pockels effect). Therefore by forming a transparent electrode on each surface of a PLZT ceramic (25 to 100 μm thick) and sandwiching this between crossed polarizing elements, the resulting cell will be opaque to incident light on the portions where a voltage is not applied, and will transmit incident light where a voltage is applied. Thus a non-volatile display function is produced, and in order to erase the display it is necessary to provide special erasing electrodes on both edge sides of the PLZT layer.

In addition to the Pockels effect described above, PLZT displays make use of the electrooptic scattering effect.

(6) Plasma displays (PDP)

A plasma display device utilizes the light produced at an intersection point of a matrix of electrodes by a plasma discharge in an inert gas. The basic device construction has a pair of glass plates provided with row electrodes and column electrodes with a discharge gap between them of approximately 0.1 mm which contains a mixture of inert gases, principally neon, under a pressure of several hundred Torr. The two principal types of plasma display are the d.c. type (direct discharge), in which the electrodes are open to the discharge gap, and the a.c. type (indirect discharge), in which the electrodes are covered by a dielectric layer. The display colour of both types is red–orange.

In the a.c. type the residual electric charge on the dielectric layer after the discharge has begun is utilized so that the discharge light emission can be maintained with a lower voltage than used for the initial discharge, and it is possible at the same time to obtain a memory effect. Therefore compared with the d.c. type the operating voltage is lower, and without reducing the luminance, the display capacity can be increased. On the other hand, the d.c. type is characterized by a relatively simple construction and scan drive circuit, and allows the luminance to be modulated easily by controlling the discharge current. Furthermore, in the d.c. type it is possible to incorporate an ultraviolet phosphor into part of the discharge cell. By causing this to luminesce with UV light from inert gases, principally xenon,* a multicolour display may be obtained.

(7) Electroluminescent display (ELD)

When a voltage is applied to a phosphor layer consisting of a matrix material such as zinc sulphide (ZnS) to which is added an activator such as manganese (yellow–orange), copper (red), or terbium fluoride TbF_3 (green), which forms luminescent centres, then a luminescence is obtained of a colour determined by the activator. An electroluminescent display is a device that utilizes this intrinsic electroluminescent effect. These displays are of two principal types: a powder ELD in which a phosphor powder is sandwiched between two electrodes and a thin-film ELD in which a thin-film phosphor is used; these may then further be divided into alternating current and direct current types.

The most common ELD is an a.c. type having a three-layer construction (with a total film thickness of approximately 1 μm), in which a zinc sulphide and manganese phosphor film is sandwiched between electrodes with dielectric layers interposed. The luminescence mechanism is thought to be as follows. When a

*This phenomenon is termed optical luminescence, as opposed to cathode ray luminescence, where the phosphor is excited by an electron beam.

voltage is applied, electrons trapped at the boundary between the phosphor layer and dielectric layer are released into the phosphor layer, and gain a high energy by being accelerated in the high electric field. Then when one of these electrons hits a luminescent centre directly, the luminescent centre is energized and when it returns from the energized state to the ground state it gives off a photon, causing luminescence. Therefore with an alternating current drive, there are two luminescences for each cycle, and also a memory function is provided by the electrons stored in the boundary between the phosphor layer and the dielectric layer.

(8) Vacuum fluorescent display (VFD)

A VFD is a display device that uses the luminescence effect when a phosphor is excited by a low-velocity electron beam. It is the same as a CRT in that the cathode ray luminescent effect is utilized, but whereas a CRT forms a display by scanning with an electron beam, a VFD has a quite different display operation in which a voltage is selectively applied to a grid and anode.

The basic construction of a VFD resembles a triode and is constructed with a front glass, a cathode (filament), a control grid, an anode on which the phosphor is applied, and a substrate that supports this anode. When a heating current is passed through the cathode which is coated with an oxide cathode material, thermal electrons are released (approximately at 650°C). These are first diffused and accelerated by a metal mesh control grid, and then next impinge on the phosphor applied on the anode, which is divided into segments, causing the phosphor to luminesce. Segments that should luminesce have a certain positive voltage applied to them, whereas a negative voltage is applied to the segments not to be displayed and the corresponding grid portions. The phosphor usually used is ZnO:Zn phosphor which is made to emit blue–green light with high efficiency by a low-speed electron beam. Additionally, although the efficiency of luminescence is low, phosphors that emit red, yellow, blue, and other colours are being developed.

(9) Light-Emitting Diodes (LED)

An LED is an optoelectronic transducer semiconductor device constructed with a junction between an *n*-type semiconductor crystal in which the charge carriers are electrons and a *p*-type semiconductor crystal in which they are positive holes. A positive voltage is applied to an anode provided on the *p*-type crystal and a negative voltage to a cathode on the *n*-type crystal, whereupon when the positive holes which are the minority charge carrier are injected from the *p*-type region to the *n*-type region (electrons are simultaneously injected from the *n*-type region to the *p*-type region), then recombine with the electrons which are the majority charge carriers, and the energy of this recombination is converted to light.

The colour of this recombination luminescence depends on the band gap of the LED semiconductor crystal; the LED crystals most widely used are group III–V compound semiconductor crystals, typically GaP (red), GaP (green), GaAsP (yellow–orange), and GaAlAs (red).

1.2.2 Display properties and characteristics

(1) Display properties and comparison

Typical display properties of both active and passive flat panel electronic displays are shown in Table 1.5. This table makes clear the comparison between properties of the various display devices, and also shows characteristics of passive versus active devices.

(a) Operating voltage and power consumption We call the applied voltage that is required to drive a display device the operating voltage V, the current flowing through the device when it is operating the operating current A, and the product of the two, the power consumption, W. Depending on the principle on which the display operates either an a.c. or a d.c. voltage may be applied, and for practical purposes a low operating voltage and low current consumption are desirable. In particular, in order to be able to use ordinary IC components in the drive circuits, the operating voltage should be held to an upper limit of about 40 V. The operating voltage of ECD, LCD, LED, SPD, and VFD display devices is low, increasing in that order in the range 0.5 to 40 V, so that using IC drive devices is easy. Other display devices need high drive voltages (50 to 250 V). With the exception of ECD devices, all of the passive display devices have a current consumption measured in $\mu A/cm^2$, which is much smaller than that of active display devices (mA/cm^2). The archetypal display device with low operating voltage and low current consumption is the LCD.

(b) Display contrast and tone The display contrast is the ratio between the luminance of on and off portions of the display. The contrast of an active display device is therefore given roughly by the luminance of light emitted, whereas for passive devices the yardstick is the degree of modulation of light by the device. In general the contrast for active displays is high (at least 30) compared with that for passive displays (10 to 30) and the apparent contrast can also easily be improved by the addition of a colour filter.

For a display in which intermediate intensities are required, tone is very important, like contrast. The measure indicating tone is the grey scale, which shows the number of stages that each corresponds to a $\sqrt{2}$ step in the display luminance.

(c) Response time The response time is normally indicated as the rise time, or the time required from applying a voltage until the display function appears, and the decay time, or the time until the display function disappears after removing the voltage. For a television picture display or similar a fast response of no more

Table 1.5 *Characteristics and performance of electronic display devices*

Display properties	Passive devices					Active devices				
	LCD	ECD	EPID	SPD	PLZT	PDP A.c.	PDP D.c.	ELD	VFD	LED
Operating voltage (V)	A.c. 2–5	D.c. 0.5–3	D.c. approx. 50	A.c. 5–10	A.c. 50–100	90–150	180–250	A.c. 160–220	D.c. 12–40	D.c. 2–5
Current consumption ($\mu A/cm^2$)	1–10	$> 5\,mC/cm^2$	approx. 10	1–10	approx. 50	1–10 mA/cm^2	1–10 mA/cm^2	1–10 mA/cm^2	Few mA/cm^2	approx $10\,mA/cm^2$
Contrast	10–20	approx. 15	10–25	10–30	approx. 10	approx. 30	approx. 35	approx. 40	approx. 50	approx. 40
Response time (ms)	30–150	approx. 500	50–200	100–300	1–10	10–20 μs	10–20 μs	approx. 100 μs	approx. 10 μs	$< \mu s$
Luminance/ brightness (fL)	Mediocre	OK	OK	OK	Mediocre	30–40	40–50	20–30	120	approx. 50
Luminous efficiency (lm/W)	—	—	—	—	—	0.3	0.1	0.3–1.5	approx. 10	0.1–1.5
Display colours	Black and white, monochrome, multicolour	Blue, red–mauve, etc.	Monochrome	Blue, etc.	Black and white	Red–orange	Red–orange etc.	Yellow–orange etc.	Green etc.	Red, orange, yellow, green
Memory function	Some	yes	yes	no	yes	yes	no	yes	no	no
Operating lifetime	OK	Moderate	Moderate	Moderate	OK	OK	OK	Fairly OK	OK	OK

than 1/30 s is required, but for a normal display the limit on response time so that the user is not generally aware of it is around 50 to 100 ms. The response time of passive display devices (10 to 500 ms) is very much slower than that for active display devices (1 to 100 μs) because, whereas the display principle of an active display is based on the movement of electrons, the passive display relies on the movement of ions, molecules, or particles.

(d) Luminance and brightness Luminance is usually used to measure the strength of light emitted by an active display device, and the units for this are fL (foot-Lamberts) or cd/m^2 (candelas per square metre).* These units are based on a photometric system of light measurement which takes into account the standard spectral luminous efficiency of the human eye (refer to Section 1.3.1). Generally, therefore, the luminance of a display device which produces green or yellow light, to which the human eye is most sensitive, is high. The luminance of active display devices generally ranges from 30 fL for a yellow–orange ELD to 120 fL for a green VFD.

The term brightness is generally used in the case of a passive display. For example, in the case of a display device that can adequately utilize ambient light, such as an ECD, a bright display can be produced, whereas a device such as an LCD which relies on a polarizer absorbing light has a lower display brightness.

(e) Display colours and light-emission efficiency The colour of light emitted by an active display device or the colour of light transmitted or reflected by a passive display device is termed the display colour. The display colour is determined by the display principle and the materials used, and may be classified into black and white, monochrome of various colours, and multicolour. In the case of a black and white display device, it is possible to obtain a single-colour display or a multicolour display by combination with a colour filter or filters. Additionally, for a single-colour display, LCD, ECD, EPID, SPD, DC-PDP, VFD, and LED devices allow a relatively wide range of colours to be obtained by the selection of the display materials used.

The luminous efficiency is defined as the product of the luminous efficiency (lm/W), which indicates the degree of perceptual response by the human eye to unit energy (W) of light emitted by an active display device and the conversion efficiency of the device from electric power consumed to optical energy produced. The vacuum fluorescent display, which is based on the cathodoluminescence phenomenon, has good light-emission efficiency (approximately 10 lm/W), but the light emission efficiency of display devices based on other principles is relatively low (0.1 to 1.5 lm/W).

(f) Memory function A display device is said to have a memory function if the display is preserved after the applied voltage is turned off. This can enable the power consumption of a display device to be reduced, and also helps to simplify the drive circuitry; it provides the crucial key to multiplex operation. Whether or

* 1 fL = 3.426 cd/m^2, 1 cd/m^2 = 1 nt (nit).

Table 1.6 *Comparison of characteristics of electronics of electronic display devices*

	Passive devices				Active devices			
Characteristic	LCD	ECD	EPID	SPD	PDP	ELD	VFD	LED
Full-colour display	yes	no	no	no	yes	no	fairly good	mediocre
Large display area (cm^2)	yes (270)	yes	yes	yes	yes (280)	yes (120)	mediocre (110)	mediocre (200)
Large display capacity (pixels)	fairly large (1.6×10^5)	no	no	mediocre	yes (2.6×10^5)	yes (1.3×10^5)	fairly large (7.7×10^4)	fairly large (7.7×10^4)
High resolution (pixel spacing, mm)	yes (0.3)	mediocre	mediocre	mediocre	quite good (0.33)	good (0.3)	quite good (0.38)	mediocre (0.5)
Display quality	mediocre	good	good	good	quite good	good	quite good	quite good
Display unit Thickness (mm)	approx. 10	—	—	—	approx. 90	approx. 30	approx. 40	—

Note: figures in ( ) are typical for commercial products.

not there is a memory function depends on the display principle; it is provided by ECD, EPID, and PLZT passive devices and EL and AC-PDP active devices.

(g) Operating lifetime The operating lifetime of a display device is affected basically by a combination of factors: the display principle and the nature of materials used, particularly the chemical stability (for example EPID, SPD), environmental durability such as resistance to water and light (for example LCD), and side reactions and the production of impurities (for example ECD). The lifetime of active ELD devices can be said to have recently reached the level of practicability, but for the passive ECD, EPID, and SPD devices an inadequate operating lifetime is the biggest obstacle to practical implementation.

(2) Comparison of other characteristics

Some characteristics of display devices not described in the previous section are compared in Table 1.6. These are all particularly important characteristics of flat panel display devices in applications for characters, graphics, and image display in direct competition with the CRT.

The display devices for which a full colour display has been achieved are the LCD using a mosaic colour filter and a DC-PDP using a phosphor excitation method. The key to a full colour version of the VFD lies in the development of phosphors of all colours which have a high luminance and long life similar to that of the green phosphor, while for LEDs the current problem is the improvement of emission luminance and efficiency of a blue LED.

Apart from LED and VFD devices, flat panel display devices are easier structurally to adapt to a large display area. The trend to a larger display area is likely to continue at least to A4 size (approximately $620\,cm^2$).

Both PDP and ELD devices are relatively easy to adapt to a large display capacity, whereas for high image resolution, the most appropriate are the LCD and ELD devices. A total of a million or so picture elements (pixels) per display can be envisaged, but it will not be easy to improve substantially on the present screen resolution (pixel spacing approximately 0.3 mm).

The display quality and visibility are greatly affected by the resolution and display colours described above and by the distinction between active and passive displays, but the best active display is the ELD device and the best passive displays are ECD, EPID, and SPD.

Even among flat panel display devices there are great differences in the overall thickness once the display is incorporated into a complete display unit or display module. The thinnest is the LCD and the thickest is the PDP.

1.3 PHOTOMETRY AND COLORIMETRY

1.3.1 Photometric quantities and radiometric quantities

The information displayed on a display device is perceived by the human eye, and therefore the light must be in the visible spectrum, with a wavelength between 380 and 780 nm; the eye corresponds to an optical detector of a certain spectral luminous efficiency. Therefore, for measuring light in this kind of application, rather than using a radiometric quantity based on the pure physical radiant flux Φ_e, a psychophysical photometric value Φ_v based on the luminous flux is used. The radiant flux is the radiant energy Q_e (J) emitted or transmitted in unit time from the light source, and its unit is the watt (W). The luminous flux, on the other hand, indicates the luminous energy Q_v (lm.s) transmitted in unit time, and the unit is the lumen (lm).*

The photometric value is an evaluation of the radiometric value through a perceptual quantity termed brightness. Therefore the relationship between the luminous flux Φ_v and the radiant flux Φ_e can be expressed using the standard spectral luminous efficiency $V(\lambda)$ (λ = wavelength), which indicates the standard human spectral luminous efficiency, as follows:

$$\Phi_v = K_m \int \Phi_e(\lambda) \ V(\lambda) \, d\lambda = K\Phi_e \tag{1.1}$$

where $\Phi_e(\lambda)$ is the spectral density of the radiant flux in the range of wavelengths

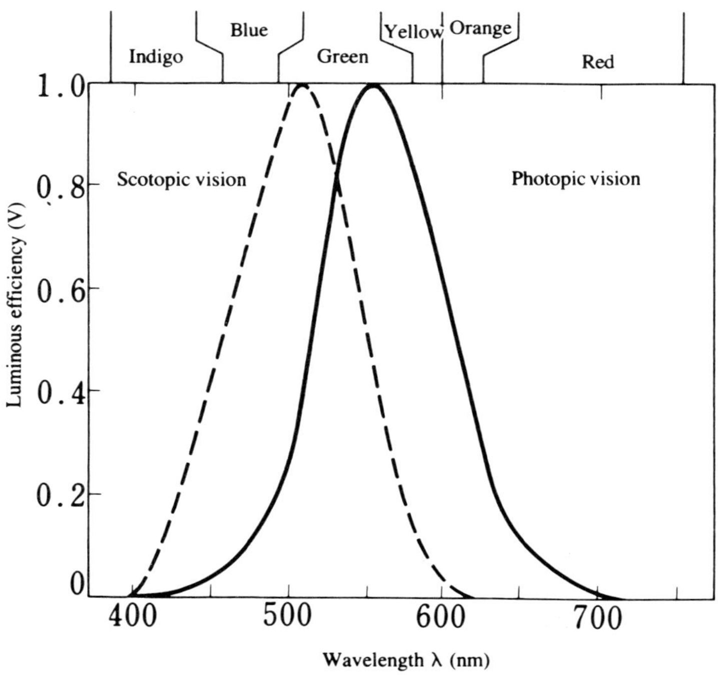

Figure 1.2 *Standard spectral luminous efficiency curves of the human eye*

*A flux of 1 lm is defined as 1/60 of the flux emitted per unit solid angle from a surface area of 1 cm^2 by a black body at the melting point of platinum (2042 K).

from λ to $(\lambda + \Delta\lambda)$ and K_m $(= 680\,\mathrm{lm/W})$ is the maximum spectral luminous efficiency in photopic vision. K indicates the average luminous efficacy over the wavelength range.

Figure 1.2 shows the standard spectral luminous efficiency curves for photopic vision and scotopic vision. Under photopic vision (a bright environment) the maximum spectral luminous efficiency ($K_m = 680\,\mathrm{lm/W}$) occurs at a wavelength $\lambda = 555\,\mathrm{nm}$, but in a dark environment (scotopic vision) the maximum spectral luminous efficiency shifts to a wavelength of $\lambda = 510\,\mathrm{nm}$, and the scotopic luminous efficiency ($K_m = 1746\,\mathrm{lm/W}$) is about three times that of photopic vision. This difference is due to the different sensitivities of the optic cells in the retina of the eye. Normally for evaluating a display device the photopic vision standard luminous efficiency characteristic is used.

In the same way that the luminous flux corresponds to the radiant flux, as shown in Table 1.7 there are a variety of other photometric values that relate to corresponding radiometric values. Among these the luminous intensity, luminance, and luminous emittance are photometric values which depend on the nature of the body emitting the light. For evaluating the characteristics of display devices the luminance is normally used. In this case the unit commonly used for the luminance is the foot-Lambert (fL).*

1.3.2 Colorimetric systems and chromaticity diagrams

Colorimetric systems for indicating colours may be broadly classified into two types: the psychophysical type based on additive colour mixing and the pure psychological type based on the three colour characteristics of hue, brightness or luminance and saturation or chroma. The CIE (Commission Internationale de l'Eclairage) colorimetric system is of the former type, and the Munsell colorimetric system is of the latter type, using standard colour cards. Below we describe principally the CIE colorimetric system. The visible spectrum ranges from approximately 380 to 780 nm.

(1) CIE colorimetric system

Under the additive colour mixing rule, any colour can be realized by mixing in suitable proportions three suitable primary colours. Therefore if the three primary colours are determined, any colour may be expressed quantitatively by the necessary amounts of the three primary colours (tristimulus values). The CIE colorimetric system uses as the primary colours R ($\lambda = 700$ nm), G ($\lambda = 546.1$ nm), and B ($\lambda = 435.8$ nm), in a red–green–blue system. By means of a mathematical transformation, this colorimetric system can be understood as an XYZ colorimetric system. Whereas the former is inconvenient in practice because it

* $1\,\mathrm{fL} = (1/\pi)\,\mathrm{cd/f^2} = 3.426\,\mathrm{cd/m^2}$, $1\,\mathrm{L} = (1/\pi)\mathrm{cd/cm^2}$. For example, the luminance of a white display on a colour television CRT is between 100 and 150 fL.

Table 1.7 *Photometric and radiometric definitions and units*

Photometric quantities				Radiometric quantities			Defining equations
Name	Symbol	Unit	Definition	Name	Symbol	Unit	
Luminous flux	Φ_v	lm	Product of luminosity and radiant flux	Radiant flux	Φ_e	W	$\Phi_v = K\Phi_e = K_m \int \Phi_e(\lambda)\, V(\lambda) d\lambda$
Luminous energy	Q_v	lm.s	Integral over time of luminous flux	Radiant energy	Q_e	J	$Q_v = KQ_e = \int Q_e(\lambda) K(\lambda) d\lambda$
Luminous intensity	I_v	cd	Luminous flux emitted per unit solid angle	Radiant intensity	I_e	W/sr	$I_v = d\Phi_v/d\Omega$
Luminance	L_v	cd/m^2 (=nt)	Luminous intensity per unit area	Radiant brightness	L_e	W/sr.m^2	$L_v = d^2\Phi_v/d\Omega.dA \cos\theta$
Luminous flux density	E_v	lx	Luminous flux impinging per unit area	Radiant flux density	E_e	W/m^2	$E_v = d\Phi_v/dA$
Luminous emittance	M_v	lm/m^2	Luminous flux emitted per unit area	Radiant emittance	M_e	W/m^2	$M_v = d\Phi_v/dA$

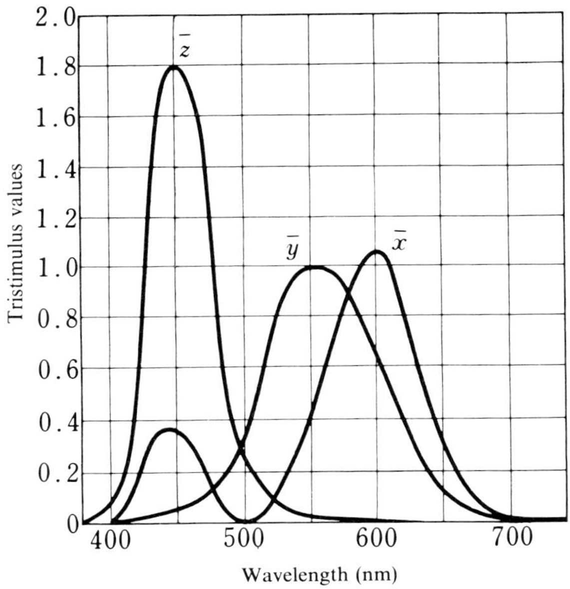

Figure 1.3 *Isochromatic functions for CIE standard colorimetric system (XYZ colorimetric system)*

involves the colour combination of negative quantities, in the latter system any colour (stimulus) can be expressed as a combination of positive values X, Y, and Z.* Thus since the XYZ colorimetric system is determined so as to be easy to use in practice, it is also known as the CIE standard colorimetric system.

In Fig. 1.3, $\bar{x}$, $\bar{y}$, and $\bar{z}$ are isochromatic functions showing the spectral tristimulus values when the CIE standard colorimetric system is used, and $\bar{y}$ is the same as the standard spectral luminous efficiency curve shown in Fig. 1.2.

(2) Colorimetry of the light source

In the XYZ colorimetric system, the tristimulus values X, Y, and Z of the colour of a light source are given by the following expressions, where $\Phi_{e\lambda}$ is the spectral intensity of the light source:

$$X = K_m \int_{380}^{780} \Phi_{e\lambda} \bar{x}(\lambda)\,d\lambda$$

$$Y = K_m \int_{380}^{780} \Phi_{e\lambda} \bar{y}(\lambda)\,d(\lambda) \qquad (1.2)$$

$$Z = K_m \int_{380}^{780} \Phi_{e\lambda} \bar{z}(\lambda)\,d\lambda$$

*The three primary colours X, Y, Z are determined for the purposes of adding together positive values, and are colour stimuli that do not actually exist.

Here $\bar{x}(\lambda)$, $\bar{y}(\lambda)$, and $\bar{z}(\lambda)$ indicate isochromatic functions, and $K_m = 680\,\text{lm/W}$. The chromatic coordinates x, y, and z in the XYZ colorimetric system are found, using expression (1.2) as follows:

$$x = \frac{X}{X + Y + Z}$$

$$y = \frac{Y}{X + Y + Z} \tag{1.3}$$

$$z = \frac{Z}{X + Y + Z}$$

From the above it will be seen that the colorimetric evaluation of the light source can be expressed as the brightness Y (photometric value) and the chromatic coordinates x, y.

(3) Colorimetry of the colour of an object

The tristimulus values of the colour of an object* are given as follows, where $\rho(\lambda)$ is the spectral reflectance or spectral transmittance of the body:

$$X = \frac{1}{K}\int_{380}^{780} \Phi_{e,\lambda}\rho(\lambda)\bar{x}(\lambda)\,d\lambda$$

$$Y = \frac{1}{K}\int_{380}^{780} \Phi_{e,\lambda}\rho(\lambda)\bar{y}(\lambda)\,d\lambda$$

$$Z = \frac{1}{K}\int_{380}^{780} \Phi_{e,\lambda}\rho(\lambda)\bar{z}(\lambda)\,d\lambda \tag{1.4}$$

where $\Phi_{e,\lambda}$ indicates the spectral density of a standard light† illuminating the object, and K is given by

$$K = \int_{380}^{780} \Phi_{e,\lambda}\bar{y}(\lambda)\,d\lambda \tag{1.5}$$

In other words, the tristimulus values for the colour of the body are given by the ratios of the tristimulus values of the light reflected or transmitted by the body and the brightness Y of the illuminating light. In expression (1.4), Y is termed the luminous reflectance (luminous transmittance).

From the above the colorimetric value for the colour of a body can be indicated

*Light reflected or transmitted by a body.

†In order to standardize the colorimetric representation of the colour of a body, standard light values A, B, and C determined by the CIE are as follows. A: light with the same spectral distribution as a black body at 2854 K (tungsten artificial light); B, C: light having a daylight spectral distribution approximating to colour temperatures of 4874 K (B) and 6774 K (C). Additionally, the value D_{65} is also used as a standard light with a colour temperature of 6504 K.

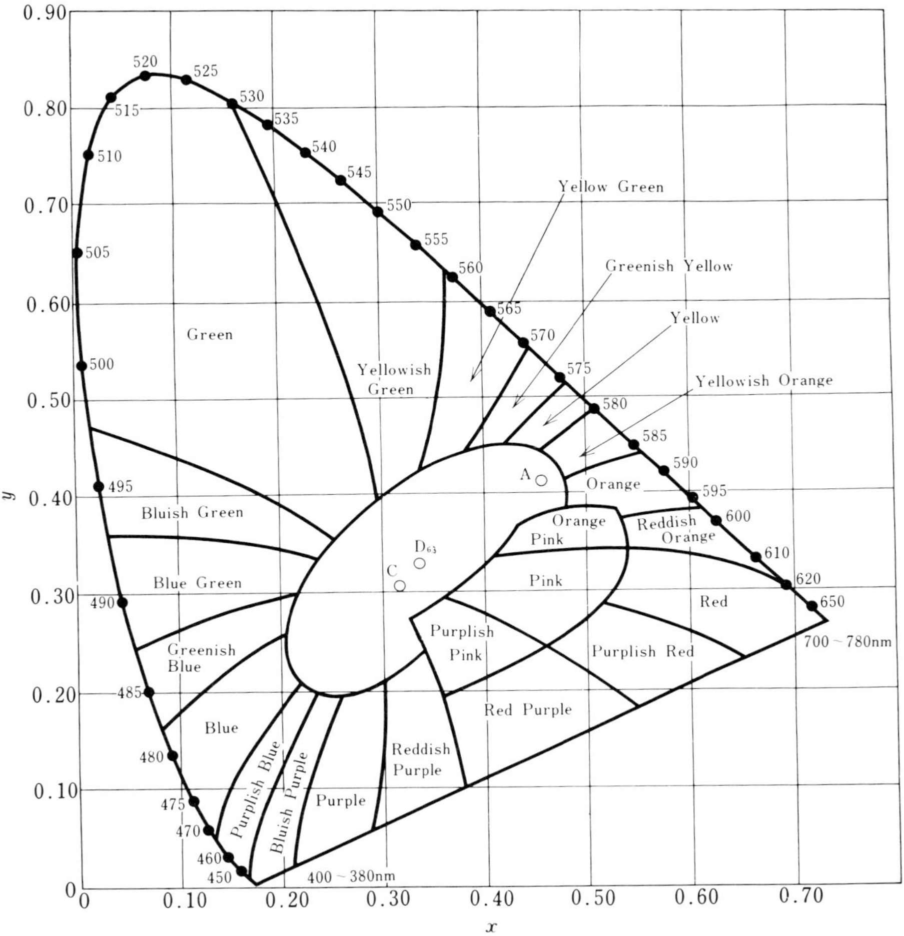

Figure 1.4 *CIE xy chromaticity diagram and colour names of chromatic stimuli*

by the combination of the value Y and the chromating coordinates x and y obtained from expressions (1.4) and (1.3).

(4) xy chromaticity diagram

Any given colour stimulus may be represented by one point on an xy chromaticity diagram.* From the colour perceptions corresponding to the colour stimuli of the light, colour names are given to each colour stimulus. Figure 1.4† shows a chromaticity diagram, with the colour names and also the

*The XYZ system is normalized so that $(x + y + z) = 1$, and therefore the chromaticity coordinates are shown on a two-dimensional chromaticity diagram.

†K.L. Kelly: *J. Opt. Soc. Am.*, **33**, 627 (1943).

coordinate locations of standard colours. The arc-shaped curve in the diagram is termed the spectrum locus linking the chromatic coordinates of monochromatic light or saturated colours of varying wavelengths, and the straight line joining the ends of this arc is termed the saturated purple line. All actually existing colour stimuli are indicated by coordinates within the area surrounded by this arc and line. In the case of a light source colour, even if the brightness (Y value) varies over a wide range, the relationship shown in Fig. 1.4 can be applied. In the case of the colour of a body, however, the colour perception changes very rapidly with the value of Y, and the relationship between a colour name and the chromatic coordinates varies depending on that value.

1.4 OUTLOOK

Various flat panel electronic display devices have been created to match the advances towards more compact and lightweight electronic devices which use low supply voltages and have low power consumptions. With the prospect of more diverse information processing systems, principally computers, they have exploited the properties of compactness, flatness, and low weight, low operating voltage, and low power consumption which are lacking in CRT displays, and have been principally used in fields where they do not compete with CRT displays.

Recently, however, a number of flat panel electronic displays have begun to offer increased display capacity and the possibility of colour displays, so that flat panel displays are beginning a new round of competition with the CRT whose traditional advantages have been large display capacity and full colour display. With this background in mind, we now look at the outlook for various electronic display devices.

The development of LCDs is continuing, with multiplex matrix displays, active matrix displays, full colour displays, video displays, and large area displays, with the object of providing the same display functions and high display quality, along with an increased display capacity to bring them in line with the CRT. The principal topics for LCD development will therefore be high-performance liquid crystal materials, inexpensive and reliable switching arrays, and red–green–blue mosaic colour filters.

For the ECD, a passive display device like the LCD, the principal problem is the improvement of operating lifetime and the realization of matrix display for a large capacity display. If only these problems can be solved the ECD can move forward to practical implementation. Further progress can be expected on the particular strengths of the ECD for a memory function, and the development of fully solid state and multicolour types.

Both a.c. and d.c. types of PDP are advancing towards large panel, high resolution, and full colour displays to match the capacity of a CRT. The most important problem is the improvement of generated light intensity and efficiency, and then the development of gas compositions and phosphors of various colours allowing higher intensity and cathode materials allowing a lower supply voltage.

For the PDP to become commonplace a more simple panel construction and drive electronics are therefore required.

There have been many problems with electroluminescent displays, such as lifetime, reliability, light intensity, and cost, but both lifetime and reliability are now at the level of practicality. Development is continuing on the search for materials and construction that will yield high light intensity and low supply voltage, and also red–green–blue phosphors are being developed; an electroluminescent display panel can be expected to be developed which is fully solid state, has an adequate memory function, and is compact, providing a low-voltage, high-intensity display with a display capacity comparable to a CRT.

VFD devices are at the stage of development of improved resolution for higher capacity displays, bigger display panels, and also higher light-intensity phosphors of various colours. In parallel with these developments, progress continues on active matrix drive displays and modules integrating the display panel and drive circuit, so that at some stage VFDs will be able to be used as colour image display panels.

The LED is different from other electronic display devices in that its principal application is for indicator lamps. This means that the principal problems to tackle are the improvement of light intensity and light generation efficiency for various colours. As the application of LEDs widens to display devices, attention will turn to the improvement of intensity of blue LEDs, and at the same time the development of large display capacities and full colour versions will continue.

In this book flat panel electronic display devices other than those described above have been treated together with the flat CRT as special displays. The majority of these special displays are still not yet at the stage of practical implementation, and still have many development problems so that their future is unclear but potentially hopeful. Of these, EPID, SPD, and PLZT passive display devices suggest strongly that they will be able to be implemented usefully as display devices because of the display quality. In the case of both EPID and SPD devices, the principal problem is stabilization of colloidal particles, and for PLZT devices the current problems are development of a suitable drive method and the establishment of device technology.

The outlook for various current electronic display devices has been considered in this chapter, but even if all of these development problems are solved there are still difficulties in the realization of an ideal electronic display device which can fulfil all requirements (high intensity, high contrast, high response, large capacity, high resolution, full colour, flat form, light weight, low voltage supply, low power consumption, building in drive circuitry, lifetime and reliability, low cost, etc.). Therefore, in order to pursue the ideal form, it is likely that a variety of new electronic display devices will continue to be invented.

BIBLIOGRAPHY

Ibuki, J.: *Solid State Display Devices*, Asakura Shoten, Tokyo (1972).
Kazan, B., and M. Knoll: *Electronic Image Storage*, Academic Press (1968).

Kmetz, A., and F. K. von Willisen: *Nonemissive Electrooptic Displays*, Plenum Press, New York and London (1975).
Kobayashi, T.: *ABC of Electronic Displays*, Nihon Hoso Shuppan Kyokai, Tokyo (1971).
Luxemberg, H., and R. Kuehn (translators [into Japanese] Wada and Shibata): *Display Engineering*, Kindai Kagaku-sha, Tokyo (1971).
Maeda, K.: *Optoelectronic Transducers*, Corona-sha, Tokyo (1975).
Oishi, Ogoe, and Nakayama: *Image Displays*, Corona-sha, Tokyo (1975).
Pankove, J.J. (ed.): *Display Devices*, Springer-Verlag, Berlin, Heidelberg, New York (1980).
Poole, H. (translator [into Japanese] Morita): *Electronic Display Systems*, Nihon Keiei Shuppan-kai, Tokyo (1968).
Richards, C. (ed.): *Electronic Display and Data Systems*, McGraw-Hill (1973).
Sherr, S.: *Fundamentals of Display System Design*, Wiley-Interscience (1970).
Shikisai Kagaku Kyokai (Colour Science Association) (ed.), *Colour Science Handbook*, Nankodo, Tokyo (1966).
Television Gakkai (Television Institute) (ed.), *Recent Display Devices*, Nihon Hoso Shuppan Kyokai, Tokyo (1974).
Television Gakkai (Television Institute) (ed.), *Television Image Engineering Handbook*, Ohm-sha, Tokyo (1980).
Wada, M. (ed.): *Electronic Displays*, Kyoritsu Shuppan, Tokyo (1975).
Wiley–Interscience Publication (ed.): *Electronic Displays*, John Wiley and Sons (1979).

Chapter 2

Liquid crystal displays(LCDs)

2.1 INTRODUCTION

2.1.1 What is a liquid crystal?

Most substances have a single melting point at which they change from a solid to a transparent liquid, but a group of materials, termed liquid crystals, first become an opaque and cloudy liquid at one temperature, and then turn to an ordinary transparent liquid at a somewhat higher temperature.* These phase changes in a liquid crystal substance with temperature can be shown externally as in Fig 2.1. The term liquid crystal has two uses: it is applied both to the liquid crystal phase, which as shown in the diagram is intermediate between the solid phase and the liquid phase, and also to any material that itself has a liquid crystal phase.

Within a certain temperature range, a liquid crystal substance behaves externally as a fluid, but at the same time has a crystalline structure which exhibits optical double refraction. This is the background to the term 'liquid crystal', and the liquid crystal phase was discovered by the Austrian biologist. F. Reinitzer in 1888, in benzoic acid and acetic acid esters of cholesterol [2].

The majority of liquid crystal substances are organic compounds which either have long thin rod-shaped molecules or flat molecules. In the liquid crystal phase, the molecular orientation takes up certain special organizations, as shown in Fig. 2.2, and based on differences in these spatial organizations liquid crystals are divided into three types: smectic, nematic, and cholesteric.†

In a smectic liquid crystal, the rod-shaped molecules are arranged in layers with the molecules parallel to each other and aligned typically perpendicular to the layers. The binding between the molecular layers is relatively weak, so they slip easily over each other, and for this reason a smectic liquid crystal behaves as a

* An ordinary liquid is optically isotropic, and is therefore termed an isotropic liquid. Liquid crystals, on the other hand, are termed anisotropic liquids.

†Specific substances showing these three variations of the liquid crystal phase are described in Section 2.4.

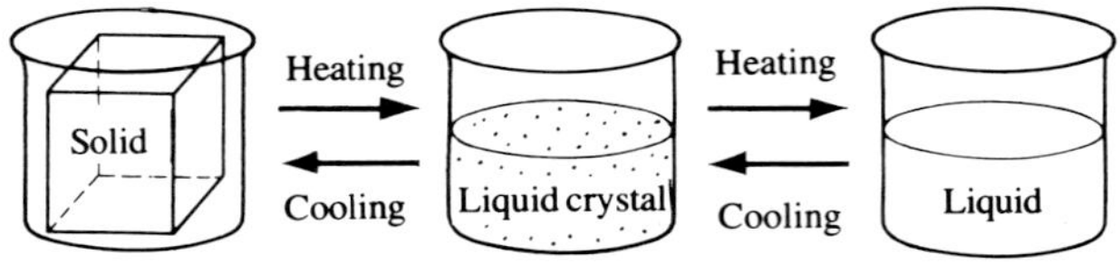

Figure 2.1 *Phase changes with temperature in a liquid crystal substance [1]*

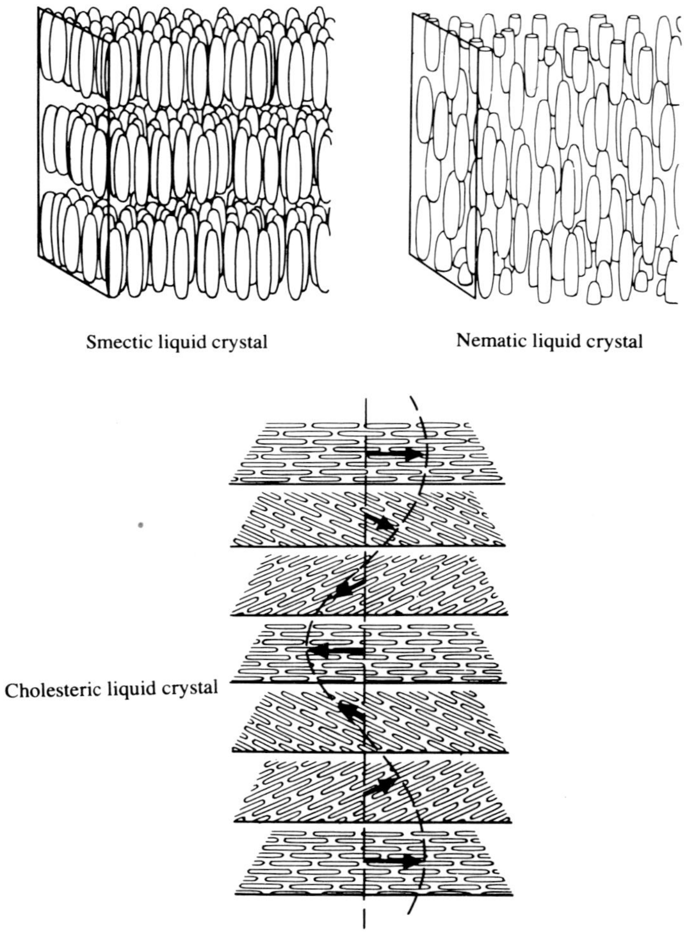

Figure 2.2 *Molecular orientations in three types of liquid crystal [3]*

two-dimensional fluid. Compared with an ordinary liquid, however, the viscosity is very high.

In a nematic liquid crystal, the rod-shaped molecules are also aligned parallel to each other, but the individual molecules move relatively easily in the direction along their axis, so that there is no layering structure. The nematic substance is therefore very fluid, with low viscosity

A cholesteric liquid crystal has a layered structure like a smectic liquid crystal, but the axis of the molecules is in the plane of each layer, with a parallel orientation similar to that of a nematic liquid crystal. The axis of alignment of molecules in neighbouring layers slightly deviate one from another, and therefore overall the liquid crystal has a helical structure. The optical characteristics of a cholesteric liquid crystal, which include optical rotatory power, selective light reflection, and circular dichroism, are due to this helical structure.

2.1.2 Liquid crystals and displays

As will be seen from the preceding section, the structural organization of molecules in a liquid crystal is not rigid as in a crystalline structure. The molecules can therefore be easily reoriented as a result of an external stimulus such as an electric field, a magnetic field, temperature or mechanical strain, and thus the optical behaviour of the liquid crystal can be changed. It is precisely this flexibility in the molecule orientation in a liquid crystal that is the basis for the applications of liquid crystals in display devices, optoelectronic devices, sensors, and so forth.

In particular, in a liquid crystal display the molecules are in a particular initial orientation; this is changed to a different orientation by the effect of a voltage or heat, and the change in optical characteristics such as the double refraction, optical rotatory power, dichroism, light scattering, or optical rotatory dispersion in the liquid crystal cell, which occurs with the change in molecular orientation, is converted to a visible change. In other words a liquid crystal display utilizes the modulation of light by the liquid crystal cell, and is thus a passive (non-emitting) display. Generally the abbreviation LCD is used for a liquid crystal displays.

2.1.3 Characteristics of liquid crystal displays

The most important characteristics of LCDs, compared with other displays, are their unparalled low power consumption and low voltage operation. They are therefore extremely suitable for use with LSI drive circuits, and are extremely valuable in practice. We now list these advantages individually:

1. Low power consumption (a few several tens of $\mu W/cm^2$) means that long-term battery drive is possible; energy saving.
2. Low drive voltage (10 V or less) enables direct IC drive, and drive circuits are both compact and simplified.
3. The device is thin and can be used for large displays (tens of square centimetres) and small displays (a few square milimetres), and so is particularly suitable for portable devices.

4. Because it is a passive display device, the display is clear even in a brightly lit position.
5. Making a colour display is easy, so the display functions can be enlarged and varied.
6. The display can be used for projection, so a large display (several square metres) is possible.

Set against these are the following disadvantages:

1. Since the display is a non-emitting type, it is not clear in a dark place.
2. The display contrast often depends on the viewing angle, which is therefore restricted.
3. The response depends on the ambient temperature, so low temperature operation is inadequate (approximately $-20°C$).

2.2 *PHYSICAL CHARACTERISTICS OF LIQUID CRYSTALS*

2.2.1 *Anisotropy of physical characteristics*

There are various principal types of liquid crystal (see Fig. 2.2), but all of these share the fact that molecules are aligned with their long axes parallel. It is for this reason that various physical characteristics such as refractive index n, dielectric constant ε, magnetic susceptibility η, conductivity σ, and viscosity of the liquid crystal substance have different values in the direction parallel to the molecular axis ($\parallel$) and orthogonal to the molecular axis ($\perp$), or in other words anisotropy ($n_{\parallel} \neq n_{\perp}$, $\varepsilon_{\parallel} \neq \varepsilon_{\perp}$, $\chi_{\parallel} \neq \chi_{\perp}$, $\sigma_{\parallel} \neq \sigma_{\perp}$, $\eta_{\parallel} \neq \eta_{\perp}$).

It is the combination of this anisotropy in physical characteristics in the liquid crystal and the ease of reorienting the molecular axes by applying a voltage or heat that opens up the possibility of applications for liquid crystals, in particular for displays.

In considering the anisotropy of the liquid crystal physical characteristics, it is very important to discover to what extent the liquid crystal molecules are properly aligned with each other. The degree of orderliness of liquid crystal molecular orientations may be represented by the molecular alignment order parameter S defined by the following expression [4]:

$$S = \tfrac{1}{2}\langle 3\cos^2\theta - 1\rangle \tag{2.1}$$

Here a unit vector in the predominant molecule axis orientation when the liquid crystal is considered as a whole is termed the director $\mathbf{n}$, the angle of deviation of the molecular axis orientation $\mathbf{a}$ for individual liquid crystal molecules from $\mathbf{n}$ is φ (Fig. 2.3), and the symbols $\langle\ \rangle$ represent averaging over the whole space.

When the molecular axis orientation is unordered, as in an isotropic liquid, then $S = 0$, whereas when all the molecules are perfectly parallel aligned, as in an

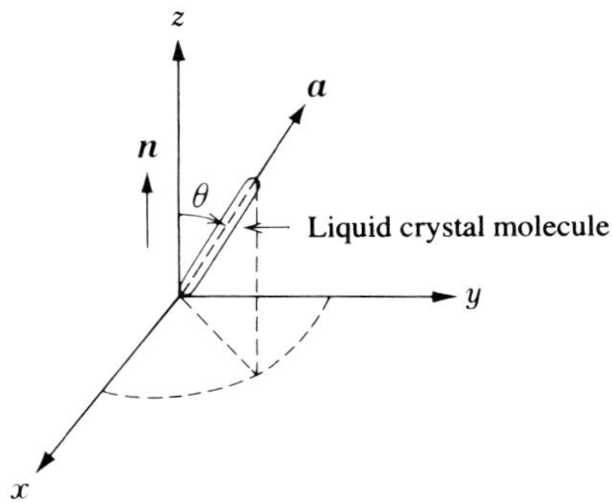

*Figure 2.3 Spatial orientation of the liquid crystal director **n** and the molecular alignment direction **a** [1]*

ideal liquid crystal, $S = 1$. In a typical nematic liquid crystal, S is in the range 0.3 to 0.8, depending on temperature.

2.2.2 Refractive index anisotropy and optical characteristics

1. Anisotropy of refractive index

A liquid crystal has an anisotropy of refractive index similar to that of an optical uniaxial crystal and exhibits double refraction. A uniaxial crystal has two different principal refractive indices n_0 and n_e. These are the refractive indices of ordinary rays in which the axis of oscillation of the electrical vector is orthogonal to the optical axis of the crystal and extraordinary ray for which it is parallel.

In nematic and smectic liquid crystals the direction of the liquid crystal director **n** corresponds to the optical axis of a uniaxial crystal, and therefore the refractive indices are $n_\perp$ and $n_\parallel$ for rays whose oscillation is respectively orthogonal and parallel to the director **n** i.e.

$$\begin{aligned} n_0 &= n_\perp \\ n_e &= n_\parallel \end{aligned} \tag{2.2}$$

Thus the refractive index anisotropy Δn is given by

$$\Delta n = n_e - n_0 = n_\parallel - n_\perp \tag{2.3}$$

Thus the refractive index of these liquid crystals in three-dimensional space is realized, as shown in Fig. 2.4(a), as a spherical surface with respect to n_0 and an ellipsoidal surface with respect to n_e. n_0 is never greater than n_e and the two coincide only on the direction of the director **n**. Therefore always $n_\parallel > n_\perp$ and Δn is positive. For this reason nematic and smectic liquid crystals are said to be optically positive.

For cholesteric liquid crystals the helical axis orthogonal to the director **n** corresponds to the optical axis, so the principal refractive indices n_0 and n_e are

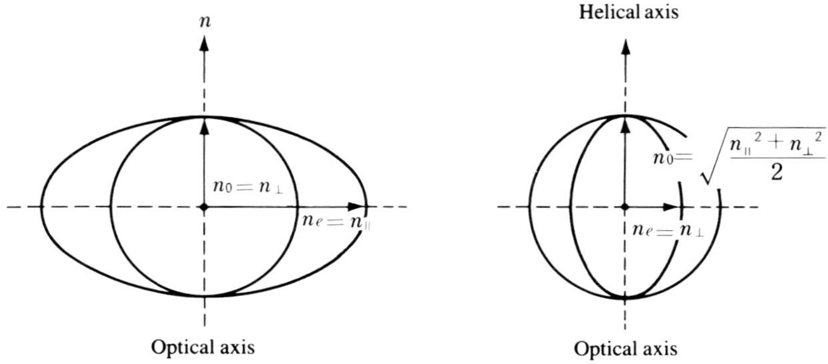

(a) Nematic and smectic liquid crystals (optically positive)

(b) Cholesteric liquid crystals (optically negative)

Figure 2.4 *Refractive index anistropy of a uniaxial liquid crystal [1]*

given by [5]

$$n_0 = [\tfrac{1}{2}(n_\parallel^2 + n_\perp^2)]^{1/2}$$
$$n_e = n_\perp \tag{2.4}$$

Since in a cholesteric liquid crystal the relation $n_\parallel > n_\perp$ holds, $\Delta n\ (= n_e - n_0) < 0$ holds. Therefore a cholesteric liquid crystal is termed optically negative. The distribution over space of the magnitude of the refractive indices of ordinary ray and extraordinary ray is shown in Fig. 2.4(b).

(2) Optical characteristics

Because of the anisotropy of refractive index, liquid crystals exhibit the following useful characteristics, which form the basis of liquid crystal displays:

1. Deflection of the direction of travel of light to the direction of the director **n** (molecular long axis)
2. A change in the polarization state (linearly polarized light, elliptically polarized light, circularly polarized light) or the direction of polarization
3. Reflection or transmission of incident polarized light according to whether the optical rotation is to right or left

The direction of travel of light incident on a liquid crystal in the direction of the director **n** is due to the fact that within the liquid crystal $n_\parallel < n_\perp$ holds, and the speeds of light $v_\parallel$ and $v_\perp$ are inversely proportional to the refractive indices $n_\perp$ and $n_\parallel$ so that $v_\parallel$ in the **n** direction is higher than $v_\perp$ in the orthogonal direction.

Next, considering a liquid crystal in which the director **n** is uniformly aligned with the x axis, we consider the case where linearly polarized light with its axis of oscillation at an angle θ to the x axis impinges in the direction of the z axis. When

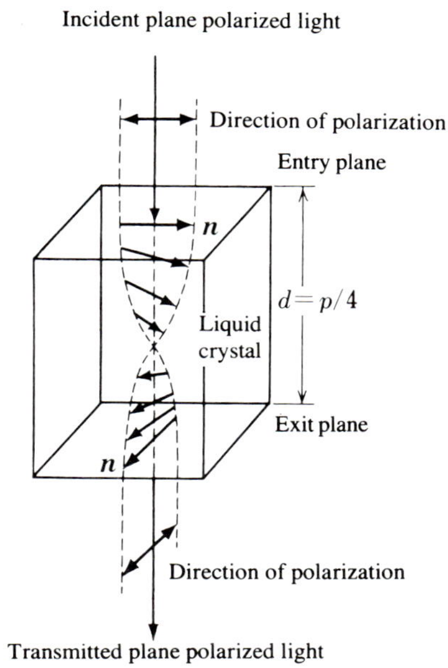

Figure 2.5 *Rotation of the plane of polarization of incident light by a liquid crystal with twisted orientation [1]*

$\theta = 0$ or $\pi/2$, the polarization of the linearly polarized light is not affected at all, but in other cases, for example when $\theta = \pi/4$, as the incident linearly polarized light proceeds along the z direction, the polarization changes in order from linear, to elliptical, to circular, to elliptical, to linear polarization, and at the same time the direction of the linear polarization is changed. What happens when linearly polarized light is incident on a liquid crystal with a twisted director **n** will be seen by considering Fig. 2.5. When the pitch p of the liquid crystal twist is sufficiently greater than the wavelength λ of the incident light, then light that is incident with its direction of oscillation parallel to **n** at the entry surface rotates together with the twist of **n**, and emerges at the opposite side with its direction of oscillation still parallel to **n**. Light that is incident with its direction of oscillation orthogonal to **n** emerges with its oscillation still orthogonal to **n** at the plane of emergence, but light incident with any other direction of oscillation will emerge with its polarization either elliptical, circular, or linear according to the phase difference $(n_{\parallel} - n_{\perp})\omega d/c$ from the incident surface to the surface of emergence, where ω is the angular frequency of the light, c is the speed of light, and d is the thickness of the liquid crystal layer.

If the pitch p of the twist is of the same order of magnitude as the wavelength λ of the light, as in an ordinary cholesteric liquid crystal, the component of incident polarized light in the same direction as the optical rotation direction of the liquid crystal is reflected and the rotation component in the opposite direction only is transmitted. This phenomenon is termed selective optical scattering.

2.2.3 *Molecular orientation by application of an electric field*

When an electric field is applied to a liquid crystal with a particular initial orientation, then with the transition to a new molecule orientation* there is an accompanying change in the optical characteristics of the liquid crystal. This transition phenomenon is due to the dielectric anisotropy $\Delta\varepsilon$ $(=\varepsilon_{\parallel}-\varepsilon_{\perp})$ of the liquid crystal.

The value of $\Delta\varepsilon$ for a liquid crystal may be positive or negative, which is determined by the molecular structure of the liquid crystal (see Section 2.4). Depending on whether $\Delta\varepsilon$ is positive or negative, as shown in Fig. 2.6 the initial molecular orientation of the liquid crystal will normally be with the liquid crystal molecular axis (director **n**) either parallel (Fig. 2.6 (a)) or orthogonal (Fig. 2.6 (b)) to the plane of the two electrode substances.

When an electric field **E** is applied to a liquid crystal with a dielectric anisotropy $\Delta\varepsilon$ then electrical energy f_e is produced, as given by the following expression:

$$f_e = -\tfrac{1}{2}\varepsilon_{\perp}\mathbf{E}^2 - \tfrac{1}{2}\Delta\varepsilon(\mathbf{n}\cdot\mathbf{E})^2 \tag{2.5}$$

As will be seen from expression (2.5), when an electric field ($\mathbf{E} > \mathbf{E}_c$†) is applied to a liquid crystal with $\Delta\varepsilon > 0$, the molecular axis (director **n**) is realigned to be parallel with the direction of the electric field **E**, whereas in the case of a liquid crystal with $\Delta\varepsilon < 0$, **n** is reoriented to be orthogonal to **E**. The reason for this is that in the reoriented state, the electrical energy f_e of the liquid crystal reaches a minimum.

However, the total free energy **F** of the liquid crystal when the molecule orientation in the initial state has undergone a deformation by the application of the electric field is the volume integral of the sum of the electrical energy f_e as given by expression (2.5) and the elastic energy f_d:‡

$$F = \int (f_e + f_d)\,\mathrm{d}\Gamma \tag{2.6}$$

The electric field threshold value $\mathbf{E}_C$ at which the transition from the initial molecule orientation of the liquid crystal begins can be determined from a computation of the value at which the total free energy F given by expression (2.6), under appropriate boundary conditions, is a minimum [6]:

$$\mathbf{E}_C = \frac{\pi}{d}\left(\frac{k_{ii}}{|\Delta\varepsilon|}\right)^{1/2} \tag{2.7}$$

*Termed the Freédericksz transition.

†The threshold electric field value where the transition from the original molecule orientation to the new molecule orientation begins.

‡A liquid crystal is an elastic body, and shows three types of basic strain under an external force: splay, twist, and bend. The elastic coefficients showing the relationship between these strains and the external force are called respectively k_{11}, k_{22}, and k_{23}, and the elastic free energy f_d of a liquid crystal under a certain deformation is given by the following [7]:

$$f_d = \tfrac{1}{2}k_{11}(\mathrm{div}\,\mathbf{n})^2 + \tfrac{1}{2}k_{22}(\mathbf{n}\cdot\mathrm{rot}\,\mathbf{n} + t_0)^2 + \tfrac{1}{2}k_{33}(\mathbf{n}\cdot\mathrm{rot}\,\mathbf{n})^2$$

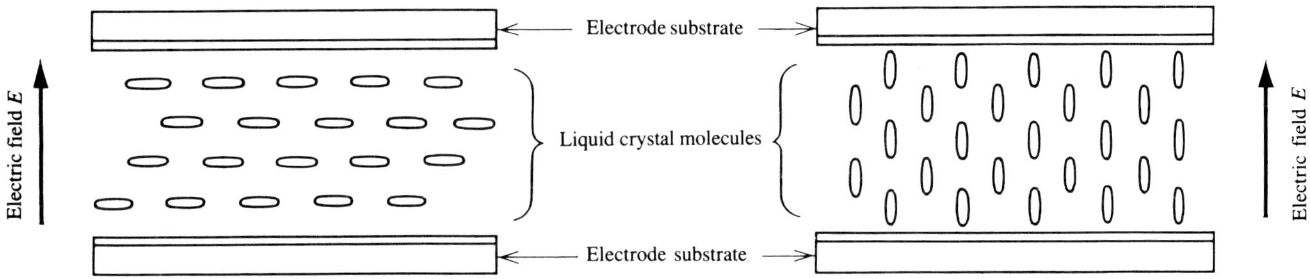

(a) Parallel orientation

(b) Perpendicular orientation

Figure 2.6 *Relationship between the sign of the dielectric anisotropy Δ_ε of a liquid crystal and the initial molecule orientation*

The threshold voltage V_{th} for the various electrooptic effects of liquid crystal described in the next section, 2.3, may be obtained as follows from expression (2.7):

$$V_{th} = \pi \left(\frac{k_{ii}}{|\Delta\varepsilon|} \right)^{1/2} \tag{2.8}$$

In expressions (2.7) and (2.8), d is the cell thickness of the liquid crystal and k_{ii} is the elastic coefficient. When the initial molecule orientation is parallel, $k_{ii} = k_{11}$, when it is the orthogonal orientation $k_{ii} = k_{33}$, and in the case of a twisted orientation which is a deformation of the parallel orientation $k_{ii} = k_{11} + (k_{33} - 2k_{22})/4$.

2.3 LIQUID CRYSTAL DISPLAY MODES

2.3.1 Liquid crystal electrooptic effects and display modes

When the molecule orientation within a liquid crystal is changed by the application of an electric field, the optical characteristics of the liquid crystal cell change; this phenomenon of an electrically induced optical modulation is termed an electrooptic effect of the liquid crystal.

Of the many electrooptic effects of liquid crystals, the most important from the point of view of application to liquid crystal displays are shown in Table 2.1. Historically, the phase change (PC) effect [8] was discovered first in 1968 by J. Wyscoki *et al.*, and then in the same year the dynamic scattering (DS) effect [9] and the guest–host (GH) effect]10] were announced by G. Heilmeier *et al.* Then in 1971 the twisted nematic (TN) effect [11] was announced by M. Schadt *et al* and the electrically controlled birefringence (ECB) effect [12, 13] was announced at almost the same time by groups represented by M. Schiekel and M. Hareng.

Table 2.1 *Classification of electrooptic effects of liquid crystals*

Liquid crystal electrooptic effect
Current effects
Dynamic scattering effect (DS)
Electric field effects
Twisted nematic effect (TN)
Guest–host effect (GH)
Electrically controlled birefringence (ECB)
Phase change effect (PC)
Thermal effects
Smectic effect
Cholesteric effect

Of the electrooptic effects in Table 2.1, the twisted nematic, guest–host, electrically controlled birefringence, and phase change effects are caused by the effect of a dielectric torque due to the anisotropy of the dielectric constant, and are therefore called electric field effects. On the other hand, the dynamic scattering effect is due to the effect of the conductivity torque, due to anisotropy of the conductivity, becoming dominant, and is therefore called a current effect. In addition to these types of effects, there are also electrooptic effects termed thermal effects which require the application of heat in addition to an electric field. These thermal effects can be classified into smectic and cholesteric effects by differences in the liquid crystal materials used.

Typical display principles based on the electrooptic effects of liquid crystal described above are shown in Table 2.2, with various characteristics. These display modes are described in more detail in the following sections.

2.3.2 Twisted nematic (TN) type

A layer of nematic liquid crystal (N_p liquid crystal) for which $\Delta\varepsilon$ is greater than 0, of thickness approximately 10 μm, is sandwiched between transparent electrode substrates, and a cell is constructed so that the molecular axis of the liquid crystal rotates continuously through 90° between the two substrates (see Section 2.5.3); this is a twisted nematic (TN) cell. Since the pitch of this twist is sufficiently large compared with the wavelength of visible light, the direction of polarization of linearly polarized light incident vertically on one of the electrode substrates is rotated through 90° by the twist of the liquid crystal molecules as it passes through the cell (see Section 2.2.2).* Therefore this twisted nematic cell shuts off light when placed between two parallel polarizers and transmits light when placed between two orthogonal polarizers.

Next, when a voltage is applied to a twisted nematic cell, from a certain threshold voltage V_{th} the molecule axes begin to align themselves with the electric field. When the voltage applied is about twice V_{th} the majority of molecules have their axes realigned in the direction of the electric field, and the 90° optical rotatory power is eliminated. In this case, in completely the opposite fashion to the case when a voltage is not applied, between parallel polarizers light will be transmitted and between orthogonal polarizers light will be shut off.

In Fig. 2.7 the principle of a twisted nematic display is illustrated: it is based on the electrooptic effect when a twisted nematic cell is placed between orthogonal polarizers. In this case light is transmitted when the voltage is not applied and light is shut off when the voltage is applied. Between parallel polarizers, however, the relationship of transmitting or shutting off light is reversed. Currently the most widely used LCD devices are based on this twisted nematic principle and provide a black display on a white background or a white display on a black background.

*The direction of polarization of incident light coincides with the direction of the liquid crystal director at the electrode substrate on the entry side.

Table 2.2 *Typical liquid crystal display modes*

Display type	Principle	Liquid crystal material: Type	Liquid crystal material: Sign of $\Delta\varepsilon$	Number of polarizers	Threshold voltage (V)	Operating voltage (V)	Power comsumption ($\mu W/cm^2$)	Liquid crystal molecular orientation change: Initial orientation → voltage (heat) applied orientation	Optical phenomenon
Twisted nematic (TN)	Optical rotation	N	Positive	2	1–3	2–5	a few to several tens	Twisted → perpendicular	Brightness change
Guest–host (GH)	Dichroism	N with dye additive	Positive Negative	1	1–6	4–10		Homogeneous → perpendicular Homeotropic → parallel	Colour intensity change
		Ch with dye additive	Positive Negative	Not necessary	2–8	3–15		Focal conic → homeotropic Homeotropic → focal conic	

Electrically controlled birefringence (ECB)	Optical interference	N	Negative Positive Positive (negative)	2	1–5	1–10		Homeotropic → parallel Homogeneous → perpendicular Hybrid → perpendicular (parallel)	Colour change
Phase change (PC)	Light scattering	Ch	Positive	Not necessary	3–10	5–20		Focal conic → perpendicular	Turbid to transparent trans- formation
Dynamic scattering (DS)	Light scattering	N with electrically conductive additive	Negative	Not necessary	5–10	7–30	few hundred	Homeotropic → random Homogeneous → random	Transparent to turbid trans- formation
Thermal effects	Light scattering	S Ch	Positive Positive	Not necessary	10–20	15–40	a few to several tens	Homeotropic → focal conic	Transparent to turbid trans- formation

Note: N = nematic liquid crystal; S = smectic liquid crystal; Ch = cholesteric liquid crystal.

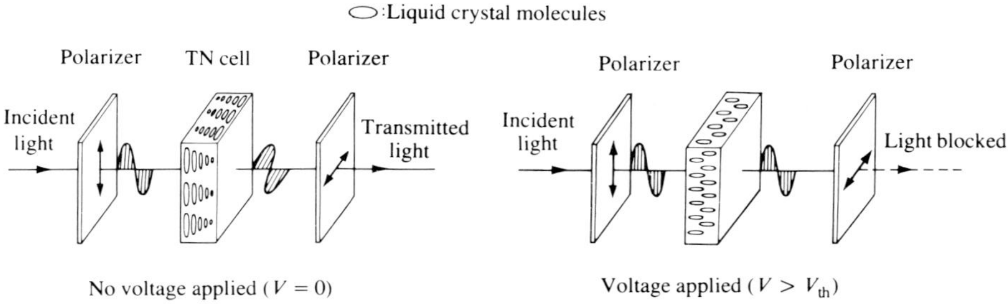

***Figure 2.7** Principle of twisted nematic (TN) display [1]*

The threshold voltage V_{th} of this type of liquid crystal display is given, following expression (2.8), by the following expression, which includes the dielectric anisotropy $\Delta\varepsilon$ of the N_p liquid crystal and the elastic moduli k_{11}, k_{22}, and k_{33}.

$$V_{th} = \pi \sqrt{\frac{k_{11} + (k_{33} - 2k_{22})/4}{\varepsilon_0 \cdot \Delta\varepsilon}} \tag{2.9}$$

From expression (2.9) it will be seen that V_{th} can be varied by choosing appropriate dielectric anisotropy and elastic moduli. Generally for twisted nematic LCDs the value of V_{th} is between 1 and 3 V.

Super-twisted nematic LCDs, called STN or SBE type LCDs, have a twisting angle larger than about 180°, and are based on the optical interference effect.

2.3.3 Guest–host (GH) type

When a dichroic dye* (the guest) that has anisotropy in the absorption of visible light in the molecular long axis and short axis directions is in solution in a liquid crystal (the host) with a certain molecule orientation, the molecules of the dichroic dye align themselves parallel to the liquid crystal molecules. Thus, if the molecule orientation of the host liquid crystal is changed by the application of an electric field, the molecule oreientation of the guest dye will change in the same way, and thus the amount of visible light absorbed by the dichroic dye can be controlled electrically. A display using this type of electrooptic effect is termed a guest–host type. This type can be subclassified into a number of types depending on the combination of molecular orientation type, liquid crystal material, and dye used, and also cell construction.

Figure 2.8 shows the principle of a guest–host display based on the electrooptic effect of a guest–host cell of homogeneous orientation (see Section 2.5.3) in which a *p*-type dichroic dye dissolved in an N_p liquid crystal is sandwiched between

*A dye that has a transition moment for light absorption in the molecular long axis direction is termed a *p* (positive) type dichroic dye and one with a transition moment in the molecular short axis direction is termed an *n* (negative) type dichroic dye.

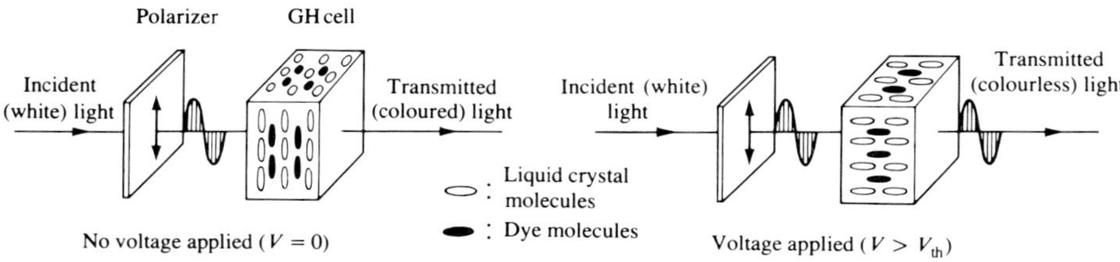

Figure 2.8 Principle of guest–host (GH) display [10]

transparent electrode substrates. When a voltage is not applied, the direction of oscillation of white linearly polarized light from the polarizer coincides with the light absorption axis of the dye, so that light is absorbed and the transmitted light is coloured. When, on the other hand, a voltage is applied, the direction of oscillation of the linearly polarized light is orthogonal to the absorption axis of the dye, and in principle there is no light absorption and the transmitted light is not coloured.

In Fig. 2.8, if a nematic liquid crystal (N_n liquid crystal) for which $\Delta\varepsilon < 0$ is used in place of the N_p liquid crystal and an n-type dichroic dye is used in place of the p-type dichroic dye, then when a voltage is not applied the transmitted light is not coloured, whereas when the voltage is applied the transmitted light is coloured. On the other hand, if a smectic liquid crystal or cholesteric liquid crystal is used in place of the nematic liquid crystal as the host, or if a two-layer construction for the nematic liquid crystal cell is used, then the polarizer can be dispensed with.

The threshold voltage V_{th} for a colour LCD using the guest–host principle is shown in expressions (2.13) and (2.12) below in the cases respectively that the host liquid crystal is an N_p liquid crystal or an N_n liquid crystal. When a cholesteric liquid crystal is used, then V_{th} is given by a transform of expression (2.8) as follows [14]:

$$V_{th} = \frac{d\pi^2}{p_o}\sqrt{\frac{k_{22}}{\varepsilon_0 \cdot |\Delta\varepsilon|}} \qquad (2.10)$$

where p_0 is the helical pitch length of the cholesteric liquid crystal, k_{22} is the twist modulus of elasticity, $\Delta\varepsilon$ is the dielectric anisotropy, and d is the cell thickness.

2.3.4 Electrically controlled birefringence (ECB) type

This display technique is based on controlling the birefringence of a liquid crystal cell by the application of an electric field, and is very important as a multicolour liquid crystal display technique. There are a number of types of this display depending on the molecule orientation of the liquid crystal cell used; these include the DAP type, homogeneous type, and HAN type.

A cell with homeotropic (DAP) orientation is made with an N_n liquid crystal sandwiched between transparent electrode substrates (see Section 2.5.3). This

DAP cell is placed between crossed polarizers, and light impinges perpendicularly on it as in the construction shown in Fig. 2.9. When an electric field is not applied, the incident linearly polarized light does not undergo birefringence when passing through the cell, so that it cannot pass through the second polarizer. On the other hand, when a voltage is applied, the molecular axis of the liquid crystal is inclined through an angle φ, so that the linearly polarized incident light becomes elliptically polarized by the effect of birefringence (see Section 2.2.2) and a portion of the light is able to pass through the second polarizer (the detector), and the light passing through is coloured.

The intensity I of the light passing through the detector is given by [12]

$$I = I_0 \sin^2 2\theta \sin^2 \frac{\pi d \, \Delta n(V)}{\lambda} \tag{2.11}$$

where I_0 is the intensity of incident light, θ is the angle between the direction of polarization of the incident light and the direction of oscillation of the ordinary light within the cell, d and $\Delta n(V)$ are respectively the cell thickness and the double refractive index at the applied voltage V, and $d\Delta n$ is the optical phase difference, and λ is the wavelength of the incident light. As will be seen from expression (2.11), the intensity of light transmitted depends on the magnitude of the voltage applied, and also on the wavelength of the incident light. Therefore, when the incident light is white light the light passing through the detector will be coloured by the interference effect variously, according to the magnitude of the applied voltage.

The threshold voltage V_{th} for a multicolour LCD using the DAP tecnique is found, following expression (2.8), as follows:

$$V_{th} = \pi \sqrt{\frac{k_{33}}{\varepsilon_0(-\Delta\varepsilon)}} \tag{2.12}$$

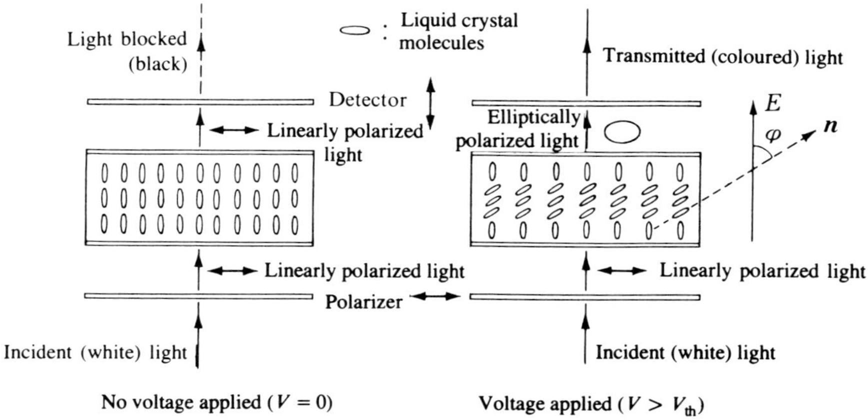

Figure 2.9 Principle of ECB (DAP) display [1]

Here $\Delta\varepsilon$ is the dielectric anisotropy of the N_n liquid crystal and k_{33} is the bending elasticity coefficient.

In the case of the homogeneous technique [15], the liquid crystal cell of Fig. 2.9 uses an N_p liquid crystal of homogeneous orientation. Therefore, with this method the value of V_{th} is obtained as follows, by replacing the k_{33} in expression (2.12) by the splay elasticity coefficient k_{11}:

$$V_{th} = \pi\sqrt{\frac{k_{11}}{\varepsilon_0 \cdot \Delta\varepsilon}} \tag{2.13}$$

There is no difference in principle between the homogeneous type and the DAP type, but compared with the DAP type, the order of the colour change is in the reverse direction when a voltage is applied, and additionally the threshold voltage value can be substantially reduced.

The HAN (hybrid aligned nematic) technique [16] is characterized by using a hybrid orientation cell (see Section 2.5.3) in which the molecules of the liquid crystal are aligned perpendicularly to one substrate, but parallel to the other substrate. Therefore for a colour LCD based on this technique, both N_p and N_n types of liquid crystals can be used, and since there is no clear threshold voltage, they have the advantage that a very low drive voltage can be used.

2.3.5 Phase change (PC) type

The phase change display technique is based on a change in molecular orientation from the helical cholesteric phase to the homeotropic nematic phase, and vice versa. For this technique cholesteric liquid crystals with long helical pitch and $\Delta\varepsilon > 0$ or $\Delta\varepsilon < 0$ (Ch_n or Ch_n liquid crystals) can be used.

Figure 2.10 illustrates the principles of the phase change type display using a Ch_p liquid crystal focal conic orientation cell (see Section 2.5.3). When a voltage is not applied, the incident light passing through the cell is optically dispersed in a focal conic orientation, and the cell appears cloudy. When, however, a voltage of at least a certain threshold value V_{th} is applied, the helical structure of the cholesteric phase disintegrates, and a nematic phase of homeotropic orientation appears. Therefore when a voltage is applied the cell becomes transparent.

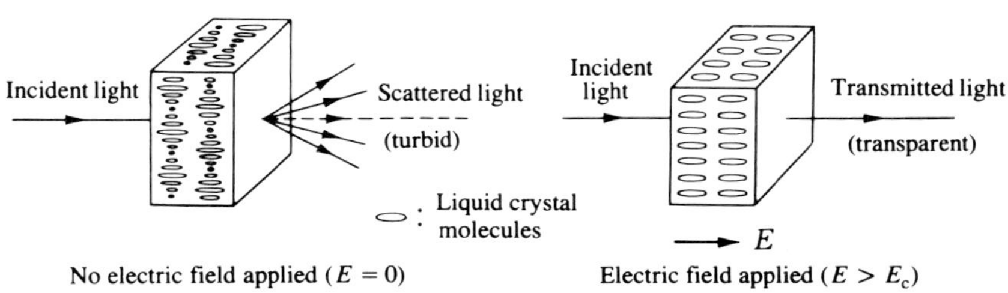

***Figure 2.10** Principle of phase change (PC) display [1]*

It is also possible to use a Ch_p liquid crystal planar orientation cell (see Section 2.5.3), but in any case the value of V_{th} is given by expression (2.10).

2.3.6 *Dynamic scattering (DS) type*

A homogeneous orientation liquid crystal cell (see Section 2.5.3) is made by sandwiching a layer of thickness approximately 10 μm of an N_n liquid crystal, which has a resistivity not more than approximately $5 \times 10^{10}\,\Omega\cdot$cm and in which is dissolved an electrically conductive substance, between transparent electrode substrates. When a voltage V, which is either d.c. or is a.c. of not more than a certain frequency f_c,* is applied to the cell, as shown in Fig. 2.11, when V is not more than a certain voltage V_W ((a) in the figure), then incident light is transmitted straight through the cell which is thus transparent. When V exceeds V_W ((b) in the figure) then eddy currents occur in the liquid crystal, and a pattern of fine lines termed Williams domains is observed.

Next, when the applied voltage increases so that $V \gg V_W$ ((c) in the figure), the eddy currents give way to turbulence, and the orientation of the liquid crystal molecules then becomes random. Thereby a very large number of very small birefringent domains are produced, with orientations different from surrounding domains, and light is strongly scattered at the boundaries of these domains in the so-called dynamic scattering phenomenon. This electrooptic effect is also found in homeotropic orientations, and is the operating principle of dynamic scattering LCD.

2.3.7 *Thermal effect types*

When the temperature of a liquid crystal is changed, phenomena whereby the optical characteristics change are called thermooptic effects, and are due to the change in molecular orientation in the liquid crystal when it is heated or cooled. We discuss an example with smectic liquid crystals.

A smectic A (S_A) liquid crystal is sandwiched between transparent electrodes that have been subject to vertical orientation processing (see Section 2.5.3) to form a liquid crystal cell. If this cell is heated until the isotropic liquid phase is reached, and is then cooled, then if the cooling is sudden the whole cell will become cloudy, whereas if the cooling is gradual the whole cell will become transparent. This cloudy state and transparent state correspond respectively to the focal conic orientation and homeotropic orientation of the S_A liquid crystal. This thermooptic effect of an S_A liquid crystal cell can be used as the principle for a large-capacity display in which a laser beam is used to write the image

*f_c is termed the cut-off frequency for the dynamic scattering effect, and is proportional to the conductivity of a liquid crystal.

† V_W indicates the threshold voltage (of the order of 10 V) for Williams domains.

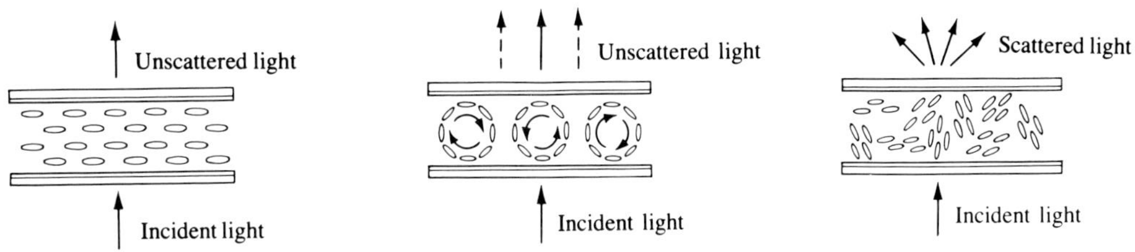

(a) Homogeneous molecule orientation $V < V_w$ (b) Williams domains $V > V_w$ (c) Dynamic scattering (DS) $V > V_w$

Figure 2.11 *Dynamic scattering (DS) phenomenon and changes in liquid crystal molecule orientation [17]*

thermally, and can also be used as a principle for a memory type display in which one of the transparent electrodes of the cell is treated as the heating electrode and the other as the signal electrode, and an electric field is applied at the same time as heating (see Section 2.6.5).

2.3.8 Response times of LCDs

For an LCD based on one of the display modes using nematic liquid crystals—twisted nematic, guest–host, electrically controlled birefringence, and dynamic scattering types—the rise time response speed τ_r and decay time response speed τ_d are given as follows [18]:

$$\tau_r = \eta_i d^2 (\varepsilon_0 \cdot |\Delta\varepsilon| V^2 - k_{ii}\pi^2)^{-1} \tag{2.14}$$

$$\tau_d = \frac{\eta_i d^2}{\pi^2 k_{ii}} \tag{2.15}$$

where η_i and k_{ii} are the appropriate viscosity parameter and elasticity parameter, $\Delta\varepsilon$ is the dielectric anistropy, d is the cell thickness, and V is the applied voltage.

For the phase change and other types of display mode using cholesteric liquid crystals, the values of τ_r and τ_d for an LCD are given as follows in the case that the

Table 2.3 Response times and parameters η_i and k_{ii} for different display modes [1]

		Response times (ms)		Parameters	
Display type	$\Delta\varepsilon$	τ_r	τ_d	k_{ii}	η_i
TN type	+ ve	20–80	60–150	$k_{11} + \frac{(k_{33} - 2k_{22})}{4}$	$\sim\eta_1$
GH type	+ ve	70–150	90–200	$\sim k_{11}$	$\sim\eta_1$
	− ve	100–200	150–300	$\sim k_{33}$	$\sim\eta$
ECB type	+ ve	10–30	30–100	k_{11}	η_1
	− ve	20–50	50–200	k_{33}	η_2
PC type	+ ve	10–40	60–200	k_{22}	η_2
DS type	− ve	10–50	30–150	k	η

helical pitch p_0 is shorter than the thickness of the cell d [18]:

$$\tau_r = \eta_i \left[\varepsilon_0 \cdot |\Delta\varepsilon| \left(\frac{V}{d} \right)^2 - \frac{K_{ii}\pi^2}{p_0^2} \right]^{-1} \tag{2.16}$$

$$\tau_d = \frac{\eta_i p_0^2}{\pi^2 k_{ii}} \tag{2.17}$$

The viscosity parameter η_i and elasticity parameter k_{ii} in expressions (2.14) to (2.17) vary depending on the display type. Table 2.3 shows the η_i and k_{ii} for individual display types, together with approximations to the response times τ_r and τ_d.

2.4 LIQUID CRYSTAL MATERIALS

2.4.1 Practical liquid crystal materials

The most important requirement at the beginning of the development of liquid crystal materials for display purposes was for a liquid crystal material that could be used at room temperature. Thereafter, as liquid crystal displays entered the tactical phase, the range of applications widened and a variety of new liquid crystal display modes were developed; in addition to the liquid crystal temperature criterion the following items were the principal criteria in making practical liquid crystal materials:

1. Exhibit the liquid crystal phase from low temperatures to high temperatures, giving a wide working temperature range.
2. Good chemical and photochemical stability, and long life.
3. Low viscosity and high response speed.
4. Large dielectric anisotropy; applicable to low-voltage operation.
5. The magnitude of the double refraction index is appropriate for increased display contrast.
6. The elasticity coefficients are balanced to allow multiplex driving.
7. The degree of ordering of molecule orientation is high, allowing increased display contrast.

These objectives are common to all the modes of liquid crystal display devices, but naturally depending on the display technique there will be variations in the type of liquid crystal used, whether the dielectric anisotropy is positive or negative, the magnitude of the double refraction index, and the necessity or not for additives (refer to Table 2.2).

Thus various characteristics, functions and qualities are required of practical liquid crystal materials, and in order to answer these requirements a wide variety of liquid crystal compounds have been synthesized to date (refer to Sections 2.4.3

Table 2.4 *General molecular structures for liquid crystal compounds*

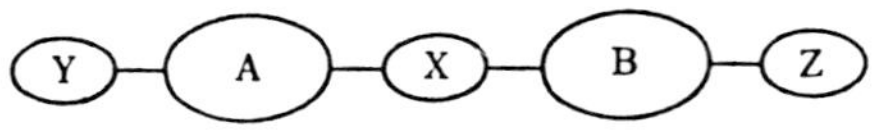

Central group X	Terminal groups Y, Z	Ring roups A, B
—CH=N—	R—	
—N=N(O)—	RO—	
—N=N—	RCO—	
—CO—O—	ROCO—	
—CO—S—	RCO—O—	
—CH=N(O)—	ROCO—O—	
—CH=C(CN)—	—COOH	
—CH=C(Cl)—	—CN	
—$(CH_2)_n$—	—NO_2	
—CH=CH—	—Cl	
—C≡C—	—Br	
——	—I	
—CO—CH_2—	—F	
—CH_2—O—	R*—	
—O—$(CH_2)_2$—O—	R*O—	
—CH=CH—CO—O—		

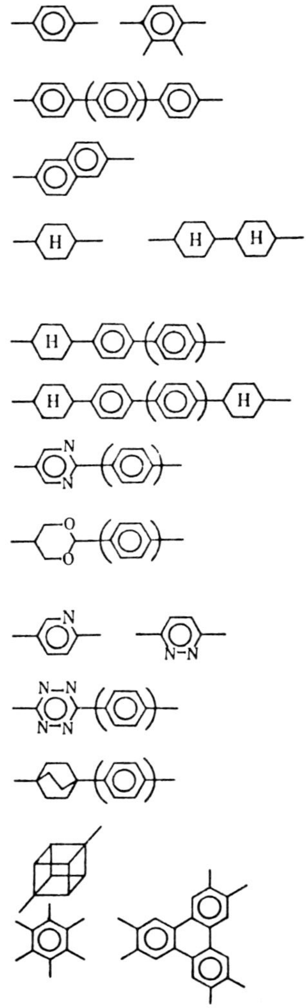

Cholesteryl ring ($C_{27}H_{45}$)

R: alkyl group, : benzene ring, 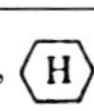: *trans*-cyclohexane ring, R*: optically active alkyl group

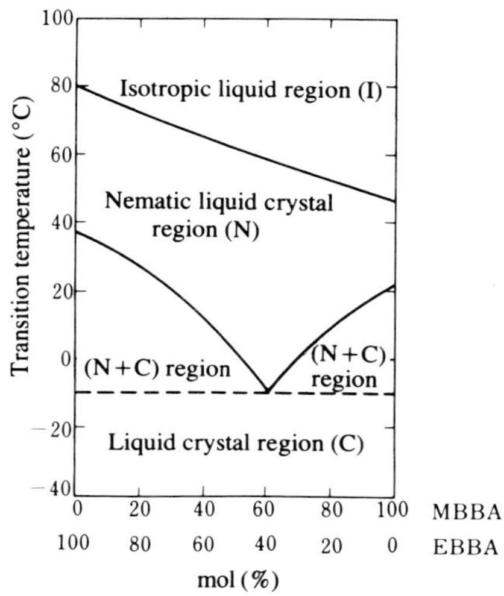

Figure 2.12 *Phase diagram for MBBA–EBBA composite liquid crystal material [19]*

to 2.4.5) [1]. The general molecular structures of these basic liquid crystal compounds are summarized in Table 2.4. By trying various combinations of the central group, terminal groups, and ring groups, practical liquid crystal compounds can be designed and synthesized.

2.4.2 Composite liquid crystal materials

In practice it is difficult to find a single liquid crystal compound that has all the characteristics required for a display. Therefore, for application purposes, rather than a single liquid crystal material, a composite blend of materials is used to provide the characteristics and qualities required for a particular display mode. Depending on the type of display, electrically conductive materials, dichroic dyes, chiral materials, and so forth may also be added.

This technique of using a blended composition is necessary not only to obtain the sum of the characteristics of individual liquid crystal materials but also in order to produce characteristics that cannot be obtained by the use of a single liquid crystal material. A good example of this is the increase in the temperature range for the liquid crystal phase that can be obtained from a composite material, which is described next.

Figure 2.12 shows the phase diagram for a composite liquid crystal material formed from MBBA* and EBBA† liquid crystals. This composite material has a eutectic point at a mol composition where MBBA/EBBA = 60/40, and for this composition the nematic liquid crystal range extends from −10 to +60°C.

* *p*-Methoxybenzylidene-*p*′-butylaniline (see Section 2.4.3).
† *p*-Ethoxybenzylidene-*p*′-butylaniline (see Section 2.4.3).

2.4.3 *Nematic liquid crystal materials*

(1) Schiff bases

MBBA (*p*-methoxybenzylidene-*p*′-butylaniline) was the first room temperature liquid crystal material; this is a typical example of the Schiff base compounds shown in Table 2.5. Liquid crystal substances of the same family such as MBBA and EBBA (*p*-ethoxybenzylidene-*p*′-butylaniline) have negative dielectric anisotropy ($\Delta\varepsilon$ approximately -0.5), large birefringence (Δn approximately 0.25), and medium viscosity (η approximately 35 cSt, and are used for dynamic scattering and electrically controlled birefringence type LCD devices. For a twisted nematic LCD, a liquid crystal material with $\Delta\varepsilon > 0$ is required, and the first liquid crystal compositions used for this were Schiff base compounds with cyano terminal groups. Although their viscosity is high, their dielectric anisotropy is large ($\Delta\varepsilon = 15$–20), so they are appropriate for low-voltage operation. However, all of the Schiff base compounds are unstable in the presence of moisture, which is an obstacle to practical implementation.

(2) Benzoate esters

These liquid crystal compounds have benzene rings joined by an ester group, and as shown in Table 2.6 encompass a large number of materials with widely varying characteristics. Since they also have good chemical stability, they are used in a range of LCD techniques as components of composite liquid crystal materials. Compounds with both terminal groups being alkyl groups or alkoxy groups show changes in the sign of the dielectric anisotropy with a small difference in these terminal groups, and the absolute value of $\Delta\varepsilon$ is less than 1. When both terminal groups are alkyl groups then the viscosity is low (η approximately 20 cSt), but in other cases the viscosity is always high (η at least 45 cSt).

The temperature dependency of physical characteristics of these liquid crystal compounds is low, so various N_p liquid crystals and mixtures thereof are useful

Table 2.5 *Schiff base liquid crystal compounds*

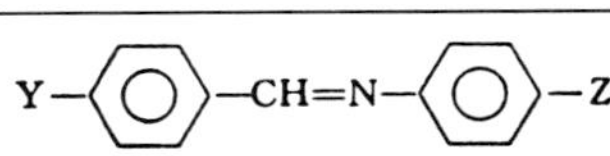

		Transition temperature (°C)		
Y	Z	C–N	N–I	$\Delta\varepsilon$
CH_3O—	C_4H_9—	22	47	−
C_2H_5O—	C_4H_9—	37	80	−
C_3H_7—	—CN	65	77	+
C_4H_9O—	—CN	65	108	+

Table 2.6 *Benzoate ester liquid crystal compounds*

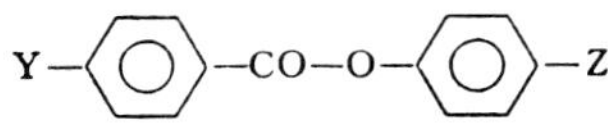

Y	Z	Transition temperature (°C) C–N	Transition temperature (°C) N–I	$\Delta\varepsilon$
CH_3O—	$C_6H_{13}O$—	55	77	−
CH_3O—	C_5H_{11}—	29	42	+
C_5H_{11}—	C_5H_{11}—	(33)	(12)	+
C_6H_{13}—	—CN	45	47	+
C_5H_{11}—  —COO—	C_5H_{11}—	78	179	+
C_4H_9—	—  —CN	108	225	+

Note: Figures in ( ) are monotropic transitions.

for multiplex driving. Liquid crystal compounds with a terminal cyano group have $\Delta\varepsilon$ positive and very large ($\Delta\varepsilon > 20$), and are used as liquid crystal material for low-voltage drive. On the other hand, liquid crystal compounds having three or more benzene rings show high viscosity, and the N–I transition temperature is very high. They are therefore useful as components in liquid crystal mixtures for increasing the liquid crystal temperature range.

(3) Biphenyl and terphenyl compounds

Cyanobiphenyl liquid crystal compounds that have an alkyl or alkoxy terminal group are colourless and have extremely good chemical stability and photochemical stability. Furthermore, they have balanced characteristics for dielectric anisotropy ($\Delta\varepsilon$ approximately 13), viscosity (η approximately 35 cSt), and birefringence (Δn approximately 0.2), and are representative N_p liquid crystal materials for use in a wide range of LCD devices (Table 2.7).

The practical use of cyanobiphenyl liquid crystal is often in combination with cyanoterphenyl liquid crystals as an additive in order to expand the nematic liquid crystal temperature range, increase the birefringence, and improve the multiplex drive characteristics.

(4) Cylohexylcarboxylic ester compounds

The cyclohexylcarboxylic ester liquid crystal compounds shown in Table 2.8 have a fairly low viscosity and a relatively wide range of liquid crystal temperatures. N_n liquid crystal compounds having a terminal group Z which is an alkyl group or alkoxy group have a very low viscosity ($\eta < 20$ cSt), so that they

Table 2.7 *Biphenyl and terphenyl liquid crystal compounds*

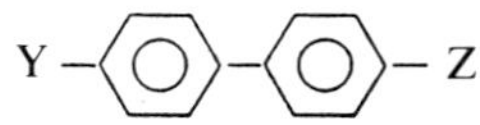

Y	Z	Transition temperature (°C) C–N	N–I	Δε
C_5H_{11}—	—CN	24	35	+
C_6H_{13}—	—CN	14	29	+
$C_5H_{11}O$—	—CN	48	68	+
$C_7H_{15}O$ –	– CN	54	74	+
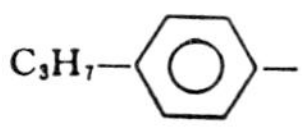	—CN	182	257	+

Table 2.8 *Cyclohexylcarboxylic ester liquid crystal compounds*

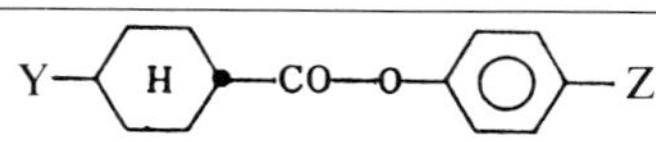

Y	Z	Transition temperature (°C) C–N	N–I	Δε
C_4H_9—	C_6H_{13}—	26	31	—
C_5H_{11}—	C_5H_{11}—	37	47	—
C_3H_7—	C_2H_5O—	47	78	—
C_4H_9—	$C_5H_{11}O$—	29	66	—
C_3H_7—	—CN	54	69	+
C_2H_5—	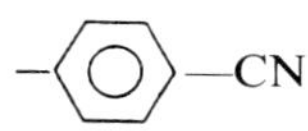	85	229	+

are used as a constituent of low viscosity liquid crystal mixtures. They can also be used as constituents to improve the multiplex drive characteristics. The N_p liquid crystal compounds having a cyano terminal group have low birefringence (Δn approximately 0.12) and low dielectric anisotropy ($\Delta\varepsilon$ approximately 8).

On the other hand, liquid crystal compounds with a terminal group that is a cyanophenyl group have a wide liquid crystal temperature range and low viscosity, and so are often used to extend the temperature range of an N_p liquid crystal material.

(5) Phenylcyclohexane and biphenylcyclohexane compounds

These liquid crystal compounds shown in Table 2.9 combine the stability shown by biphenyl liquid crystals with the low viscosity shown by cyclohexylcarboxylic

Table 2.9 *Phenylcyclohexane and biphenylcyclohexane liquid crystal compounds*

Y	Z	Transition temperature (°C)		Δε
		C–N	N–I	
C_3H_7—	—CN	43	45	+
C_5H_{11}—	—CN	30	55	+
C_5H_{11}—	 —CN	95	219	+
C_3H_7—	C_2H_5O—	(41)	(37)	—
C_3H_7—	C_2H_5—	(0)	(− 70)	+

Note: figures in ( ) are monotropic transitions.

ester liquid crystal and are therefore extremely useful as materials for liquid crystal displays. The series of cyanophenylcyclohexane liquid crystal compounds have reasonable dielectric anisotropy and extremely low viscosity, and so are useful to liquid crystal displays requiring high response speed. Cyanobiphenylcyclohexane liquid crystals have a high N–I transition temperature, and are therefore used to extend the liquid crystal temperature range. Phenylcyclohexane liquid crystal compounds which have both end groups that are alkyl or alkoxy groups are monotropic transition liquid crystals, but have extremely low viscosity, and are therefore very important as viscosity-lowering additives for composite liquid crystal materials.

(6) Pyrimidine and dioxane compounds

Typical examples of these liquid crystal compounds are shown in Table 2.10. The cyanophenylpyrimidine liquid crystal compounds having a cyano group at one end have a high dielectric anisotropy (Δε approximately 25) and a wide liquid crystal temperature range, so are used as liquid crystal materials for wide temperature range and low-voltage operation. On the other hand, phenylpyrimidine liquid crystal compounds which have both terminal groups that are alkyl groups or alkoxy groups have a very low elastic coefficient ratio (k_{33}/k_{11}), and therefore are used as principal components of liquid crystal materials for multiplex drive. Cyanophenyldioxane liquid crystal compounds generally have a monotropic transition and high viscosity (η approximately 70 cSt), but show high dielectric anisotropy (Δε approximately 18) and relatively small elastic coefficient ratio (k_{33}/k_{11}); again they are useful as liquid crystal materials for multiplex drive.

Table 2.10 *Pyrimidine and dioxane liquid crystal compounds*

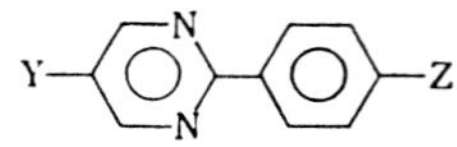

		Transition temperature (°C)		
Y	Z	C–N	N–I	Δε
C_7H_{15}—	—CN	44	50	+
C_4H_9— 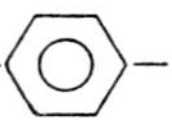 —	—CN	94	246	+
C_6H_{13}—	$C_6H_{13}O$—	31	60	+
C_6H_{13}—	$C_9H_{19}O$—	37	61	+

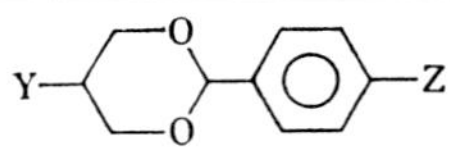

C_4H_9—	—CN	(42)	(35)	+
C_6H_{13}—	—CN	(47)	(40)	+
C_6H_{13}— 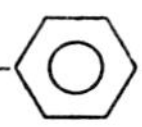 —	—CN	84	151	+

Note: figures in ( ) are monotropic transitions.

2.4.4 Cholesteric liquid crystal materials

Cholesteric liquid crystal compounds are divided, as seen from Table 2.11, into two main groups: cholesterol derivative compounds, which have a cholesteryl group ($C_{27}H_{45}-$), and chiral nematic liquid crystals, which are ordinary nematic liquid crystal compounds with a terminal group replaced by an optically active branching alkyl or alkoxy group such as 2-methylbutyl (2MB) or 2-methylbutoxy (2MBO) groups. These cholesteric liquid crystals have characteristic optical properties such as optical rotatory power, selective light reflection, and circular dichroism, and these optical properties are very sensitive to changes in temperature and applied voltages. Such phenomena have therefore been used to develop various applications such as colour displays.

The following two purposes are important applications which make direct use of the helical structure of cholesteric liquid crystals:

1. By adding a very small amount of cholesteric liquid crystal to a nematic liquid crystal in a twisted nematic LCD, reverse twist molecular orientation defects are prevented.

Table 2.11 *Typical cholesteric liquid crystal compounds*

Molecualr structure	Transition temperature (°C)	
	C–Ch	Ch–I
Cholesterol acetate $CH_3COO—(C_{27}H_{45})$	114	(90)
Cholesterol palmitate $C_{15}H_{31}COO—(C_{27}H_{45})$	77	146
Cholesterol benzoate $—COO—(C_{27}H_{45})$	150	182
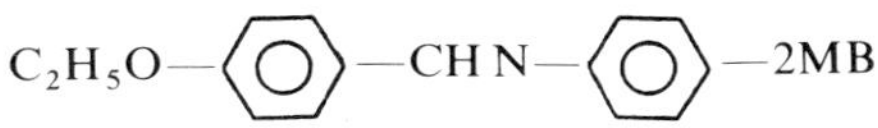	25	58
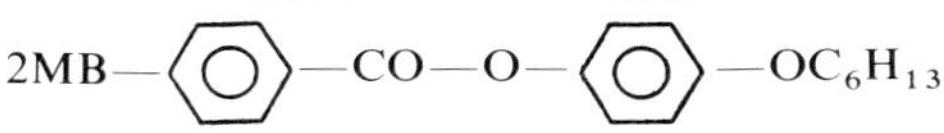	24	30
	53	(9)

Note: figures in ( ) are monotropic transitions; 2MB = 2-methylbutyl group; 3MBO = 3-methylbutoxy group.

2. By adding a small quantity of cholesteric liquid crystal to a normal nematic liquid crystal, a liquid crystal material with long helical pitch can be fabricated for use in a phase change type LCD.

2.4.5 Smectic liquid crystal materials

Smectic liquid crystals are classified on the basis of very fine differences in molecular orientation structure as A, B, C, ..., H variants. Of these, smectic A liquid crystals and smectic C liquid crystals are relatively susceptible to external fields; the former is particularly used as an LCD based on a thermooptic effect. Table 2.12 shows examples of smectic A (S_A) liquid crystal compounds.

Some kinds of chiral smectic liquid crystals, typically chiral smectic C materials, show a spontaneous polarization phenomenon, so they are useful as ferroelectric LCDs based on the interactive force between spontaneous polarization and electric field.

2.5 STRUCTURE AND MANUFACTURE OF LIQUID CRYSTAL DISPLAYS

2.5.1 LCD structure

Liquid crystal displays have a wide range of dimensions depending on the applications; these range from an area of a few square millimetres or a few square

Table 2.12 *Smectic liquid crystal compounds*

Molecular structure	Transition temperature (°C)		
	C–S_A	S_A–N	S_A[N]–I
	21	33	(40)
	44	—	50
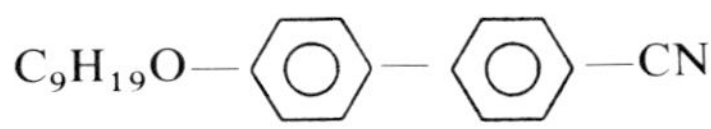	64	77	(80)
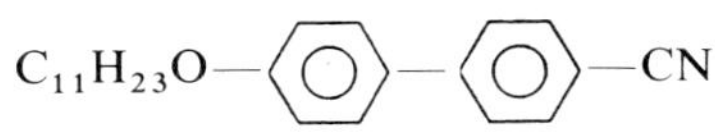	71	—	87

centimetres for wrist watches or pocket calculators to hundreds of square centimetres for vehicle instrumentation panels or computer terminals. Figure 2.13 is an external view of a typical reflection type twisted nematic LCD for use as a numeric display in an industrial meter or other measuring instruments. The external dimensions of this LCD are overall length 71 mm, overall width 31 mm, and overall thickness 3 mm, with an effective display area of $16 \times 63\ mm^2$.

Figure 2.14 shows the sectional construction of this LCD. A liquid crystal layer of thickness about 10 μm is sandwiched between two glass substrates which have transparent electrodes and molecular orientation layers (see Section 2.5.3), and

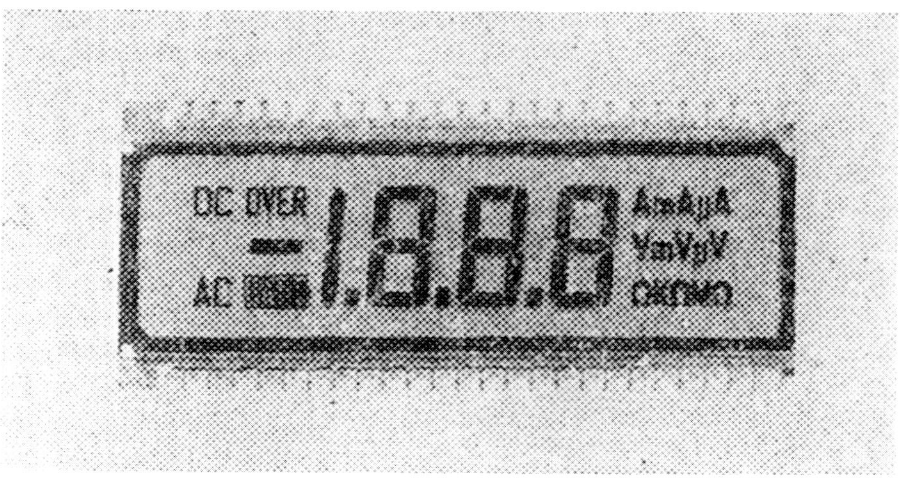

Figure 2.13 *Exterior view of reflective TN type LCD*

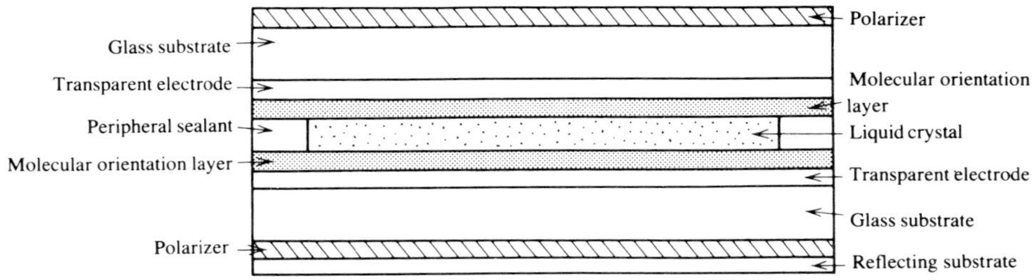

Figure 2.14 *Sectional structure of reflective TN type LCD*

the cell construction is completed by an air-tight sealing material around the periphery of the liquid crystal layer. Then a polarizer film is adhered to both sides of the cell, and a thin reflective layer is added to one side.

In displays using techniques that make polarizers unnecessary, or in LCDs for use with transmitted light, the polarizers or reflective layers can be omitted from the construction of Fig. 2.14.

2.5.2 Material and parts for LCDs

The principal constituents of a LCD are the liquid crystal material, the molecular aligning material, the transparent electrodes, the sealant, the spacer, the polarizers, the reflecting layer, and then various filters and connectors.

Liquid crystal materials have already been discussed in detail in Section 2.4 above. Molecular alignment materials needed to form the molecular alignment layers are described in the next section 2.5.3.

The transparent electrode substrates are usually glass plates with a film of indium oxide or tin oxide. Generally their optical transmission coefficient is at least 90 per cent and the surface resistivity is some tens or hundreds of ohms. The formation of the display pattern electrodes is mostly done by photoetching the whole transparent electrode substrate.

Thermosetting epoxy adhesive is widely used as a sealant, and in cases where reliability is particularly important glass soldering, for example, can be used. For spacing purposes, in order to control the cell gap, glass or plastic in granule or fibre form can be used.

For the polarizers that are necessary for the majority of LCD devices, a stretched polyvinyl alcohol film dyed with iodine or a dichroic dye is used, in sheet form and sandwiched between protecting layers of cellulose acetate. The optical transmission coefficient is between 40 and 50 per cent and the degree of polarization is approximately 90 per cent. The reflecting layer is often made integral with the polarizer.

The filters used are ultraviolet cut-off filters and coloured filters for colour displays. The electrical connections between the liquid crystal display and the drive circuits are provided by pin connectors, elastic connectors, or flexible connectors.

Table 2.13 *Classification of liquid crystal molecular orientations*

MOLECULAR ORIENTATION TYPE	HOMEOTROPIC	HOMOGENEOUS	TILT	HYBRID	TWIST	PLANAR	FOCAL CONIC
Orientation (diagrammatic)	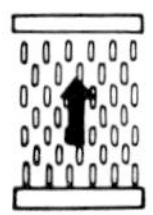	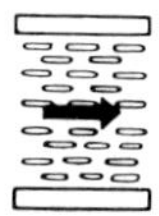			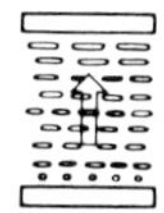	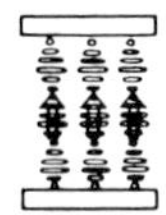	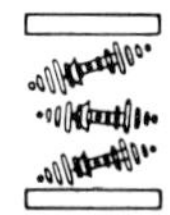
Substrate arrangement processing combination	$\perp/\perp$	$\parallel/\parallel$	$\angle/\angle$	$\perp/\parallel$	$\parallel/=$	$\parallel/\parallel$	$\perp/\perp$

Note: ⇔ liquid crystal molecules, ➡ director orientation, ⇨ helical axis orientation

2.5.3 *Liquid crystal molecular orientations and molecular alignment techniques*

(1) Types of molecular orientation and their roles

A uniform and stable liquid crystal molecular orientation is essential for an LCD. This is because all LCD techniques depend on the liquid crystal having an initial molecular orientation which is changed to a new orientation under the effect of an external field such as an electric field or heat.

There is a wide range of liquid crystal molecular orientations, but seven typical examples are shown in Table 2.13, and we list them here:

1. Homeotropic: the liquid crystal molecules are arranged perpendicular to both substrates.
2. Homogeneous: the liquid crystal molecules are parallel to both substrates and aligned in one fixed direction.
3. Tilt: the liquid crystal molecules are tilted at a certain angle to both substrates, and are aligned in one fixed direction.
4. Hybrid: the liquid crystal molecules are perpendicular to one substrate, and are parallel to the other substrate, all in the same direction; the overall molecular orientation curves continuously through 90° between the two substrates.
5. Twist: the liquid crystal molecules are arranged parallel to both substrates, but the alignment orientations on the two substrates are 90° apart, so that overall the molecular orientation twists continuously through 90° between the substrates.
6. Planar: the liquid crystals are aligned such that the helical axis is perpendicular to both substrates.
7. Focal conic: the orientation is such that the liquid crystal molecule helical axis is parallel to both substrates.

(2) Techniques for molecular alignment

A cell with any of the molecular orientations described above may be formed by using a suitable combination of two substrates and the three-molecule alignment processing techniques described next (see Table 2.13):

1. Perpendicular processing: the substrate surface is processed so that the liquid crystal molecule long axis is aligned perpendicular to the substrate ($\perp$).
2. Parallel processing: the substrate surface is processed so that the liquid crystal molecule long axis is aligned parallel to the substrate ($\parallel$).
3. Inclined alignment processing: the substrate surface are processed so that the liquid crystal molecule long axis is inclined at a constant angle to the substrate surface (V, IV)

Table 2.14 shows actual techniques for carrying out these three types of alignment processing [1]. In direct methods of substrate processing, the substrate

Table 2.14 *Molecular alignment processing methods*

Type	Process	Alighment material	Orientation type
Direct processing of substrate surface	Coating	Chromium perfluorononanoate complex, organic silane DMOAP, lecithin	Perpendicular
		Chromium brassylate complex, organic silane MAP, polyvinyl alcohol, polyimide	Parallel
Indirect processing of substrate surface	Solution	CTAB ammonium salt, octadecyl malonic acid	Perpendicular
		Dicarboxylic acid, varsamide 125	
Substrate surface transformation	Rubbing	Parallel alignment material	Parallel
—	Oblique evaporation	Silicon monoxide SiO, metals Au, Pt	Tilted (small evaporation angle) Parallel (large evaporation angle)

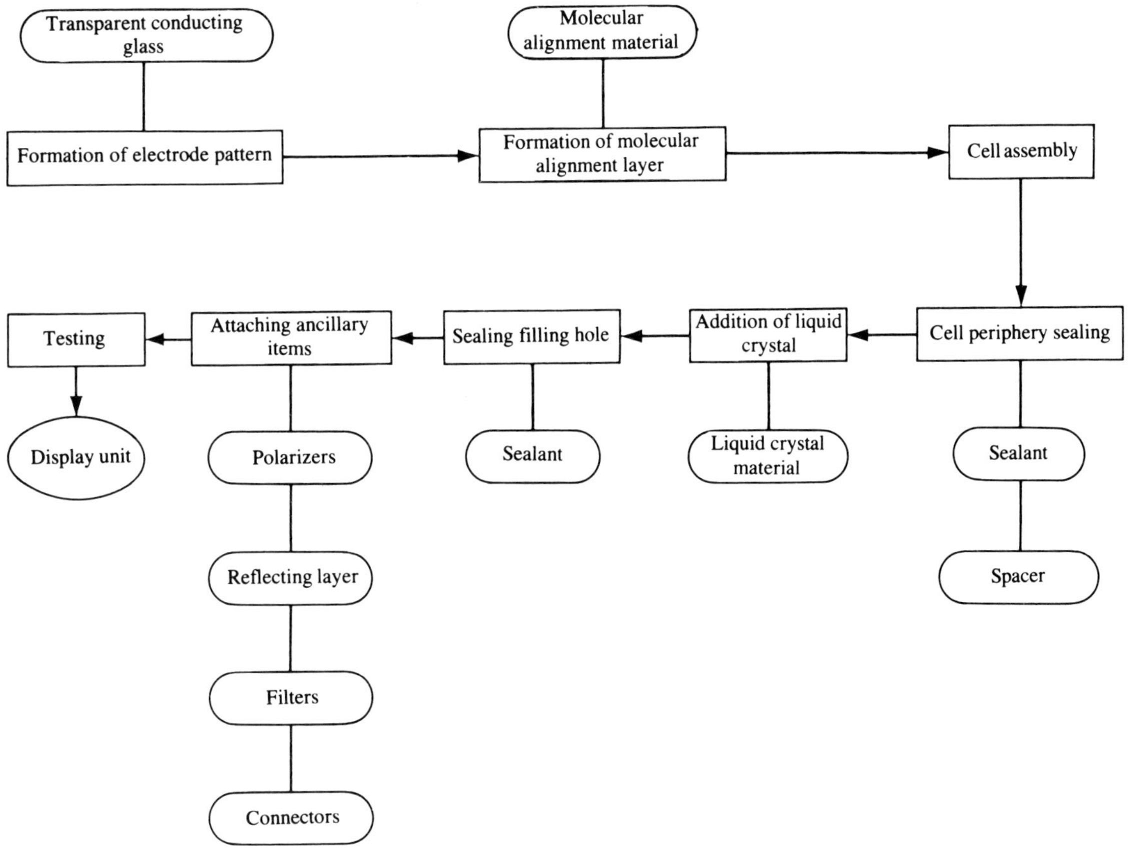

Figure 2.15 *Stages in manufacture of an LCD*

surface is coated with a solution of an aligning agent, the solvent is removed by heat drying or similar, and a thin film of the aligning agent is formed on the surface. In indirect methods of processing, the aligning agent is dissolved in the liquid crystal. When the liquid crystal is introduced into the cell the aligning agent is adsorbed onto the substrate surface from within the liquid crystal, thus forming the alignment layer.

One substrate surface transformation processing method consists of rubbing the surface of the substrate lightly in a single direction with a cotton cloth. Generally, after direct processing of the substrate surface with a parallel aligning agent, rubbing processing is carried out. This technique ensures that the parallel orientation of the liquid crystal molecules is definitely in a fixed direction. Another substrate surface transformation processing method consists of oblique evaporation of an aligning agent such as silicon monoxide on the substrate surface. When the angle of vapour evaporation θ* is shallow (θ from 5° to 20°) inclined orientation is obtained, while when the angle of vapour deposition is deep (θ between 20° and 45°) parallel orientation is obtained.

2.5.4 LCD manufacture

Figure 2.15 shows an outline of the stages in manufacture of a LCD. Of these stages, that which affects most greatly the device characteristics and quality is the formation of the liquid crystal molecular alignment layer. As described in Sections 2.3 and 2.5.3, different molecular orientations and molecule aligning methods are needed for different display types, and the surface processing for a particular orientation is carried out on the patterned electrode substrate surface.

The cell gap between electrodes is normally set to be approximately 10 μm, and this cell gap can be controlled either by dispersing some spacer material between the electrodes during the assembly of the cell or by mixing a spacer material in with the sealant. The introduction of the liquid crystal material to the empty cell is done under reduced pressure in a bell jar, and either the filling aperture of the empty cell is immersed in the liquid crystal or the liquid crystal material is introduced by a dropping method. After the various ancillary parts are fitted, the liquid crystal display device will undergo external inspection, display operation testing, reliability testing, mechanical strength testing, and other checks before becoming a finished product.

2.6 DRIVE TECHNIQUES FOR LIQUID CRYSTAL DISPLAYS

2.6.1 Drive electrode structure

The electrode constructions used in liquid crystal displays are classified in Table 2.15, together with their principal applications. The most common

*Angle of the vapour deposition beam to the surface of the substrate.

Table 2.15 LCD electrode constructions and their principal applications

LCD electrode constructions
Segmented electrodes
(numeric displays, analogue bar graph displays)
Fixed pattern electrodes
(symbol and pattern displays)
Matrix electrodes
(character, graphics, and video displays)

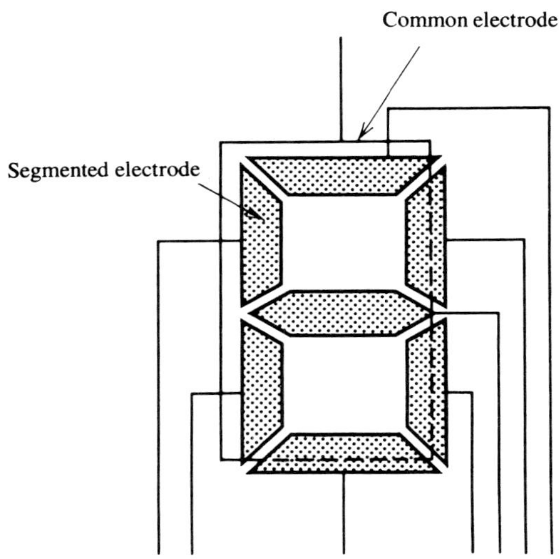

Figure 2.16 Seven-segment electrode construction

segmented electrode is the seven-segment pattern shown in Fig. 2.16, which has seven individual segment electrodes on the front substrate and a common electrode on the back substrate. A voltage is applied selectively to each of the segments so that any of the digits 0 to 9 can be displayed. For an analogue bar graph display a number of segment electrodes are arranged in a line. When a special pattern or symbol is to be displayed an electrode of the appropriate shape is used.

Next, Fig. 2.17 shows the matrix electrode construction which is used when the display is to be capable of an arbitrary pattern. One substrate has strip (scan) electrodes X_n and the other substrate strip (signal) electrodes Y_m, so by means of selectively applying a voltage to one scan electrode and one signal electrode, any intersection point (pixel) can be activated, and character displays, graphic displays, and video displays can be obtained.

The principal drive techniques actually used for liquid crystal displays using

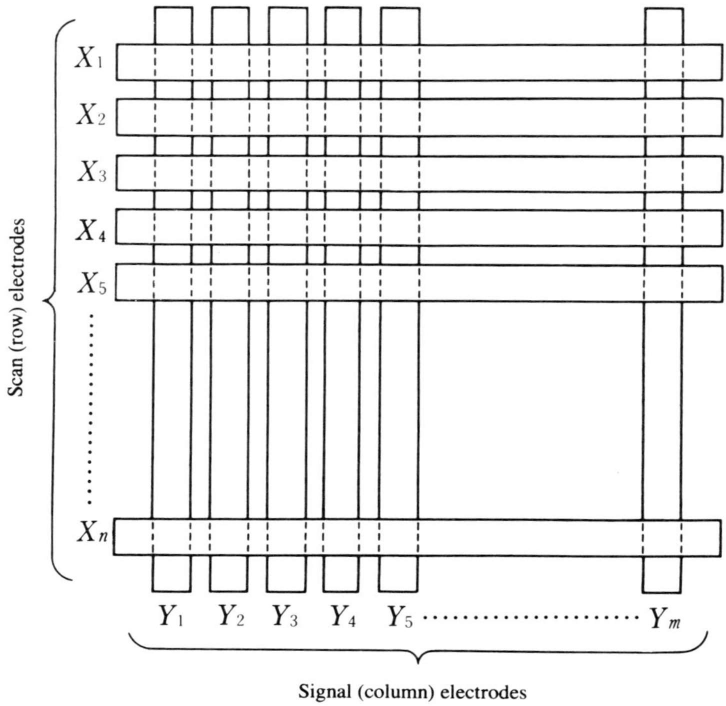

Figure 2.17 *Matrix electrode construction*

Table 2.16 *Principal LCD drive techniques*

LCD drive techniques
Static drive
Multiplex drive
Active matrix drive
Thermal addressing drive
Optical addressing drive

the above electrode constructions are listed in Table 2.16, and we then discuss these individually.

2.6.2 Static drive

In the static drive method each electrode segment to be displayed is driven separately and simultaneously. Therefore each segment requires an independent drive circuit element and a voltage is applied continuously to the display segment

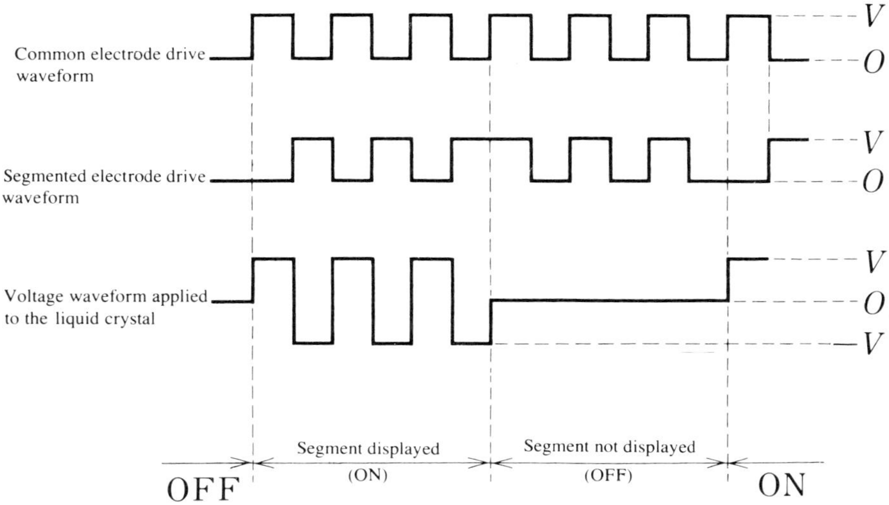

Figure 2.18 *Static drive waveform examples*

and the common electrode while the segment is to be displayed. Figure 2.18 shows the waveforms of a typical static drive. Both the segment electrode and the common electrode are supplied with a pulse wave form of the same polarity, and when the display is to be on these are put out of phase by 180°. The voltage applied to the liquid crystal is therefore actually an a.c. waveform, and the deterioration of the liquid crystal due to direct current operation is avoided.

Since in a static drive separate drive circuit elements are required for each

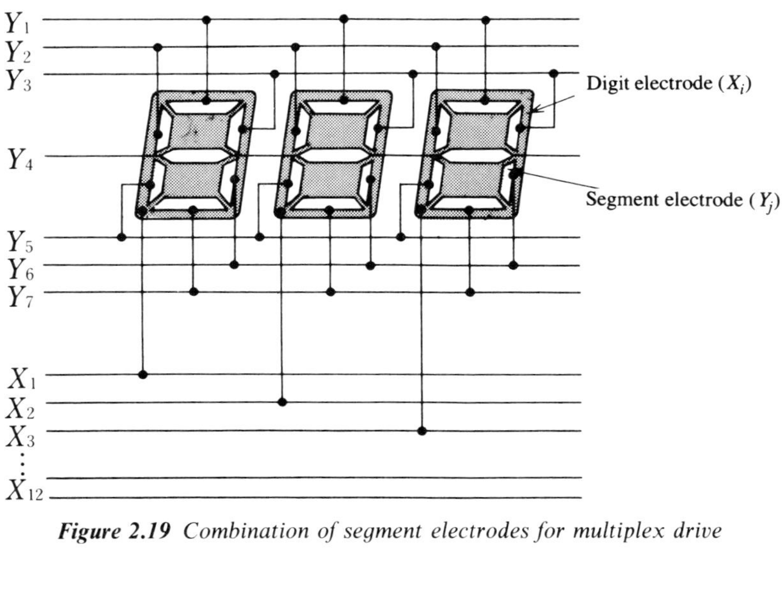

Figure 2.19 *Combination of segment electrodes for multiplex drive*

electrode segment, for a numeric display with a large number of digits the number of drive circuit elements and the number of lead connections is inconveniently large. The multiplex drive method described next alleviates this problem.

2.6.3 Multiplex drive

This technique is used for cases with a relatively large number of segments, such as multidigit numeric displays, and also for matrix displays; it is also referred to as time-division drive or dynamic drive.

Consider, for example, a twelve-digit seven-segment numeric display that is using a multiplex drive; as shown in Fig. 2.19, there will be twelve common

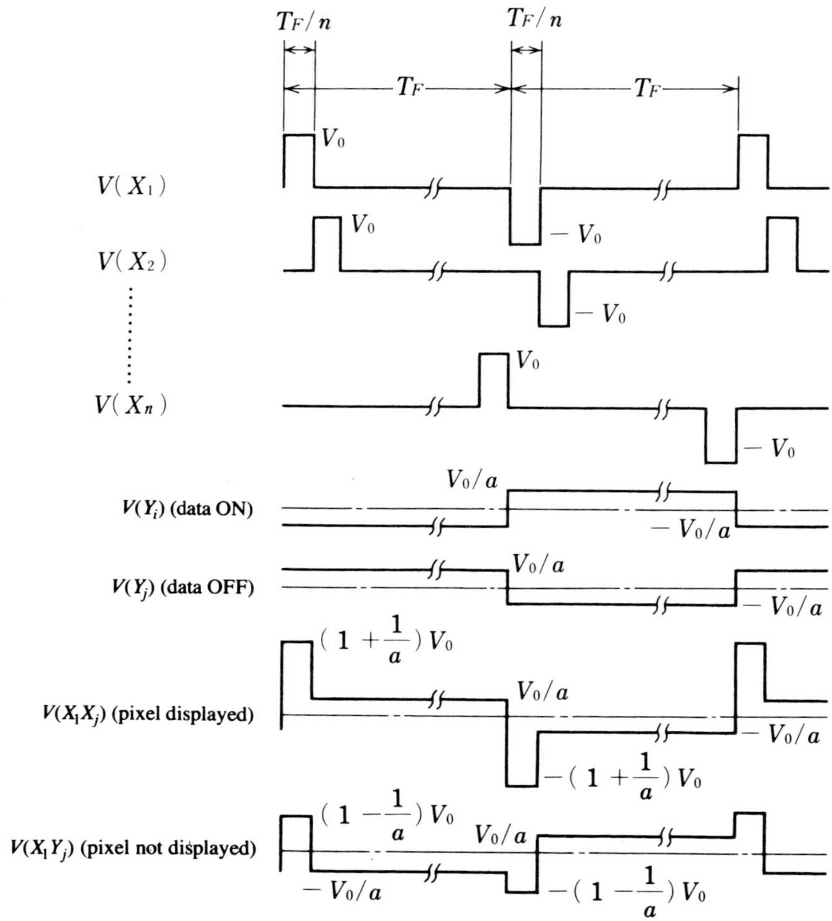

Figure 2.20 Drive waveforms using the voltage-averaging method (T_F: frame time)

electrodes, one for each digit, and all of the segment electrodes will be connected together in seven sets. Thus the drive operates by operating the twelve-digit electrodes $X_1, X_2, \ldots, X_{12}$ in time sequence and selectively driving the displayed segment electrodes Y_1, $Y_2, \ldots$, Y_7 for each digit in turn. Thus, in this example the number of necessary connections and drive elements is nineteen, which is very much less than if the static drive technique is used.

Since, however, the segment electrodes are connected in common, various partial voltages will be applied to segments other than those to be displayed, so that the display contrast will be reduced. This phenomenon is called crosstalk, and arises because the response of the liquid crystal to voltage is bipolar and because the electrooptic response (voltage to display contrast) characteristics are not steep. In order to avoid this crosstalk phenomenon the following technique, termed voltage averaging, is usually used with a liquid crystal multiplex drive.

In a multiplex drive using a voltage-averaging method by line sequential drive and the XY matrix electrode construction* shown in Fig. 2.17, the drive waveform is designed to satisfy the following conditions (Fig. 2.20):

1. In order to eliminate the crosstalk phenomenon, the effective voltages V_s and V_{us} applied to the liquid crystal of a group of pixels for which the display is on and a group of pixels for which the display is off are the same for the respective groups of pixels.
2. In order to increase the display contrast, V_s is set to be a maximum limit exceeding the operating threshold voltage of the liquid crystal.
3. V_{us} is set to be a size that does not exceed the threshold voltage.

As shown in Fig. 2.20, the scanning pulse for time-division driving and the signal pulse for data input are set to V_0 and V_0/a respectively, and if the number of scanning electrodes is n, then the effective voltage values V_s and V_{us} are given as follows:

$$V_s = \frac{V_0}{a}\sqrt{\frac{a^2 + 2a + n}{n}} \tag{2.18}$$

$$V_{us} = \frac{V_0}{a}\sqrt{\frac{a^2 - 2a + n}{n}} \tag{2.19}$$

The value a in expressions (2.18) and (2.19) is found from the condition that the display contrast should be maximum or, in other words, that the ratio α of V_s/V_{us} (the voltage operating margin) should be maximum, and thus $a = \sqrt{n}$. This means that in Fig. 2.20, when the scanning pulse has the value V_0, if the signal pulse is set to a value of $V_0/\sqrt{n}$ the maximum display contrast will be obtained.

At this time the margin α_{max} of the operating voltage is given by the

*The segmented electrode construction of the twelve-digit numeric display shown in Fig. 2.19 is exactly the same electrically as the matrix electrode construction of Fig. 2.17, with $n = 12$ and $m = 7$.

following expression:

$$\alpha_{max}\left(\frac{V_s}{V_{us}}\right)_{max} = \sqrt{\frac{\sqrt{n}+1}{\sqrt{n}-1}} \tag{2.20}$$

Expression (2.20) shows that as the number of scanning electrodes n is increased, it is necessary to have a liquid crystal that can produce adequate display contrast for a smaller voltage difference ($V_s - V_{us}$), and for a particular liquid crystal the display contrast will fall as n increases.

A current practical limit on the number of scanning electrodes appears to be about 200.

2.6.4 Active matrix drive

Using the multiplex drive technique described in the previous section, when a voltage averaging technique is used to prevent crosstalk the number of the scanning electrodes that will allow time-division driving is restricted, and therefore the technique cannot be applied to applications such as a video display where more than 200 scanning electrodes are required. The active matrix drive technique is applicable to large display capacity applications such as this.

In this technique, a switching element and if necessary a capacitor is added for the pixel corresponding to each matrix intersection point of the scanning electrodes and signal electrodes, in order to improve display characteristics such as contrast and response. The switching element and capacitor for each pixel have the roles of preventing crosstalk and storing the signal charge. In this way restrictions on the number of scanning electrodes are practically removed, and in principle a liquid crystal display by a static drive with nearly 100 per cent duty factor can be obtained.

As illustrated in Table 2.17, there are two main types of this drive, depending

Table 2.17 *Classification of active matrix drive methods*

Active matrix drive methods
Field effect transistor (FET) type
MOS transistor type
TFT transistor type
a-Si type
p-Si type
CdSe type
Te type
Non-linear two-terminal device type
Varistor type
MIM type
Diode type

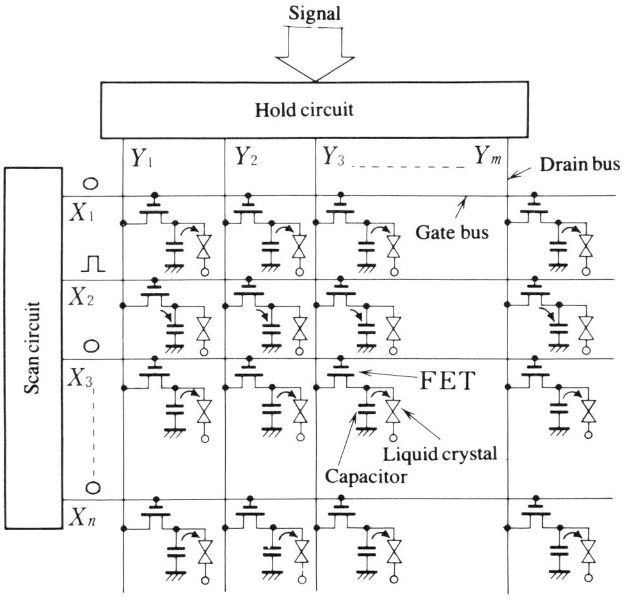

Figure 2.21 *Operating principle of active matrix drive LCD*

on whether a field effect transistor (FET) is used as the switching element or non-linear two-terminal element. Figure 2.21 shows the operating principle of an active matrix drive LCD using FETs. Using a line scan technique, the row electrodes (gate buses) X_1, X_2,.., X_n are scanned sequentially, so that all of the FETs on one gate bus are turned on (conducting) simultaneously, and the signals for the current row are supplied from the hold circuit through the drain buses Y_1, Y_2,.., Y_m through the conducting FETs to the capacitors. The liquid crystal elements are sustained in their current state by the charge on the capacitor until the next scan.

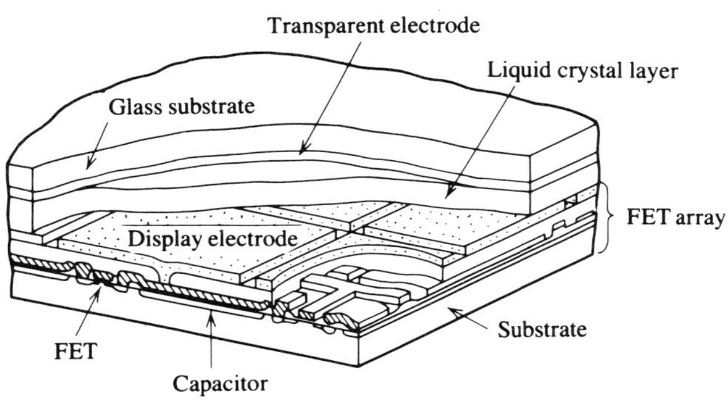

Figure 2.22 *Construction of an active matrix drive LCD [20]*

There are two types of FET drive: the MOS transistors type, which uses a silicon single-crystal substrate, and the thin-film transistor (TFT) type, which uses a thin film of α-Si, β-SI, CdSe, or Te formed on a transparent substrate of glass or quartz. The construction of an active matrix drive LCD using either type of FET is basically as shown in Fig. 2.22. For some of these drive techniques, the peripheral drive circuitry can be integrated on the same substrate as that for the FET array for the display, which simplifies the connection between the drive circuit and the display. Since the TFT type allows a transmissive display, a multicolour display or a full-colour display can easily be implemented.

Active matrix drive techniques using non-linear two-terminal elements as the switching elements are all designed with the objective of simplifying construction and reducing cost. The MIM type uses a three-layer construction for the switching array, consisting of metal–insulator–metal layers (M–I–M), typically $Ta/Ta_2O_5/Cr$. The varistor type uses a sintered ZnO varistor as the switch element and capacitor element.

2.6.5 Thermal and optical addressing

The thermal addressing drive method uses a thermooptic effect (see Section 2.3.7) of a smectic or cholesteric liquid crystal. The liquid crystal may be heated either by an infrared laser beam or by an electrical resistive heater. In the former case a

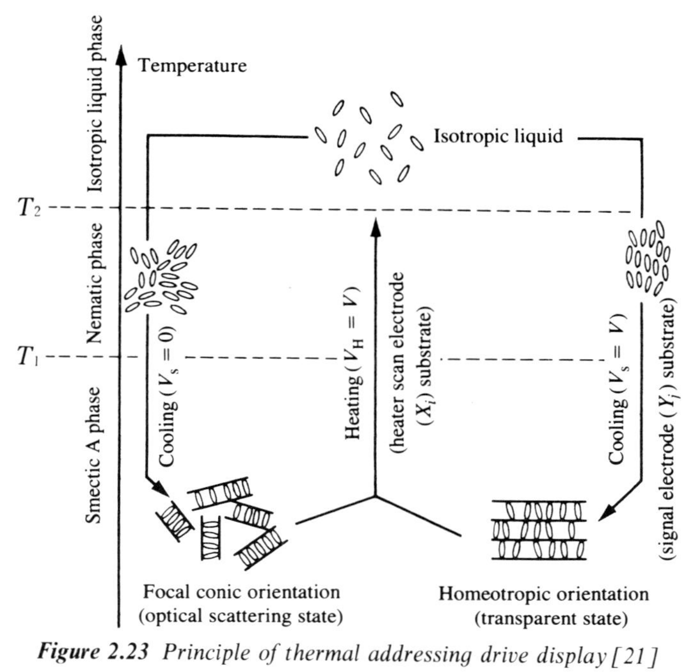

Figure 2.23 *Principle of thermal addressing drive display [21]*

liquid crystal cell is used which has an infrared absorbing layer and the display image is written directly by scanning with the laser beam. This laser beam technique is principally used for large projection displays (see Section 2.7.5) and is characterized by a large-display capacity, high resolution, and a memory function.

In the latter case, an XY matrix electrode construction is used, and the scanning electrodes (X) are the heater resistance electrodes. In the display principle illustrated in Fig. 2.23 a smectic A liquid crystal (S_A) is used which has a nematic phase in a high-temperature region. The S_A liquid crystal is progressively heated ($V_H = V$) by the heater scan electrode (X_i) to the isotropic (I) phase, and then as it cools through the N phase the presence or absence of a voltage signal ($V_s = V$ or $V_s = 0$) on the signal electrode (Y_j) determines whether the clear state or the cloudy state (scattering texture) is reached. These two states are strongly retained by the smectic A phase. Therefore this technique gives in principle no crosstalk, and no limit to the number of scanning electrodes. It can therefore be applied to large-capacity displays which cannot be implemented using the normal multiplex drive technique (see Section 2.7.2). The difficulty is, however, that a fast write speed is not possible.

The optical addressing drive method uses a layer of a photoconductive material such as CdSe on one substrate and a liquid crystal cell construction in which the other substrate is a whole surface electrode. The display image is projected on the liquid crystal cell as an optical image, and an electric field corresponding to the optical intensity of the image is applied to the liquid crystal layer. This technique is principally used for large-scale projection displays (see Section 2.7.5).

2.7 *APPLICATIONS OF LIQUID CRYSTAL DISPLAYS*

The first stage in the development of liquid crystal displays as practical devices came in 1973 with the implementation of the numeric display panel for watches and calculators. These applications exploited to the full the advantages of liquid crystal displays, which are incomparably low operating voltages and power consumption, together with compact, lightweight, and thin construction. Building on the success of these first implementations, the second stage of applications came in about 1977 when liquid crystal displays began to appear in a wide range of applications such as measuring instruments, domestic electrical and audio equipment, office equipment, game machines, and vehicle instruments. These became established by 1980 or 1981. In this second stage the application broadened from digital displays to analogue displays, small character displays, and colour displays.

In order to reach the third and fourth stages of applications the development of

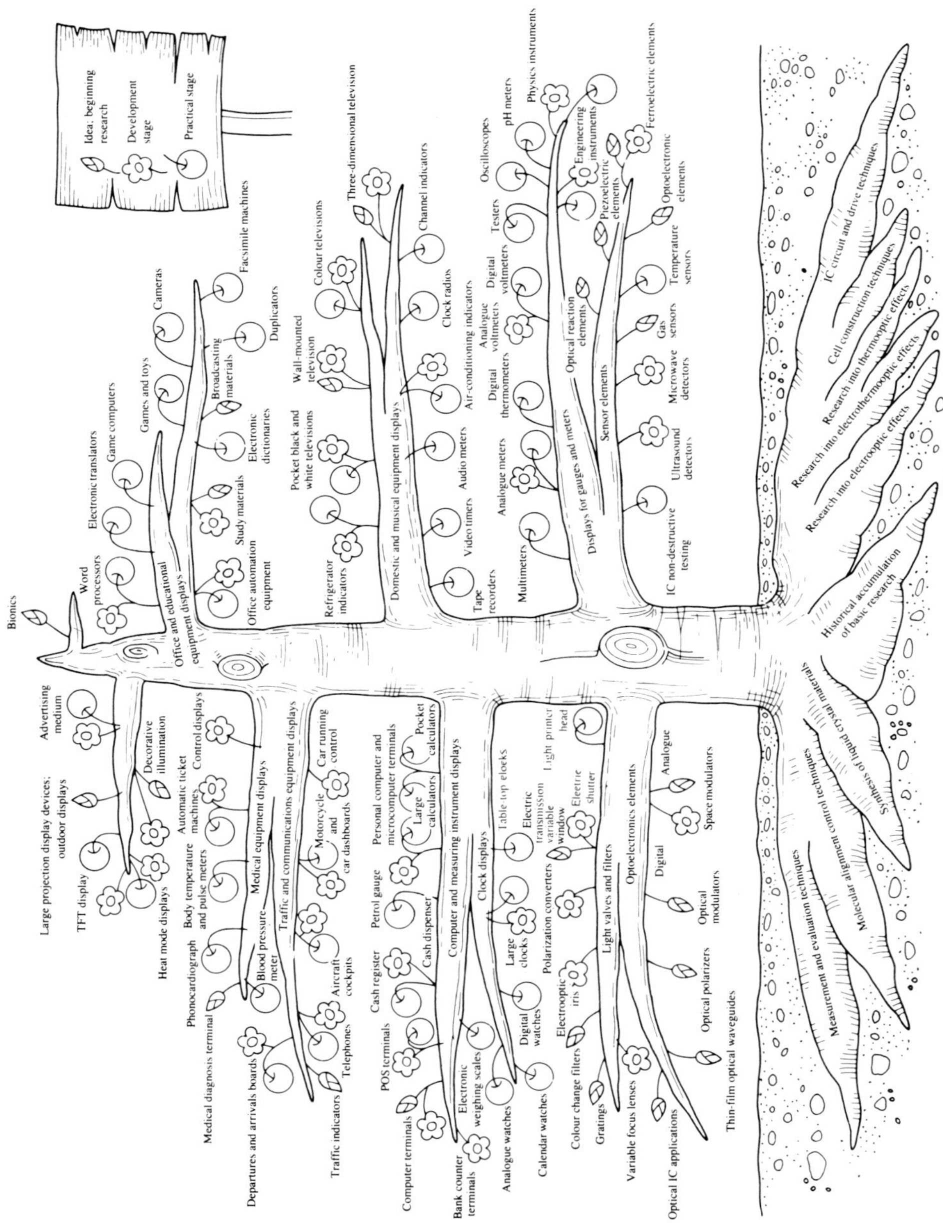
Idea: beginning research
Development stage
Practical stage
Bionics
Word processors
Electronic translators
Game computers
Games and toys
Cameras
Facsimile machines
Duplicators
Broadcasting materials
Electronic dictionaries
Study materials
Office automation equipment
Office and educational equipment displays
Colour televisions
Three-dimensional television
Channel indicators
Clock radios
Wall-mounted television
Pocket black and white televisions
Refrigerator indicators
Domestic and musical equipment displays
Air-conditioning indicators
Audio meters
Video timers
Tape recorders
Physics instruments
pH meters
Oscilloscopes
Testers
Digital voltmeters
Analogue voltmeters
Digital thermometers
Analogue meters
Multimeters
Displays for gauges and meters
Optical reaction elements
Engineering instruments
Piezoelectric elements
Ferroelectric elements
Optoelectronic elements
Temperature sensors
Gas sensors
Microwave detectors
Ultrasound detectors
IC non-destructive testing
Sensor elements
IC circuit and drive techniques
Cell construction techniques
Research into thermooptic effects
Research into electrothermooptic effects
Research into electrooptic effects
Historical accumulation of basic research
Synthesis of liquid crystal materials
Molecular alignment control techniques
Measurement and evaluation techniques
Large projection display devices; outdoor displays
TFT display
Heat mode displays
Advertising medium
Decorative illumination
Control displays
Automatic ticket machines
Body temperature and pulse meters
Phonocardiograph
Medical diagnosis terminal
Blood pressure meter
Medical equipment displays
Departures and arrivals boards
Traffic indicators
Telephones
Aircraft cockpits
Motorcycle and car dashboards
Car running control
Traffic and communications equipment displays
Personal computer and microcomputer terminals
Pocket calculators
Large calculators
Petrol gauge
Cash register
Cash dispenser
POS terminals
Computer terminals
Bank counter terminals
Electronic weighing scales
Computer and measuring instrument displays
Table-top clocks
Light printer head
Electric transmission variable window
Electric shutter
Clock displays
Large clocks
Digital watches
Analogue watches
Calendar watches
Polarization converters
Electrooptic iris
Colour change filters
Gratings
Light valves and filters
Optoelectronics elements
Analogue
Space modulators
Digital
Optical modulators
Optical polarizers
Variable focus lenses
Optical IC applications
Thin-film optical waveguides

liquid crystal displays in the 1980s has aimed next at large-capacity character displays, graphic displays, large-area displays, and then colour video displays, and in 1982 medium-capacity character displays and black and white television displays entered the stage of practical implementation. Following on from this the implementation of LCDs as the large-capacity character displays and colour television displays which will compete with CRT displays is approaching.

Figure 2.24 shows the liquid crystal display application 'tree' which shows the current development states of the liquid crystal displays described above; these are now described in more detail.

2.7.1 Digital displays and analogue displays

The majority of domestic devices such as watches clocks, calculators, household electrical and audio equipment, and also industrial equipment such as measuring instruments, to which liquid crystal displays are applicable are digital, but analogue displays are also needed. In this type of application the display capacity is relatively small, and most of the demand is for displays of about ten digits, so that a segment electrode drive can be adopted. For applications such as clocks where the number of segments is small a static drive technique can be used, but

Figure 2.25 *Colour liquid crystal display digital clock (Toshiba)*

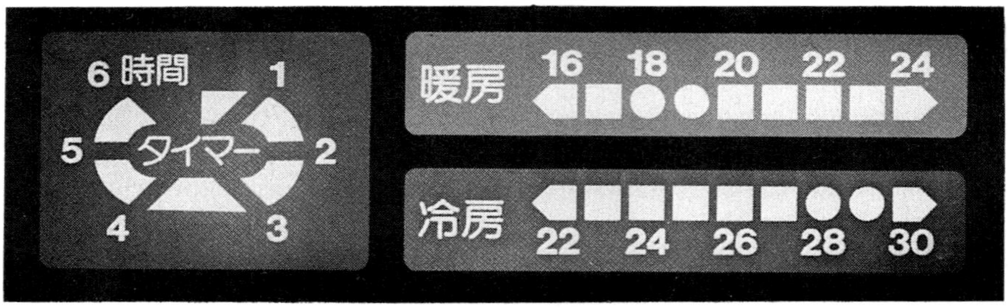

Figure 2.26 *Multicolour liquid crystal display air conditioning indicator (Stanley)*

when the number of segments increases as in, for example, calculator displays a multiplex drive technique is used. Figure 2.25 shows a digital clock using a guest–host type color liquid crystal display with static drive, and Fig. 2.26 shows an air conditioner indicator panel using a multicolour liquid crystal display combining digital and analogue displays.

2.7.2 *Character displays and graphic displays*

Character and graphic displays for computer terminals, word processors, and so forth require a large display capacity, and therefore a matrix electrode drive is

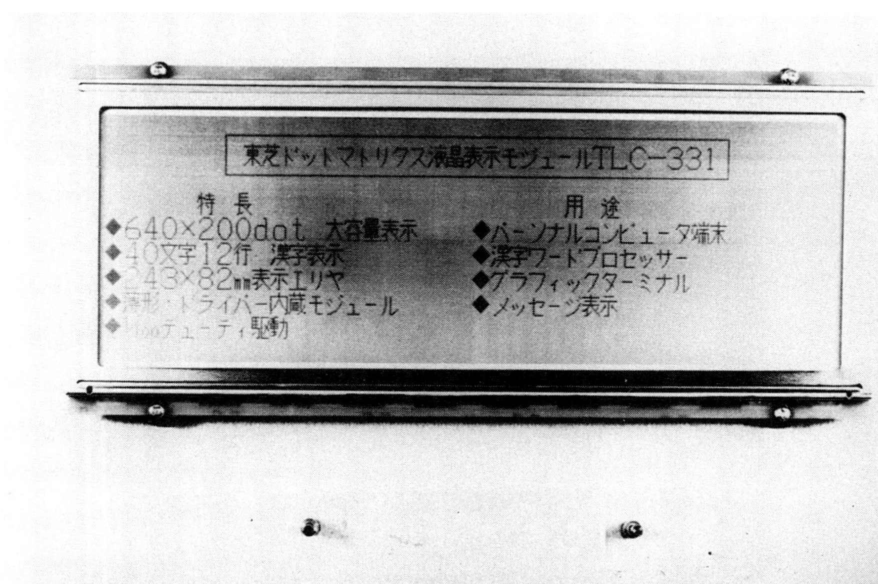

Figure 2.27 *Character display using multiplex matrix drive twisted nematic LCD (Toshiba)*

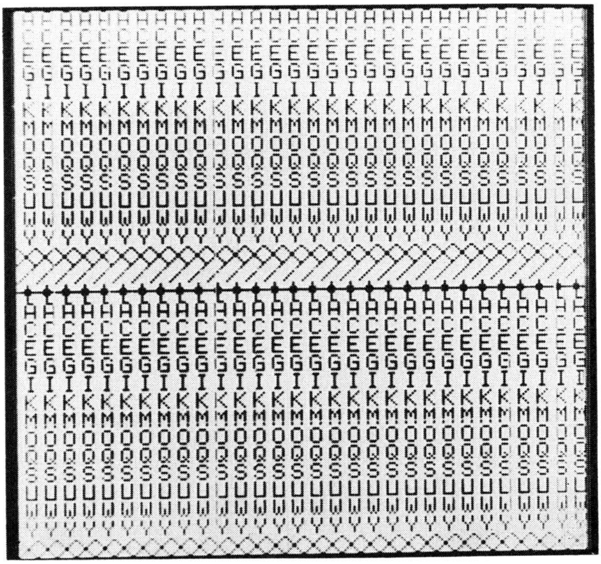

Figure 2.28 *Character display using MOS type active matrix drive guest–host LCD (Toshiba)*

adopted. For a number of scan electrodes N up to approximately 200 a multiplex matrix drive technique which relies on the multiplex capability of the liquid crystal material itself can be adopted, but for a higher number of scan lines an active matrix drive technique is used in which switching elements are added. For the former drive technique twisted nematic, super-twisted nematic and dynamic scattering displays are used, and for the latter technique also a guest–host type.

Figure 2.27 shows a twisted nematic liquid crystal display module of full-dot type which uses a multiplex matrix drive. This has a multiplexing number $N = 100$ and can display alphanumeric characters, Chinese characters and graphics.

A high-density large-capacity character display that uses an active matrix drive and guest–host LCD is exemplified in Fig. 2.28. It uses a MOS matrix array (480×480 pixels/3 inch-diagonal screen), and displays 30 rows of 30 characters. As well as the MOS type, the active matrix drive displays of the varistor type, MIM type, diode type, and TFT type have been developed, and of these the first three, which use two-terminal switching elements, appear to be appropriate as simple and inexpensive drive methods for character and graphic displays that do not require grey tones.

For large-capacity displays the thermally-addressed multiplex matrix/drive method using smectic A liquid crystals is also used (see Section 2.3.7). In Fig. 2.29 a character display with black characters on a white background is shown which uses a smectic guest–host matrix liquid crystal cell using a black dissolved dye.

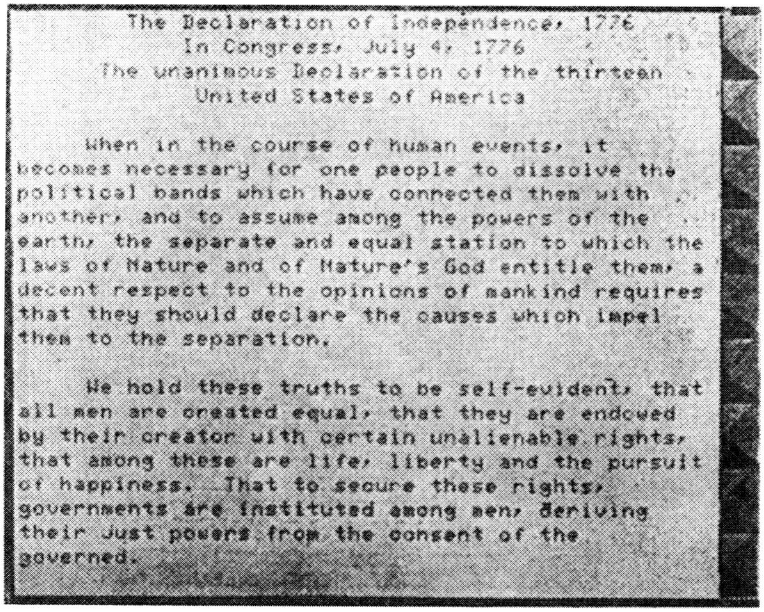

Figure 2.29 *Thermal drive multiplex matrix guest–host LCD (Kylex) [22]*

2.7.3 Video displays

A television display requires high resolution, good half-tone resolution, fast response, high contrast, and other high-grade characteristics. The principal display types that have been used are dynamic scattering, twisted nematic, and guest–host types, and the drive techniques are active matrix drive and multiplex matrix drive.

Figure 2.30 shows a liquid crystal TV set of dynamic scattering type using a MOS active matrix drive. It has a fast response time ($< 30\,\text{ms}$), high contrast ($> 20{:}1$), high resolution (220×240 pixels/2 inch-diagonal screen, and is conveniently small, thin, and light. A wristwatch including a guest–host liquid crystal TV using the same MOS drive method has been marketed by Seiko. The TV screen size is 2 inches diagonal and the number of picture elements is 210×152.

Figure 2.31 shows an α-Si type TFT active matrix drive colour liquid crystal TV set. The display type is twisted nematic, and a red, green, and blue mosaic colour polarizer film is incorporated into the cell so that a colour image is generated by light transmission. A colour television using p-Si type TFT active matrix drive has been marketed by Seiko. It has a resolution of $240 \times$ pixels/$43 \times 32\,\text{mm}^2$ and uses the same twisted nematic display technique, but the colour image is produced by a mosaic of red, green, and blue colour filters.

A means that has been proposed to increase the number of scanning electrodes

Figure 2.30 *MOS type active matrix drive dynamic scattering liquid crystal TV set (Toshiba)*

in the multiplex matrix drive arrangement is the multiple electrode matrix technique which has a number of signal electrodes scanned simultaneously. A test version of a twisted nematic liquid crystal TV of this type has been made [23], with a quadruple electrode matrix drive, 120 × 160 picture elements, and 3 inch-diagonal size, and a double-electrode matrix type black and white display liquid crystal TV set has been marketed by Casio.

2.7.4 Colour displays

Single-colour liquid crystal displays have been marketed for clocks, calculators, measuring instruments, and so forth, but in these cases the principal object is simply novelty of design. The first example of using a colour display where different colours discriminated functions was a peak programme meter made by

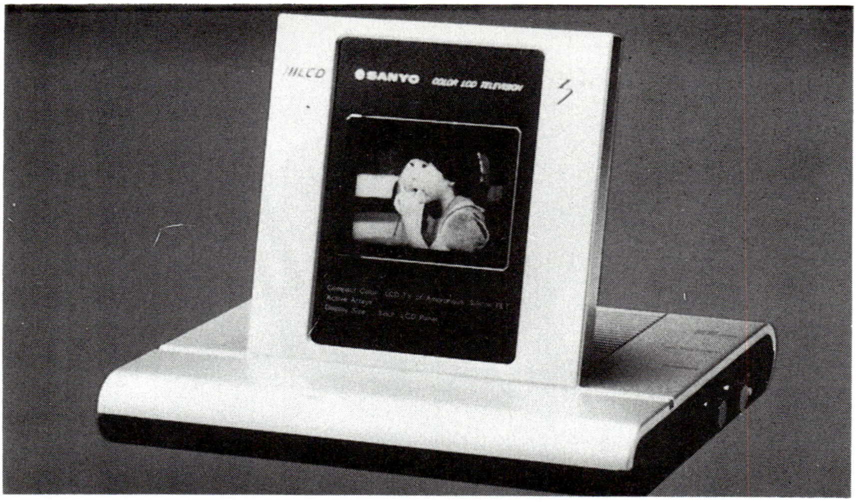

Figure 2.31 α-Si type TFT active matrix drive twisted nematic liquid crystal colour TV set (Sanyo)

Sony provided in a stereo cassette deck that used a two-colour LCD of coloured polarizers twisted nematic type.

One recent example where colour distinction has begun to be seriously used is in coloured LCD instrument panels for cars. Figure 2.32 shows a multicolour

Figure 2.32 Multicolour guest–host type liquid crystal vehicle instrumentation panel (Toshiba)

LCD instrument panel in which all of the instruments from the dashboard have been integrated into one cell. This uses a colour filter type black guest–host liquid crystal cell, and has separate functions for blue, green, red, and yellow colour displays. In this case the black guest–host liquid crystal cell acts as a light shutter and the display colour is determined by a colour filter. A multicolour liquid crystal vehicle instrument panel has also been implemented using a twisted nematic liquid crystal cell as an optical shutter.

As a means of providing a full-colour display, the method of laminating a coloured layer with a mosaic of the primary colours red, green, and blue onto the electrode surface of a black and white liquid crystal cell and effecting a natural colour display by additive colour light mixing has been attracting attention [24]. A colour TV display based on this method has already been described above (see Section 2.7.3).

Another technique is to place a two-colour polarizer type twisted nematic colour liquid crystal display panel in front of a monochrome CRT for field sequential colour display and to utilize the integrating effect of vision to provide a three-colour image display [25]. In an application of this method, a three-colour liquid crystal display oscilloscope, providing blue, green, and orange images, has been marketed by Tektronix.

2.7.5 Large-scale displays

There are basically two types of large-scale display. In the first a small liquid crystal cell, playing the same role as a slide, has a display image drawn on it by thermal, optical, or electric field methods, and the image is projected onto a large screen by a projector. In the other method, a large array of liquid crystal cells of a fixed size corresponding to an image element are integrated together to make a large screen, and this integrated display image is viewed directly by back lighting.

In an optical addressing projection type of display one electrode of the liquid crystal cell is replaced by the photoconductive layer of, for example, CdSe, and by directing a beam of light onto this layer a corresponding electric field will be applied to the liquid crystal layer. A projection colour television system has been marketed by the Hughes Corporation which uses three monochrome CRTs to provide the image source separated into red, green, and blue images and three light bulbs each combined with a liquid crystal cell of the above type. The three independently modulated images are projected in superimposition on a large screen in the system.

In the thermal addressing projection method, a smectic A liquid crystal is sandwiched between electrode substrates, one of which has a thermal absorption layer, and the image is produced by a laser beam (see Section 2.3.7). IBM, for example, have built a projection system in which a semiconductor laser array head writes an image of 8000×8000 pixels on a liquid crystal cell $10 \times 10\ mm^2$ in 62.5 seconds, and this image is then projected on a $1 \times 1\ m^2$ screen with a display contrast of 10:1 [26].

One example from the many large display systems in which small liquid crystal

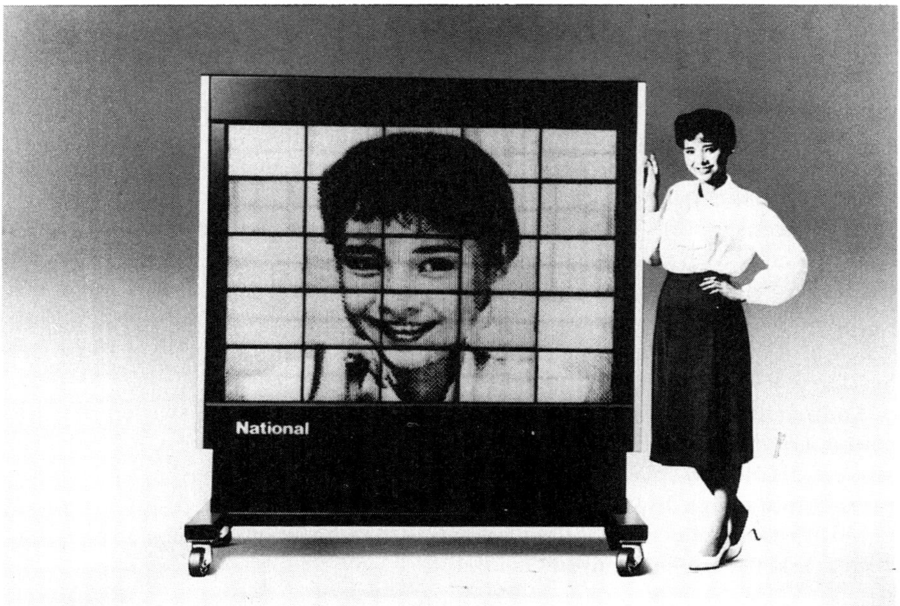

Figure 2.33 *Large-scale full-colour liquid crystal video display system using liquid crystal cell integration (Matsushita)*

cells are integrated together is the full-colour video system made by Mitsubishi. With the three primary colours of red, green, and blue as the pixel units, the basic element is a transmission type of twisted nematic liquid crystal cell of 8×46 pixels and size $6 \times 46\,cm^2$, and by combining these in an array, a full-colour video display of area up to $2.9 \times 4.6\,m^2$ is obtained. The Matsushita Group has also marketed a full-colour display device of this type (Fig. 2.33) with a display area of $1 \times 1.2\,m^2$ which consists of units that combine black and white guest–host type liquid crystal cells with red, green, and blue colour filters and have good half-tone resolution and wide viewing angle.

2.8 OUTLOOK

The development of liquid crystal display applications has been startling, and although these applications only began about ten years ago LCD production is approaching 1 billion panels per year. The range of application of liquid crystal displays is still widening rapidly (see the liquid crystal display application 'tree' in Fig. 2.24).

Even so, LCDs for the new applications other than watches, clocks and calculators have still not reached the 50 per cent level overall. The widening of the market for LCDs will of course continue, but to enable LCDs to expand their contribution to the field of electronic display devices development is still needed for high performance and high quality LCDs that can meet the new applications.

In Sections 2.6 and 2.7 developments were described in multiplex displays, active matrix displays, colour displays, video displays, large-scale displays, and so forth to enable the display capacity to be increased, to improve the display functions, and to give higher quality, and performance so that we can see that these requirements are being worked on.

BIBLIOGRAPHY

Chandrasekhar, S: *Liquid Crystals*, Cambridge University Press, Cambridge (1977).

de Gennes, P. G.: *The Physics of Liquid Crystals*, Clarendon, Oxford (1973).

de Jeu, W. H.: *Physical Properties of Liquid Crystalline Materials*, Gordon and Breach, New York (1980).

Gray, G. W.: *Molecular Structure and the Properties of Liquid Crystals*, Academic, New York (1962).

Kelker, H., and R. Hatz: *Handbook of Liquid Crystals*, Verlag Chemie, Weinheim (1980).

Kobayashi, S. (ed.): *Liquid Crystals—Nature and Applications*, Nikkan Kogyo Shinbun-sha, Tokyo (1970).

Matsumoto, S., and I. Tsunoda: *Latest Liquid Crystal Technologies—Physical Properties, Materials and Applications*, Kogyo Chosakai, Tokyo (1983).

Meier, G., E. Sackmann, and J. G. Grabmaier: *Applications of Liquid Crystals*, Springer, Berlin (1975).

Nakada, I., and F. Hori: *Synthesis and Applications of Liquid Crystals*, Sachishobo, Tokyo (1974).

Sasaki, A. (ed.): *Foundations and Applications of Liquid Crystal Electronics*, Ohm-sha, Tokyo (1979).

Tachibana, T., *et al.*: *Liquid Crystals*, Kyoritsu Shuppan, Tokyo (1972).

REFERENCES

[1] S. Matsumoto and I. Tsunoda: *Latest Liquid Crystal Technologies—Physical Properties Materials and Applications*, Kogyo Chosakai, Tokyo (1983).

[2] F. Reinitzer: *Z. physik. Chem.*, **4**, 462 (1889).

[3] J. L. Fergason: *Sci. American*, **211**, 77 (1964).

[4] V. Tsvetkov: *Acta Physicochim. (USSR)*, **16**, 132 (1942).

[5] W. H. de Jeu: *Physical Properties of Liquid Crystalline Materials*, Gordon and Breach, New York (1980).

[6] P. G. de Gennes: *The Physics of Liquid Crystals*, Claredon, Oxford (1973).

[7] K. Kobayashi: Ch. 2 in *Liquid Crystals*, T. Tachibana *et al.* (eds.), Kyoritsu Shuppan, Tokyo (1972).

[8] J. J. Wysoki, A. Adams, and W. Haas: *Phys. Rev. Lett.*, **20**, 1024 (1968).

[9] G. H. Heilmeier, L. A. Zanoni, and L. A. Barton: *Proc. IEEE.*, **56**, 1162 (1968).

[10] G. H. Heilmeier and L. A. Zanoni: *Appl. Lett.*, **13**, 91 (1968).

[11] M. Schadt and W. Helfrich: *Appl. Phys. Lett.*, **18**, 127 (1971).

[12] M. F. Schiekel and K. Fahrenschon: *Appl. Phys. Lett.*, **19**, 391 (1971).

[13] M. Hareng, G. Assouline, and E. Leiba: *Electron Lett.*, **7**, 699 (1971).

[14] R. B. Meyer: *Appl. Phys. Lett.*, **12**, 281 (1968).

[15] J. Borel, J. Robert, and F. Gharadjedaghi: The International Conference on *Alphanumerical Display Device and Systems*, B-3, p. 61 (1973).
[16] S. Matsumoto, M. Kawamoto, and K. Mizunoya: *J. Appl. Phys.*, **47**, 3842 (1976)'
[17] P. A. Penz and G. W. Ford: *Phys. Rev.*, **A6**, 414 (1972).
[18] E. Jakeman and E. P. Raynes: *Phys. Lett.*, **39A**, 69 (1972).
[19] D. L. Fishel and Y. Y. Hsu: *J. Chem. Soc.*, **23**, 1557 (1971).
[20] F. Hori and Kasahara: *Television Gakkai-shi (Journal of the Television Institute)*, **37**, 32 (1983).
[21] M. Hareng, S. Le Berre, R. Hehlen, and J. Perbet: *1981 SID International Symposium Digest*, p. 106 (1981).
[22] S. Lu, D. H. Davies, R. Albert, D. Chung, A. Hochbaum, and C. Chung: *1982 SID International Symposium Digest*, p. 238 (1982).
[23] E. Kaneko, H. Kawakami, H. Hanmura, and S. Saito: *Proc. SID*, **23**, 3 (1982).
[24] Uchida, TV Gakkai Giho (Report of the Television Institute), IPD 63–2, p. 7 (1981).
[25] M. G. Clark and I. A. Shanks: *1982 SID International Symposium Digest*, p. 172 (1982).
[26] A. G. Dewey *et al.*: *1982 International Symposium Digest*, p. 36 (1983).

Chapter 3

Electrochemical displays (ECDs)

3.1 INTRODUCTION

3.1.1 Electrochemical phenomena and displays

Some substances undergo a colour change as a result of various external stimuli. For example, coloration may be caused by the application of heat, light, or an electron beam, or by a current flowing through the substance. These phenomena are called thermochromism, photochromism, cathodochromism, and electrochromism. Of these, electrochromism, whereby the light absorption characteristics of a material are changed by turning on and off a current flowing through a pair of electrodes attached to the material is ideal for an electronic display.

Many materials have been known for a long time that exhibit this phenomenon. One example, based on a chemical phenomenon, is when an electrode is inserted into a pH indicator and the concentration of hydrogen ions is varied locally by passing a current through, after which a colour change appears. (reported in 1953 [1]); a colour change due to the electroplating reaction of silver ions has also been reported (1962) [2]. Based on physical phenomena, there is coloration due to the generation of colour centres in an alkali halide (1932) [3] and the absorption of light by KCl or NaCl crystals that have been heated and then subjected to an electric field. There are various other forms of electrochromism which have been known for many years, but the change in light absorption tends to be small compared with the energy consumption, and a high temperature is usually necessary for the reaction to proceed rapidly, so that the application of these techniques for display elements has not been very successful.

In order to use a light absorption variation mediated by electrical energy in a display device, the following points must be satisfied:

1. There must be a sufficient light absorption band in the visible spectrum.
2. Coloration and bleaching (or colour change) must have good reversibility without any accompanying unwanted reactions.

(1) Transition metal oxides (WO_3, MoO_3, TiO_2, V_2O_5, etc.)

$$\underset{\text{transparent}}{WO_3 + xM^+ + xe^-} \rightleftarrows \underset{\text{blue}}{M_x^+ WO_3e_x^-} \quad (0<x<1)$$

M: H, Li, Na, K, Ag, etc.

(2) Viologen derivatives (4, 4′-dipyridium compounds)

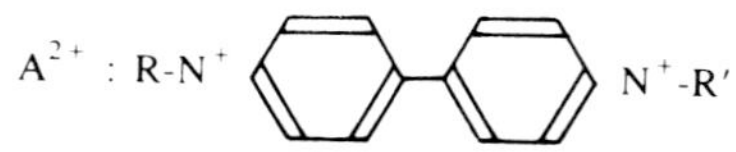

$$A^{2+} + e^- \rightleftarrows \dot{A}^+, \dot{A}^+ + X^- \leftrightarrow \dot{A}^+X^- \downarrow \quad \text{(display)}$$
$$\dot{A}^+ + e^- \rightleftarrows A\downarrow, A + A^{2+} \rightarrow 2\dot{A}^+$$

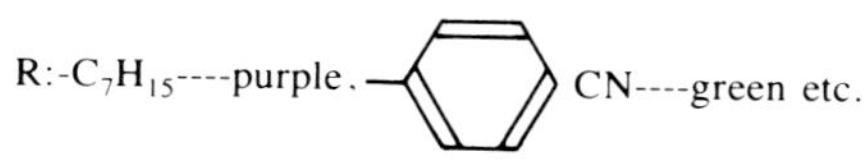

X: Br, Cl, etc.

Figure 3.1 *Typical ECD reactions*

3. A satisfactory display contrast must be obtained at room temperature with low voltage.

It is only relatively recently, around 1970, that new materials have been found that meet these conditions and lend themselves to application in electronic displays. Dividing these into inorganic and organic materials, representative examples are those based on transition metal oxides and those using a viologen derivative material. These were announced in 1969 and 1973 respectively as display elements based on the principles shown in Fig. 3.1 [4, 5], and were the first real electrochemical displays (ECD).

In tungsten oxide film a blue coloration (termed tungsten bronze) which is obtained by the double injection of electrons and cations is used for the display, and then the coloration is removed by oxidation. In a viologen system, radical cations of a reduction product are reversibly deposited on the electrode, forming a coloured film which can be used for the display. More recently, as shown in Table 3.1, many materials have been developed, including improvements to these materials.

The optical absorption changes of these materials are not based on a simple physical phenomenon or chemical reaction, but the basis is commonly an electrically controlled chemical reaction, or in other words an electrochemical reduction–oxidation reaction. These are all collectively known as electrochemical displays (ECD).

3.1.2 Characteristics of electrochemical displays

Like LCD devices, electrochemical displays are non-emissive and based on changes in light absorption. This means that under normal ambient light they are

Table 3.1 *Coloration materials for electrochemical displays*

Display material state		State of ion supply source	Typical materials	
Bleached state	Coloured state		Organic materials	Inorganic materials
Dissolved	Dissolved	Liquid	(a) pH indicators, chinone type, styryl type, fluorene type, beldazyl, viologen type (non-aqueous solution)	(b) Polytungsten ions (PTA)
Dissolved	Surface film	Liquid	(c) Viologen bipyridine mixtures	(d) AgI, I_2 derivatives,
Solid	Solid	Liquid	(e) Rare earth metal diphthalocyanin, polymerized tetrathiafulvalene (TTF)	(f) WO_3, MoO_3, TiO_2, V_2O_5, Rh_2O_3, NiO, $Ir(OH)_n$, $Zr(WO_4)_2$, InN intercalation materials (C_6Li)
		Solid	Prussian blue modified electrode	
				(g) Alkali halides $RbAg_4I_5$

easy on the eye, like printed matter, and are not adversely affected by external light sources as a CRT or LED is: moreover, they have a wide viewing angle. A second characteristic is the low drive voltage of between 0.5 and 5 V, which makes them suitable for direct battery drive and connection to integrated circuits. A third characteristic is the memory function: unlike an LCD or LED which uses a physical effect, the cause of the coloration is an electrochemical reaction, so that even when the power is turned off the display is preserved.*

*Nevertheless, to prevent the display from remaining, or for dynamic drive operation, it is also possible to remove this facility.

Table 3.2 Typical characteristics of ECDs

Drive voltage	0.5–3 V
Charge amount	5–10 mC/cm²
Response time	500 ms approx.
Operating temperature	−20–80°C
Memory function	Yes
Viewing angle dependency	No
Visibility in bright surroundings	Good
Multiplexing drive	Difficult

Fourthly, since the construction is basically simple, and there is no need for accurate control of the spacing between electrodes as in an LCD, this type is suitable for large display areas. Lastly, the operating temperature range is wide, and multicoloured displays are possible.

Typical ECD characteristic values are shown in Table 3.2.

To set against the above advantages of electrochemical displays, there are some disadvantages. One is the response time. Since coloration and bleaching rely on an electrochemical reaction, accompanied by physical movement of the materials and oxidation and reduction, the limit in practice is of the order of 0.1 second, and fast displays are not possible. Because of the memory property, a current is also required for bleaching, so that the power consumption depends on the frequency of the coloration–bleaching cycle, and a rapidly changing display will have a higher power consumption. Additionally, since it is difficult to carry out the coloration–bleaching reaction reversibly and maintain the long-term stability of the material, there are problems in reaching a practical level of repetition lifetime.

3.1.3 Construction of electrochemical displays

The general structure of an ECD is that of an electrochemical cell containing a pair of electrodes, one the display electrode and the other a counterelectrode. Figure 3.2 shows the simplest experiment to demonstrate an electrochemical coloration effect, in which electrodes are inserted into an ECD material consisting of a solution such as viologen. Ions within the solution cause a coloured layer to be formed on the electrode; the coloration and bleaching effects in a tungsten oxide type cell can also be investigated in the same type of experiment. When an electric current flows between the electrodes a change occurs in the absorption of visible light, and this is restored when the polarity of the electrodes is reversed.

It is very important for an ECD that the electrochemical reaction is smoothly reversible. In the simplest case the two electrodes may be arranged so that the reaction between them is symmetrical. In the example shown in Fig. 3.1, the rightward reaction ($WO_3 + xM^+ + xe^- \rightarrow M_x^+ WO_3 e_x^-$) proceeds on the display electrode, while the leftward reaction ($WO_3 + xM^+ + xe^- \leftarrow M_x^+ WO_3 e_x^-$) occurs on the counterelectrode. When repeated deposition and redissolving of the solution onto an electrode is used then the symmetry of the reaction may

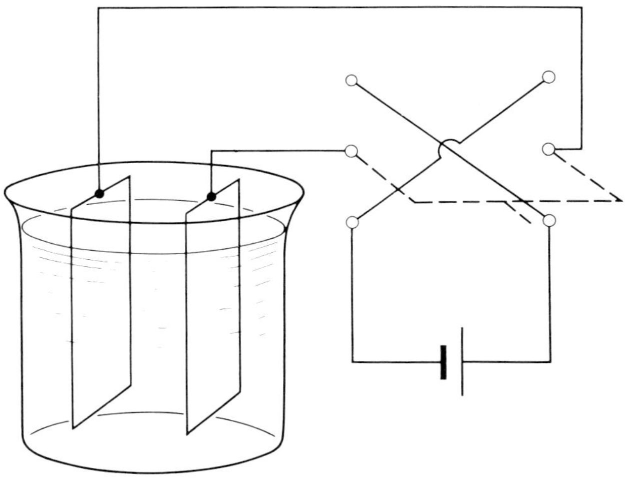

Figure 3.2 Experiment for electrochemical coloration phenomenon

be utilized with symmetrical electrodes, but in the case of a solid coloration layer, the following structure is adopted in order to facilitate easy charge transfer:

$$E_1, M_1 | I | M_2, E_2$$

Where E is an electron conductor, M is an electron–ion mixture conductor, and I is an ion conductor.

Generally M_1 is a coloration material and E_1 is a transparent electrode. I may be liquid or solid provided it contains the ions used in the electrochemical reaction. The effect of M_2 as a counterelectrode is very important, and as well as using the same materials as M_1 various other constructions have been considered.

Differences in cell construction stemming from the differences in coloration materials are shown in Fig. 3.3 under the classifying symbols shown in Table 3.1. For types (a) and (b) both the oxidation and reduction materials are soluble in the solvent, and consideration must be made of the possibility that a display occurring in the vicinity of the electrode may be dispersed by diffusion; however there is a large number of materials and a wide choice of colours. Type (d) uses a plating phenomenon of either a metal or a non-metal, one example being a silver–iodine system. This type works by the deposition from the solution onto an electrode, and has extremely good adhesion but high power consumption. The viologen system of type (c) was developed to overcome the defects of types (a), (b), and (d). These four types, (a), (b), (c), and (d), are structurally straightforward, but have a liquid material between the electrodes. Types (e) and (f), on the other hand, have the colouring material fixed on the electrode. These types have a slower response speed compared with the types that use deposition from a solution, but because of the stability involved in a transformation in the solid state, recently much research has been carried out into a range of these materials. As well as a

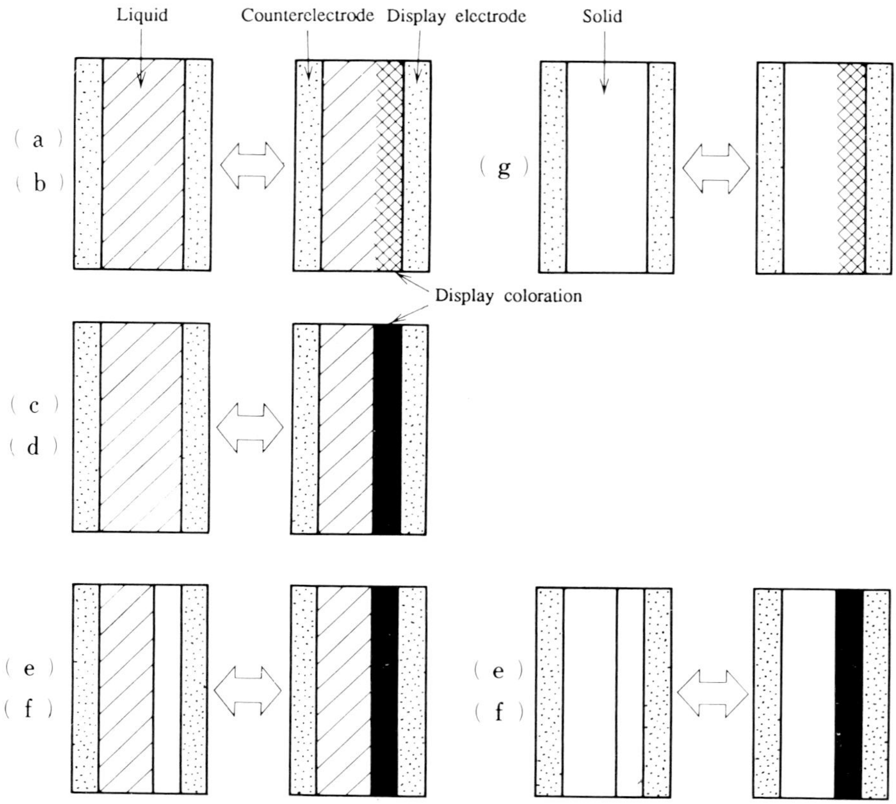

Figure 3.3 *ECD cell constructions and coloration and bleaching states*

number of metal oxides, organic materials developed include metal diphthalocyanin and modified macromolecules which have coloration and bleaching materials added to a macromolecular carrier. In types (e) and (f) it is also possible to use the solid phase as an ion source. In type (g) the coloration occurs in the solid phase adjacent to the electrode, and the coloration material and ion source are combined.

3.2 STRUCTURE AND CHARACTERISTICS OF ELECTROCHEMICAL DISPLAYS

3.2.1 Electrooptic characteristics

(1) Optical density and the Lambert–Beer law

The absorption of light by a material is shown by the Lambert–Beer law. The optical transmission ratio T of an absorption layer of thickness d is given by the following expression:

$$T = \frac{I}{I_0} = 10^{-\varepsilon Cd} \tag{3.1}$$

where I_0 is the incident light intensity, I is the intensity of light passing through the layer of thickness d, C is the concentration (mol/l) of the light absorbing material, and ε is the mol absorption coefficient.

The optical density (OD) is defined as follows:

$$\mathrm{OD} = \log\frac{I_0}{I} = \varepsilon Cd \tag{3.2}$$

If the surface density* of the colouring material is C_B, then

$$\mathrm{OD} = \varepsilon C_B \tag{3.3}$$

According to the Lambert–Beer law, if the light transmission of a material layer thickness d while bleached is I_1 and that while during coloration is I_2, then the change in optical density due to the coloration is given as follows:

$$\Delta\mathrm{OD} = \Delta\mathrm{OD}_1 - \Delta\mathrm{OD}_2 = \log\frac{I_2}{I_1} \tag{3.4}$$

(2) Faraday's rule

'The amount of a product deposited by electrolysis is proportional to the amount of electric charge which has flowed, and the amount of electric charge required to deposit a chemical equivalent weight of a product is independent of the material and is constant.' That is Faraday's law of electrolysis which was formulated in 1833.

In an ECD, the amount of the colouring material and the amount of electric charge have been found by experience to be proportional, and so if the surface charge density is $Q^{\dagger}$ and the coloration material production coefficient is $\beta^{\ddagger}$, then the area density C_B of the coloration material is

$$C_B = \frac{\beta Q}{nF} \tag{3.5}$$

Therefore, from expression (3.3), the following holds:

$$\mathrm{OD} = \frac{\varepsilon\beta Q}{nF} = \eta Q \tag{3.6}$$

*This is the product of d and c ignoring the dimension in the thickness direction, and is the density of the coloration material per unit area.

†The integral of current flowing in unit area, $Q = \int_0^t i\,dt$ (i = current density).

‡Proportion of the total electric charge that contributes to production of a coloration centre.

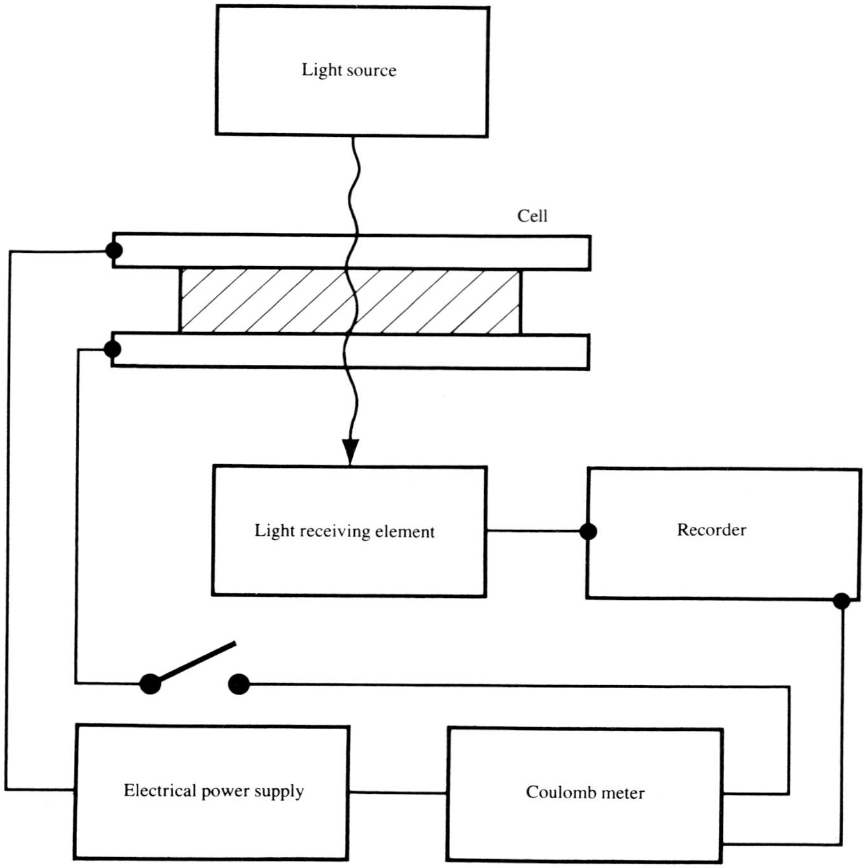

Figure 3.4 *Electrooptic measurement of an ECD cell*

where F is the Faraday constant (96 500 C per equivalent weight), η is the optical density per unit electrical charge density (electrical charge efficiency), and n is the number of electrons necessary to produce a coloration centre.

From the above, provided that the Lambert–Beer law and the Faraday rule hold, the change in optical density ΔOD is proportional to the amount of electric charge ΔQ. When measurements are made with the system shown in Fig. 3.4, the relationships between OD and Q shown in Fig. 3.5 can be demonstrated.

3.2.2 Display device structures

The principal elements making up the display construction are glass substrates, a spacer, a sealant material, a background, electrodes, and insulating materials. Particularly the material for the electrodes must be chosen with care in order to

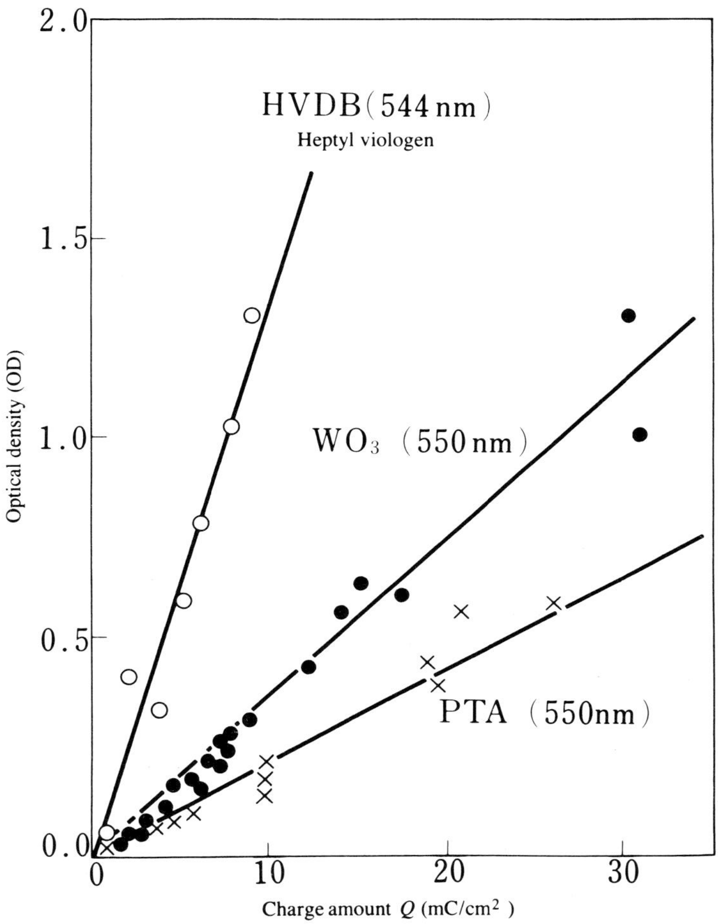

Figure 3.5 *Relationship between optical density and injected charge amount for different coloration materials*

provide (a) transparency (display side), (b) conductivity, and (c) electrochemical stability. Additionally, in cases where a liquid is used, the sealing of the container is very important, and when a solid material is used, a protective layer to prevent the influence of the atmosphere is important.

The arrangement of the electrodes is an important factor in determining the construction, and the two main arrangements are the opposing type, in which a pair of electrodes are placed opposite each other, and the planar type, in which the electrodes are all on one surface. Then depending on whether or not there is a background between the electrodes, there are reflection and transmission types.

We now describe characteristics of these various constructions.

In Fig. 3.6, the opposing type and planar type are compared. For viologen

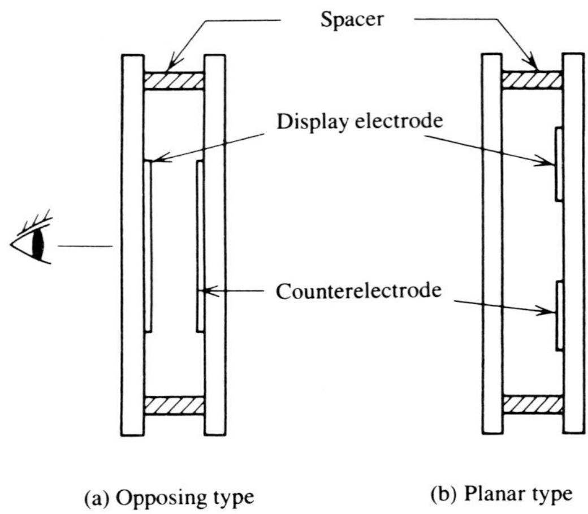

Figure 3.6 *Disposition of electrodes in an ECD cell*

systems the planar type has been much researched, but the opposing type is the mainstream in practical terms. These each have counterelectrodes as well as the display electrodes, and in some cases a third reference electrode may be added to control the electrochemical reaction by referencing the electromotive force produced between the electrodes.

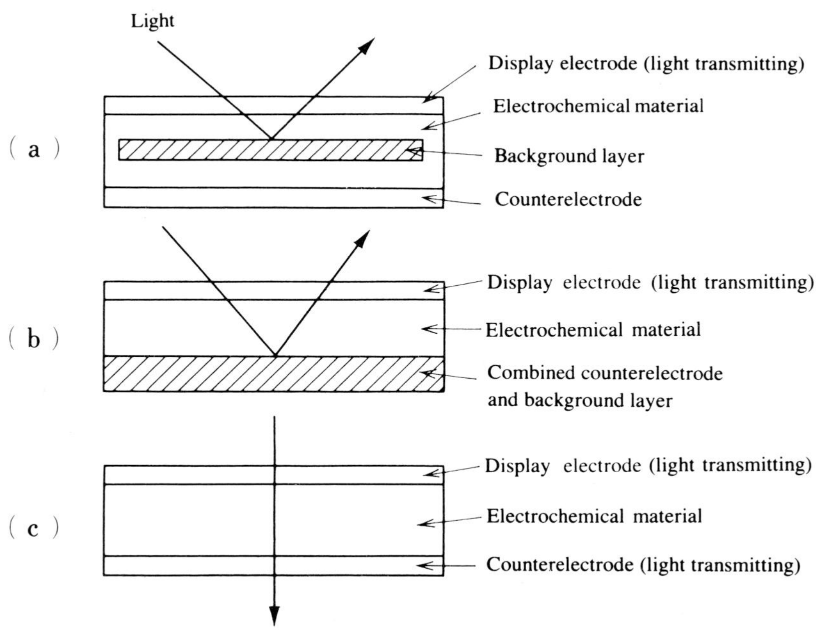

Figure 3.7 *Reflecting and transmitting type cells*

In Fig. 3.7, (a) and (b) are reflecting constructions and (c) is a transmission type of cell. In construction (a) a background is inserted between the electrodes, whereas in (b) a reflective counterelectrode is used, or a background placed behind the counterelectrode which is transparent. Since in this type of cell light that passes through the transparent layer passes through the absorbing layer twice, it has the advantage of high coloration density. In the transmission type (c), both electrodes are transparent, and therefore this type is useful for projection displays and overlapped displays such as scale readings.

Examples of ECD cells are shown in Fig. 3.8. In this drawing, (a) and (b) are solid types which do not require a glass container, and for example, in (a) the construction is laminated with in turn a display electrode (SnO_2/glass), a coloration layer (WO_x), a dielectric layer (SiO), and a counterelectrode (Au). The example in (c) uses a solid coloration layer and a liquid dielectric, while (d) shows a liquid coloration material used with the planar electrode construction.

3.2.3 Basic operating characteristics

We now describe basic characteristics, taking as examples the most well developed tungsten oxide and viologen types.

(1) Electrical charge amount and display contrast

The display contrast ratio (CR) is given by the following expression:

$$\mathrm{CR} = \frac{I_0}{I} = \frac{1}{T} \tag{3.7}$$

A practicable display contrast (CR 1.5) is obtained with a charge of 10 mC/cm^2 in a tungsten oxide system and 5 mC/cm^2 in a viologen system. The reason for the difference between the two is the difference in coloration efficiency shown in Table 3.3. The coloration centre density means the attainable limit of the optical density OD, and the charge mobility efficiency is the number of coloration centres produced by one electron.

It is desirable for the display device to have an optical absorption that coincides with luminance. As shown in Fig. 3.9, in a viologen system there is a wide peak at around 500 nm, whereas for tungsten oxide there is a peak at 900 nm and the absorption level in the visible wavelength is low. It should be noted that the contrast also depends on the absorption spectrum.

(2) Electrooptic response

Response characteristics for a reflection type cell using tungsten oxide systems and viologen systems are shown in Figs. 3.10 and 3.11. In the electrochemical reaction accompanying charge transfer a time of at least 10 or 100 ms is needed in order to obtain a satisfactory contrast. In practice in a viologen system, a figure

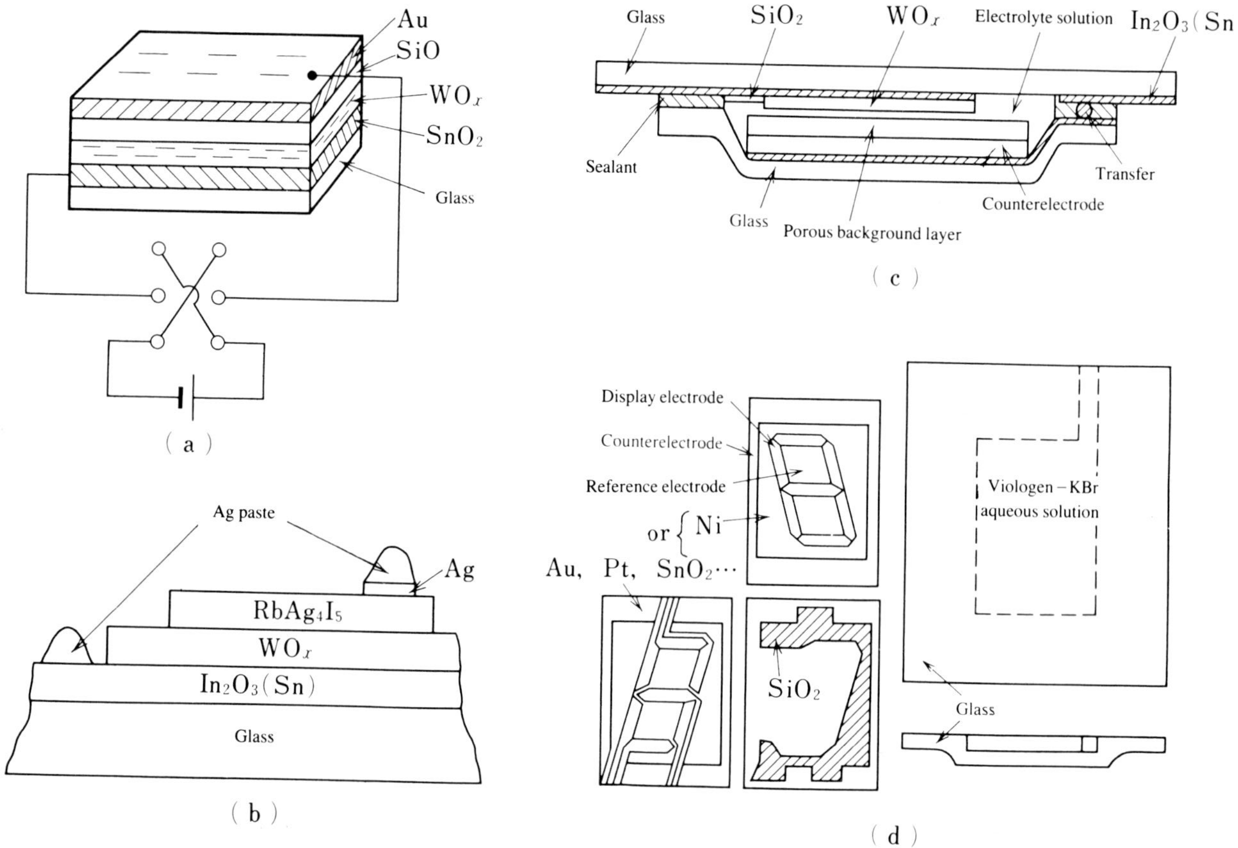

Figure 3.8 Examples of ECD cells

Table 3.3 Coloration differences between ECD materials

	Solid material	Liquid material
Asborption coefficient (cm^{-1})	10^2–10^4	10^3–10^5
Coloration centre concentration (cm^{-3})	10^{16}–10^{19}	10^{22}
Charge movement efficiency	10^{-2}	0.5–1

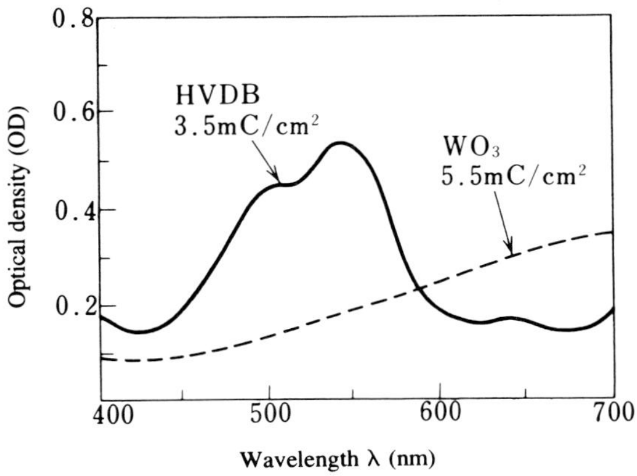

Figure 3.9 *Absorption spectra of viologen type (HVDB) and tungsten oxide (WO_3)*

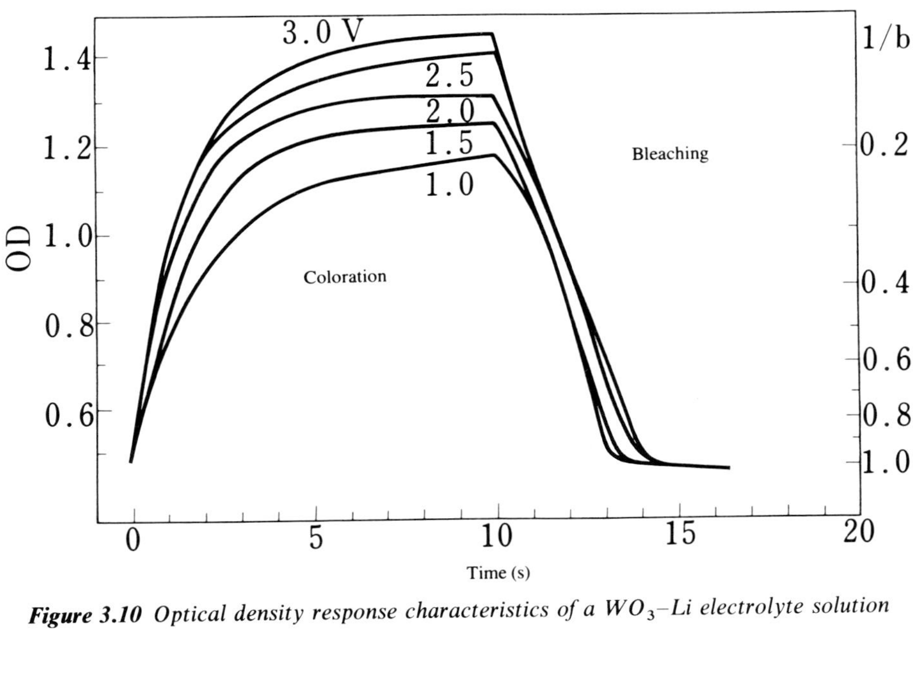

Figure 3.10 *Optical density response characteristics of a WO_3–Li electrolyte solution*

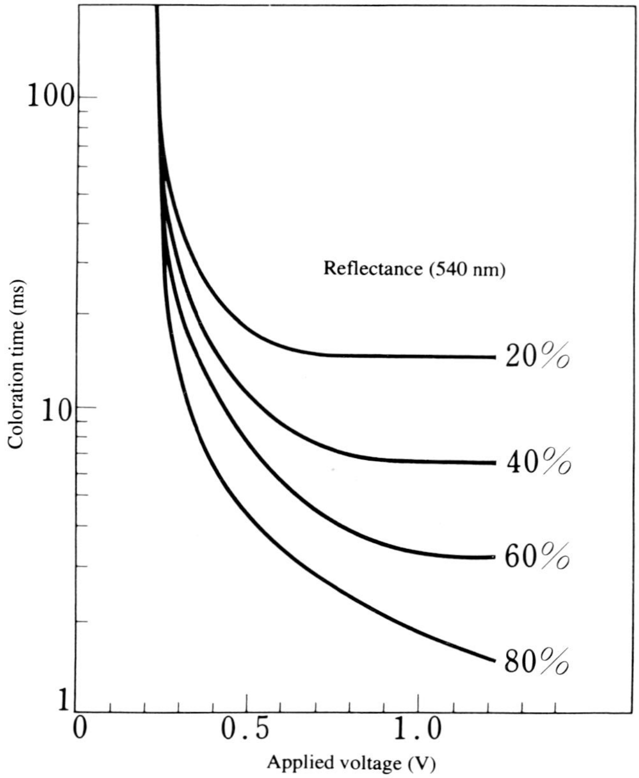

Figure 3.11 *Response characteristics of a viologen–KBr cell*

close to this is indicated as the response time, but for a tungsten oxide system it is much slower at some hundreds of milliseconds. The higher the applied voltage the faster the response, but in view of the oxidation–reduction potential of the material itself and side reactions, it should preferably not be more than one or two volts. The response characteristics also depend on the cell construction, the electrode materials and the dielectric materials.

In the case of a tungsten oxide system, in the process of coloration four factors must be considered: (a) the mobility of charge carriers within the thin film, (b) the ease of electron injection from the electrode into the film, (c) the ease of injection of ions into the film, and (d) the ease of ions reaching the film by dispersion through the electrolyte; of these (c) is the dominant factor [6]. On the other hand, during bleaching, current fluctuations [7] occur that are limited by a space charge limiting layer within the solid, and the bleaching time is inversely proportional to the product of the cation mobility μ and the square of the applied voltage V (μV^2). In an actual display device, the response speed is often determined by the bleaching characteristics, so that the cation mobility in the film is very important [7–11].

In an oxidation–reduction reaction, as in a viologen system in which insoluble salts are deposited from a solution, the charge transfer stage is very much faster than the material transfer stage, and the response speed is controlled more by the diffusion speed than by the electrochemical reaction speed and is determined by the diffusion rate-determining current which is proportional to the square root of time. Furthermore, in actual display elements there are resistance limitations; particularly the resistance of the transparent electrodes in a display cannot be ignored [12].

(3) Memory characteristics

The memory function which is an important characteristic of electrochemical displays depends on the method of formation of the coloration layer, the solvent, the cell construction, and the presence of oxygen. In principle, once a coloration layer is formed, unless the colour is removed or some external factor acts to cause the covering film to be redissolved then the display will be permanently maintained, but in practice various factors act against this. In particular chemical oxidation and reverse reactions caused by the electromotive force generated by non-uniformities in the distribution of electric charge are such factors.

(4) Reliability and lifetime

Various factors come together to degrade an electrochemical display cell. Common factors which may be mentioned are imperfections in the packaging, degradation of the electrodes, and side reaction products caused by inappropriate operating conditions or impurities in the material. The following are particular degradation phenomena of viologen systems [13, 14]:

1. If the applied voltage is too high then the deposited coloration material will undergo an irreversible secondary reduction reaction and the electrodes will be contaminated.
2. If the memory time is long the crystalline state of the deposited film will change and the erasability will be impaired.

Factors causing deterioration of tungsten oxide systems include: (a) electrochemical dissolution of the film into the electrolyte and (b) reduction of the display transparent electrodes [15].

In an attempt to overcome the dissolution phenomenon [15, 16] of the tungsten oxide film, the use of a non-aqueous solvent or a solid electrolyte has been considered, but in this case the lithium motile ions and sodium impurities are stored within the film, and a loss of reversibility of the coloration is indicated [17, 18]. The basic causes of degradation are therefore still being investigated.

The lifetime of an electrochemical display device is determined by the number of coloration and bleaching cycles. The current level is a maximum of approximately 10^7 on/off cycles [16, 17, 19], and in a viologen system, using the coloured state as the base line, the so-called diffusion systems in which the memory function is controlled allow a lifetime of 10^8 cycles [20].

3.3 INORGANIC ELECTROCHEMICAL DISPLAY TECHNIQUES

3.3.1 Transition metal oxide types

The most highly developed inorganic ECD materials are transition metal oxides such as those of tungsten, molybdenum, titanium, and vanadium. In particular, systems using a combination of tungsten oxide and a liquid electrolyte are close to the level of practicability.

(1) Amorphous tungsten oxide materials

Tungsten oxide used in electrochemical displays is normally made by the vacuum evaporation method, but may also be formed by sputtering [21] or spray [22] methods or the method of anodic oxidation [23]. In the case of vacuum evaporation using resistive heating, a powder or sintered form of WO_3 is put in a tantalum or tungsten boat in a vacuum of approximately 1×10^{-5} Torr, is heated to 1000 to 1300°C and caused to sublime, and then a substrate of glass or similar material on which the electrodes are formed and which has been heated to 80 to 120°C has a transparent film of 0.3 to 1 μm thickness deposited on it. Depending on the boat temperature, the vacuum, and the speed of vapour deposition, a film of WO_x $(2 < x < 3)$ is formed with an elementary composition intermediate between WO_3 and WO_2. This film has been shown by X-ray and electron diffraction analysis to be amorphous.

The crystalline structure of WO_3 is shown in Fig. 3.12, but it is thought that in the amorphous structure there are partial oxygen atom deficiencies distributed randomly in three dimensions, so that large tunnel structures are created within the crystal lattice and the mobility of the coloration ions is increased. The film has a density of 5.5 to 6.5 g/cm^3 and is porous, while water adsorption has also been observed. Films produced by the sputtering method are often polycrystalline because of the rise in temperature of the substrate or the sputtering rate, and this type of film has poorer coloration properties. Also, if heated in air to 340 to 400°C crystallization occurs [24], and the composition approaches WO_3. Therefore the absorption spectrum is shifted to the low-energy end and in practice the contrast may be worsened.

(2) Tungsten oxide film/liquid electrolyte cell

An example cell construction is shown in Fig. 3.8(c). On a glass substrate display electrodes formed from a transparent conductive film such as In_2O_3 (Sn) or SnO_2 (Sb) are formed in a pattern (Fig. 3.13), and a tungsten oxide film is formed by

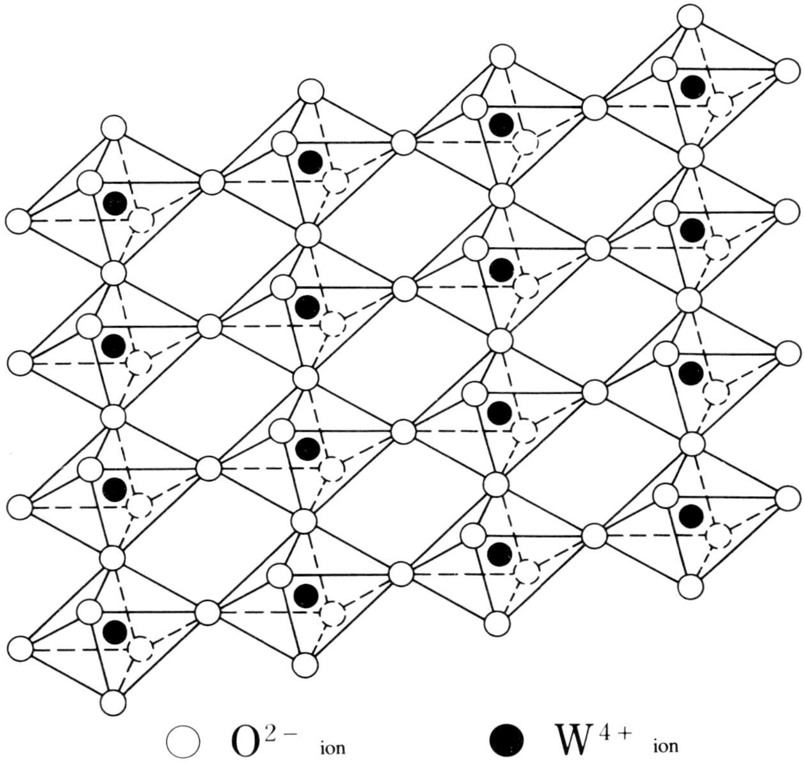

Figure 3.12 *Octahedral structure of* WO_3 *crystals*

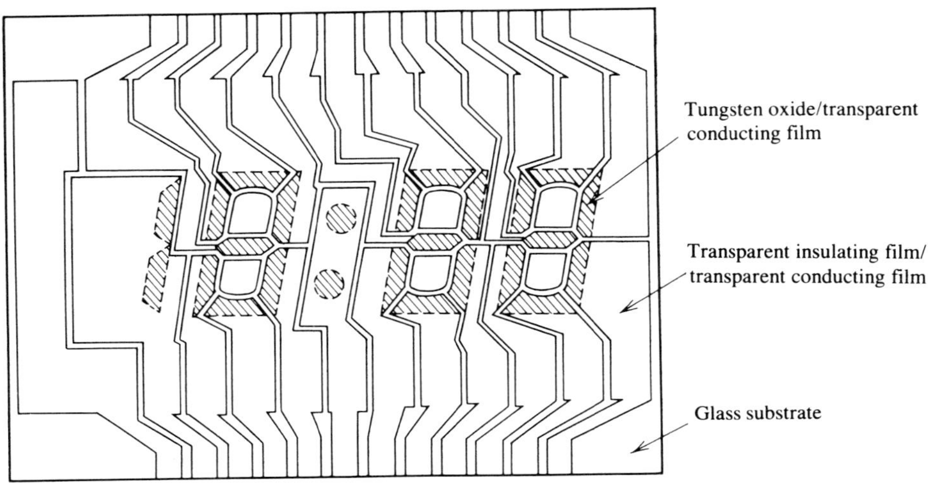

Figure 3.13 *Display electrode pattern*

vapour deposition or other method on these display portions. Those parts of the electrodes outside the display portions are covered with a transparent insulating protective film.

One key to stable operation is the counterelectrode, and reversibility of the reaction is very important; that is, as shown in Fig. 3.1, when electrons and cations are injected on the display side, the charge balance is maintained if both are emitted on the counterelectrode. It has therefore been shown that the best response is obtained with an electrode of graphite or graphite that has been mixed with tungsten oxide powder [25], or in which the counterelectrode is provided with a vacuum evaporation film that is the same as the display electrodes [10, 26], so that the counterelectrode is bronzed with a coloured tungsten oxide. There are also indications [27] that the voltage actually applied to the coloration layer fluctuates because of fluctuations in the counterelectrode potential during operation, and it has been considered to stabilize the voltage by using for the counterelectrode an iron complex [28] or a mixture of carbon with iron compounds [29]; the method has also been proposed whereby a third electrode is provided and used to counter fluctuations in the potential [30]. There is also an example of constructing a transmission type cell in which charge exchange is carried out at the counterelectrode, using crystalline porous WO_3 which has no optical absorption change in the visible spectrum region [31].

Between the display electrode and the counterelectrode a reflecting layer of a porous ceramic plate or a porous organic sheet is inserted and used as a background. An electrolyte solution which includes the ions used for coloration is also disposed between the two electrodes. Initially, inorganic aqueous solutions such as sulphuric acid were used so that the high mobility hydrogen ions could be used to improve response. Subsequently, to improve the memory characteristics and prevent the electrolysis of water during operation, with the object of improving reliability, experiments were made with ECD cells using the electrolyte diluted by glycerol or ethylene glycol. On the other hand, in long-term operation, there is still some electrochemical dissolution of the film and degradation of the electrodes, and most recently a liquid electrolyte has been used with, for example, lithium perchlorate dissolved in a non-aqueous solvent such as propylene carbonate, γ-butyrolactone, or acetonitrile. A non-aqueous electrolyte has a lower conductivity than an aqueous electrolyte, and Li^+ and Na^+ cations of relatively large diameter are used, so that the response is worsened but the reliability is very good; the response time is of the order of 0.1 to 0.5 second and the lifetime of a cell has reached 10^7 cycles. This type of display is already suitable for a clock or watch display (minute display), but further improvements in long-term stability and response are needed.

(3) Tungsten oxide thin-film/semi-solid electrolyte cell

A semi-solid macromolecular electrolyte that contains a certain amount of water is used as a hydrogen ion electrolyte, within an air-tight seal. In this way the dissolution of the tungsten oxide layer and degradation of the electrodes can be reduced, which yields a high quality cell.

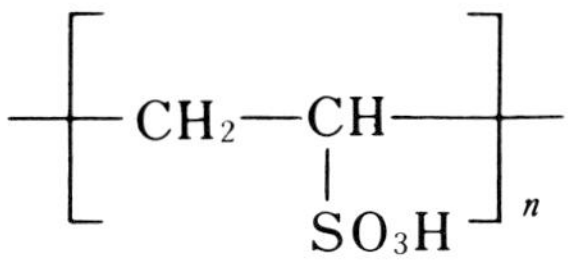

Polyvinyl sulphonic acid

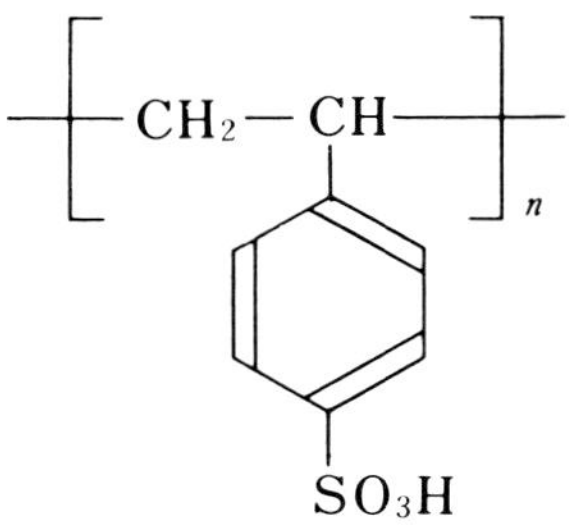

Polystyrene sulphonic acid

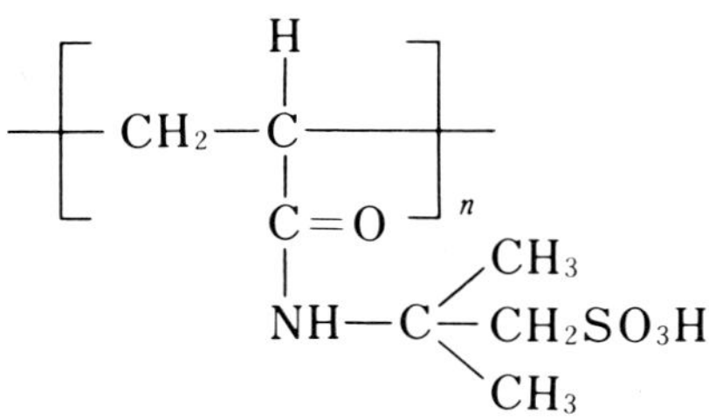

Polyacrylamide methyl propane sulphonic acid

Figure 3.14 *Macromolecular electrolytes [32]*

Various macromolecules having sulphone radicals were tested as in Fig. 3.14, and acrylamide methyl propane sulphonic acid (AMPS) polymer was chosen as having the best chemical and electrochemical stability, the best adhesion with the electrode, and the best miscibility with the background colouring element [32]. For the electrolyte itself, the AMPS polymer has mixed with it a background colour element and, to improve the condition, polyethylene oxide (PEO); water is added and it is used in a sheet form.

Between the tungsten oxide film and the electrolyte itself to prevent hydration of the tungsten oxide by the water within the AMPS a protective layer is formed from a thin film of HEM (hydroxy ethyl methacrylate)/AMPS copolymer. For the counterelectrode, graphite or a mixture of graphite and MnO_2 designed to have a maximal surface area is used as an electrode solidified with acrylic fibre and a flocculating agent. As seen in the following equation, the surface carbon atoms

Table 3.4 *Characteristics of tungsten oxide/semi-solid electrolyte ECD elements [32]*

Response time	0.9 s
Contrast	1.7
Applied voltage	1.0 V maximum
Charge amount	3.7 mC/cm^2
Lifetime	10^7 cycles minimum
Shelf life	3 years minimum
Memory function	48 hours minimum

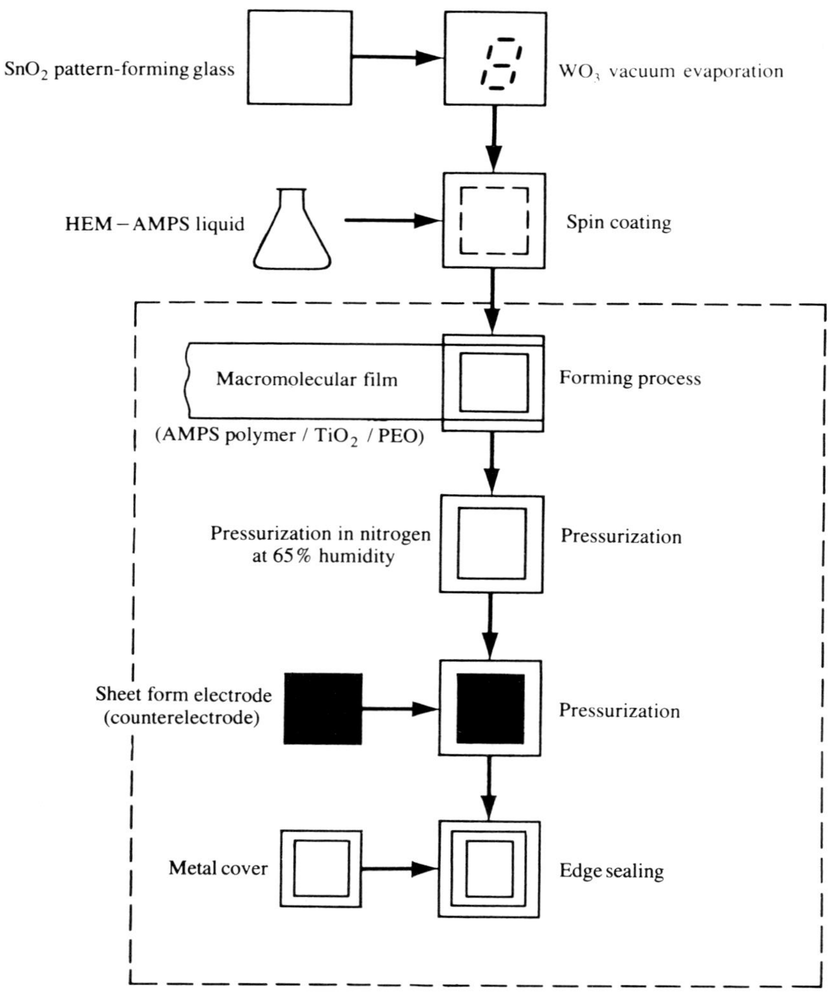

Figure 3.15 *Stages in manufacture of an ECD using a macromolecular electrolyte [32]*

provide good reversibility for releasing and receiving hydrogen ions:

$$(C{=}O) + H^+ + e^- \rightleftarrows (C{-}OH) \tag{3.8}$$

If a mixture with the oxidizing agent MnO_2 is used, the counterelectrode electromotive force is increased to 0.95 V and the display can be erased without applying an external voltage simply by shorting. An ECD element with a perfect seal designed to eliminate water evaporation has characteristics as shown in Table 3.4. Compared with the fully solid state element described next, the electrolyte layer is somewhat complicated, but compared with a liquid electrolyte, as seen in Fig. 3.15, the production process is simplified and future prospects for this type are good.

(4) Tungsten oxide thin-film/thin-film dielectric cell

The attraction of ECDs lies in the possibility of a fully solid state element. A solid state system has already been seen in leading work by Deb [4, 33] and also his patent [34], and by means of thin-film formation techniques a simple cell construction can be achieved by laminating the electrodes, coloration layers, and ion conduction layers on a substrate. One fully solid state type uses a dielectric film as an ion conduction layer and achieves the coloration cycle by means of hydrogen ions derived from water incorporated into the film from the atmosphere.

Thin films tested to date include oxides such as SiO [33], ZrO_2 [34], Ta_2O_5 [35], and Cr_2O_3 [36], and fluorides such as MgF_2 [33], CaF_2 [33], and LiF [37]. Of this type, an ECD element using chromium oxide [38] gives good characteristics, as shown in Table 3.5. As a *p*-type semiconductor, chrome oxide has a very high adsorbed water retention ability, and as shown in the following equations, gives rise to favourable results for carrying out the reversible coloration and bleaching reaction in the opposite sense to tungsten oxide, which is a *n*-type semiconductor:

$$WO_3 + xH^+ + xe^- \rightleftarrows H_xWO_3 \tag{3.9}$$

$$Cr_2O_3 + xOH^- + xH^+ \rightleftarrows (OH)_xCr_2O_3 \tag{3.10}$$

A device that is close to practical applicability has a layer that is a mixture of chromium oxide and vanadium pentoxide between the tungsten oxide and chromium oxide, and this provides low power consumption and improved memory characteristics (Table 3.5).

Another possibility which has been considered is to provide materials that are coloured in the oxidized state and materials that are coloured in the reduced state in first and second coloration layers, and construct an element with the structure: transparent electrode/first coloration layer/insulating layer/second coloration layer/transparent electrode/glass substrate [39–41]. When a voltage is applied from outside, both the first and second layer become coloured, and the display is formed from a synthesis of these colours; when the polarity is reversed, the reactions go in the opposite direction and both are bleached.

Table 3.5 Characteristics of WO_3/Cr_2O_3 and V_2O_5/Cr_2O_3 solid state ECD elements [38]

Charge amount consumed	5 mC/cm^2 approx.
Lifetime	10^6 cycles minimum (1 Hz operation)
Response time	0.5 s maximum
Memory function	1–2 hours

(5) Tungsten oxide thin-film/solid electrolyte cell

Another all solid state system uses ions within a solid electrolyte for coloration. ECD elements have been developed using super-ion conduction solid electrolytes such as $RbAg_4I_5$ [21], Na-β-Al_2O_3 [42], Li-β-Al_2O_3 [42], and $NO_3Zr_2Si_2PO_2$ [43]; of these the closest approach to the point of practical application is the β-alumina element shown in Fig. 3.16. This is a solid state system, and in view of the need for ensuring insulation and air-tightness, the design is such as to prevent reaction between ions within the solid and the atmosphere. Unfortunately response is slow, but by incorporating a heater a response time of one second or less is possible; a stable blue display on a white background is obtained, and a lifetime commensurate with the liquid electrolyte types is achieved.

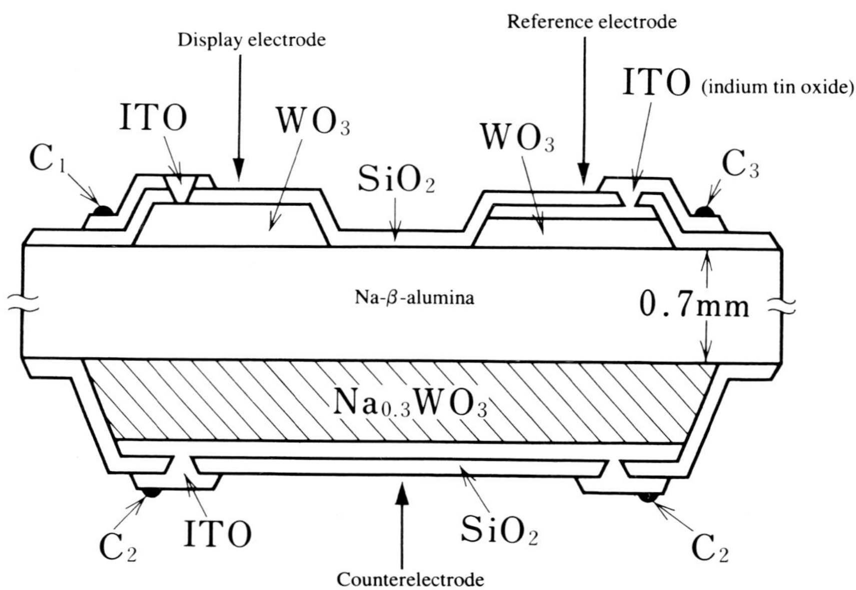

Figure 3.16 *ECD using Na-β-alumina solid electrolyte [42]*

Table 3.6 *Coloration ancillary layers used in solid ECDs*

Type	Motile ions	Material	Reported by	Reference
Dielectric film layers	H^+	MgF_2	Deb	33
	H^+	CaF_2	Deb	33
	H^+	SiO	Deb	33
	H^+	ZrO_2	Hajimoto *et al.*	35
	H^+	Ta_2O_5	Hajimoto *et al.*	35
	H^+	Cr_2O_3	Shimizu *et al.*	36
	H^+	$Cr_2O_3 + V_2O_5$	Shizukuishi *et al.*	38
	H^+	LiF	Stocker *et al.*	37
Solid electrolytes	Na^+	Na-β-Al_2O_3	Green and Kang	42
	Na^+	$Na_{1+x}Zr_2Si_xP_{3-x}O_{12}$	Barna	43
	Li^+	Li_3N	Miyamura *et al.*	44
	Li^+	LiI	Taira *et al.*	45
	Li^+	Li_2WO_4	Yoshimura *et al.*	46
	Ag^+	$RbAg_4I_5$	Green *et al.*	21

In this type of device, forming a solid electrolyte with higher conductivity into thin-film form is essential to realize an ECD with good response, and ECD devices have been reported using thin films of Li_3N [44], LiI [45], and Li_2WO_4 [46]; exploiting recent advances in solid state ionics, the similarities with battery technology will make various materials applicable, and future developments are awaited. Table 3.6 lists solid electrolytes and thin-film dielectrics applied to electrochemical displays.

(6) Integrated coloration layer and electrolyte cell

Generally electrochemical displays using a solid electrolyte have the disadvantage of slow response, but if the coloration material and the hydrogen ion conductor are integrated into a single layer then the response can be improved. A simple construction has been proposed for this type, with PWA($H_3PO_4(WO_3)_{12}\cdot 29H_2O$) or PMA($H_3PO_4(MoO_3)_{12}\cdot 29H_2O$) sandwiched between electrodes [47]. Initially, with an Ag (or Cu or Li)/PWA/SnO_2 construction, a blue coloration with an optical density of approximately 0.5 (ΔOD) was achieved by applying 2.5 V for 100 ms; later substituting graphite electrodes a response speed of 100 ms or less was achieved with an applied voltage not exceeding 2 V [48, 49]. The memory function is not so good with a half-life time of one hour for a reduction in the optical density. The reliability of the cell depends on the contact between the electrodes and the PWA, and as the coloration proceeds the overall electron conductivity increases so that the bleaching characteristics and memory function worsen; overcoming this is the key to practical application.

(7) *Materials other than tungsten oxide*

A number of materials exhibiting a similar coloration phenomenon to tungsten oxide has also been discovered. Table 3.7 shows some materials that undergo a coloration or colour change in a reversible reduction reaction by hydrogen ions or alkali ions of small ion radius. Materials that undergo coloration in an oxidizing reaction are described in the next section, 3.3.2, 'ECD anodic materials'.

Materials such as TiO_2 [50] and MoO_3 [51] have been studied for as long as

Table 3.7 *Cathodic materials other than WO_3*

MATERIAL	COLOUR CHANGE	REFERENCE
TiO_2	Colourless↔black	50
MoO_3	Colourless↔blue–black	51
V_2O_5	Yellow↔green	54
Nb_2O_5	Colourless↔dark blue	55
WO_3 (Au doping)	Blue↔red	53

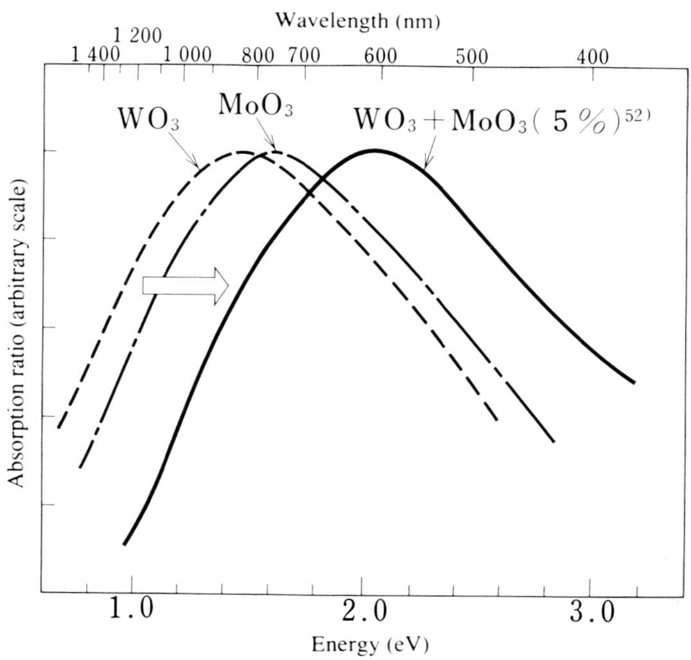

Figure 3.17 *Spectral shift of tungsten oxide caused by molybdenum additive*

WO_3, but to date tungsten oxide has the best set of properties that have been found. The object of using tungsten oxide is (a) to bring the absorption spectrum closer to the luminosity curve and improve the coloration efficiency, and (b) to change the absorption spectrum to produce colours other than blue and allow multicolour displays. The first objective has been achieved by combining MoO_3 in the WO_3 [52]. The absorption peak of WO_3 alone is 900 nm and that of MoO_3 is 800 nm; in a mixture of WO_3/MoO_3 obtained by simultaneous vapour deposition the spectrum changes with the content of molybdenum, and a content of 5% MoO_3 gave the film the shortest wavelength absorption. The absorption film is shifted to 600 nm (the luminosity curve peak is at 550 nm) and the coloration efficiency is improved (Fig. 3.17). The multicolour display of (b) has been attempted by forming a metal ceramic with a metal colloid of gold or platinum dispersed in a film of WO_3 by the sputtering method, and the coloration tested [53]. In this case the film is coloured initially by the colloid, but the WO_3/Au (5 to 20%) cermet changes colour between blue and red. Other transition metal oxides such as V_2O_5 [54] and Nb_2O_5 [55] are known, and V_2O_5 undergoes colour changes between yellow and blue–green.

3.3.2 Anodic coloration materials

(1) Iridium oxide

An iridium oxide film formed within a 0.5 M sulphuric acid, subject to an applied voltage of about 1 V, undergoes a reversible coloration reaction between colourless and blue–black as a result of anodic oxidation* [56]. In this case the coloration occurs at the anode, the reverse of the case with tungsten oxide; this type of material is termed an anodic material.

The anodically formed iridium oxide film is porous and even in the earliest cell, to achieve ΔOD approximately 0.4, a fast response time of 40 ms was obtained. Because, however, of dissolution in the solution the lifetime is short at 10^5 cycles, and as a method of reinforcing the film, heat treatment (85°C for 15 minutes) has been proposed. This improves the stability and increases the lifetime to some 6×10^5 cycles, and with an applied voltage of 1 V and ΔOD approximately 0.3, the response is improved to 20 ms (coloration) and 10 ms (bleaching) [57].

Since, however, an oxidized metal plate is not optimal for constructing a display element, trials were next made with anodic oxidation using an SnO_2 electrode on which iridium had been deposited by electron-beam vacuum evaporation method. Because of the higher resistance of SnO_2 the response was worsened at 0.2 s, but as with a tungsten oxide system an element could be constructed with a freely chosen background. Next a method was developed

*The material is steeped in an electrolyte solution and a current is passed with this as the anode; by means of an electrochemical oxidation reaction the anode itself is oxidized and a film is formed covering the surface of the anode.

whereby an iridium oxide film was formed directly on an ITO electrode of low resistance by the reactive sputtering method, and fast response, 0.1 s or less, and a lifetime of 2×10^7 cycles was obtained [59]. As in tungsten oxide systems, non-aqueous electrolytes have been considered for the cell construction, and trials have been made of completely solid state devices [60] using solid sheet form fluoride ($PbF_2/PbSnF_4$) or Nafion macromolecules with hydrogen ion conduction; practical development is continuing, concentrating on a fast response rate.

(2) Other anodic materials

Apart from iridium oxide, other anodic materials include oxides of Rh [61], Ni, and Co. These are all formed by anodic oxidation. Rhodium oxide is formed in an alkali bath and changes between yellow and dark brown, and like IrO_2 it exhibits high response speeds (coloration 40 ms, bleaching 20 ms), but has only been investigated up to a cycle life of approximately 30 hours and 10^5 cycles. Colour changes have been confirmed between black and dark blue for nickel oxide and purplish red to black or grey for cobalt.

These anodic materials are useful in themselves as coloration materials, but they are seen as offering most promise in combination with cathodic materials

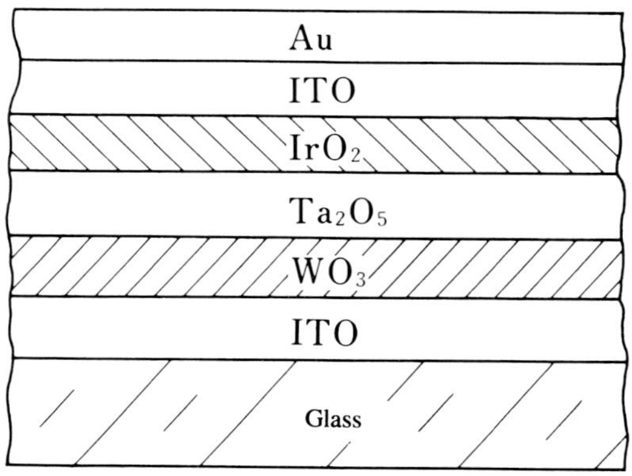

Figure 3.18 *Solid laminated ECD element*

⟨Bleaching⟩ ⟨Coloration⟩

$$WO_3 + xH^+ + xe^- \rightleftarrows HxWO_3 \text{ (within } WO_3)$$

$$xH_2O \rightleftarrows xH^+ + xOH^- \text{ (within } Ta_2O_5)$$

$$Ir(OH)_2 + xOH^- \rightleftarrows Ir(OH)_{2+x} + xe^- \text{ (within (} Ir(OH)_2$$

Figure 3.19 *Reactions within a solid laminated ECD element*

such as WO_3, as the second coloration layer in the solid laminated construction described in Section 3.3.1[4]. An example of such an ECD element is shown in Fig. 3.18, and the charge exchange reactions in each of the layers can be modelled as shown in Fig. 3.19.

3.3.3 Electroplating type ECDs

It has long been seen as possible to utilize silver plating on an electrode for a display, but this has recently reached the stage of a practical device [62]. As shown in Fig. 3.20, silver iodide or sodium iodide is dissoved in a non-aqueous solvent such as acetonitrile or methanol, and for a silver film to be deposited on a transparent electrode the response speed is relatively high.

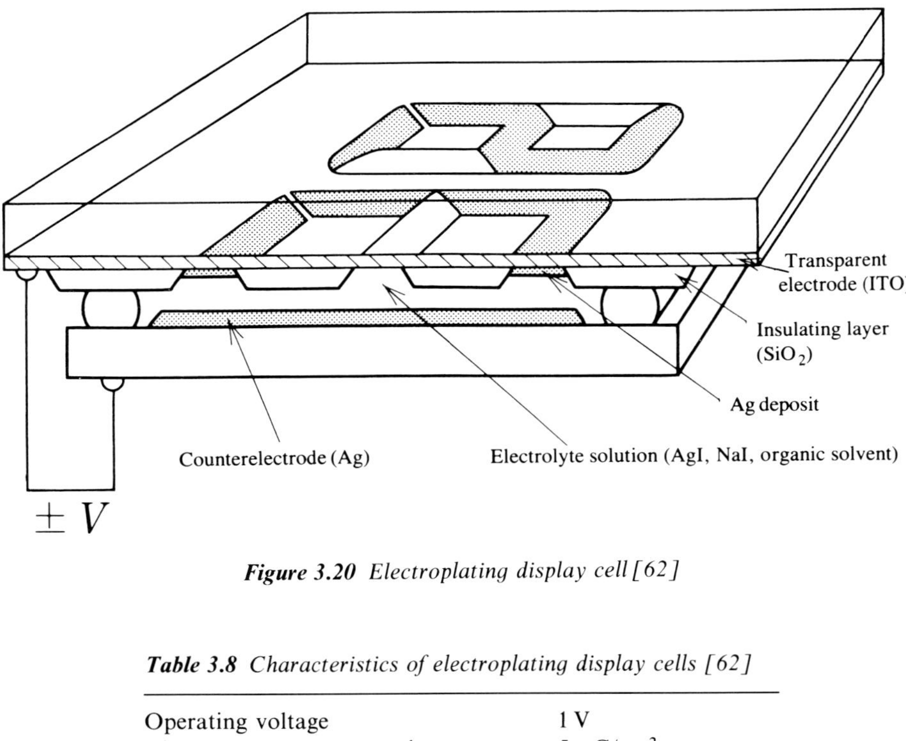

Figure 3.20 *Electroplating display cell [62]*

Table 3.8 *Characteristics of electroplating display cells [62]*

Operating voltage	1 V
Charge amount consumed	5 mC/cm^2
Average current (watch minute display—10 minute memory)	4 μA
Write time	50–200 ms
Coloration density (OD)	0.3
Lifetime	10^7 cycles minimum

When the polarity is reversed the silver on the electrode is redissolved and the colour disappears, but since in any case side reactions cause a natural elution of the silver, it is necessary in order to preserve the display to apply a refresh pulse at intervals at a number of minutes. The operating temperature range is wide (−40 to 80°C) and a lifetime of 10^7 cycles can be reached, so that application is being considered. Other characteristics are shown in Table 3.8. There are also reports [63] of a method using tetraalkyl ammonium iodide (NR_4I), ammonium iodide (NH_4I), and silver iodide, in which an electroplating display of $Ag^0 \cdot (AgI)_3$ is obtained by applying a one second pulse of 2 V.

3.3.4 Graphite laminar compound type ECDs

A laminar compound (C_6Li, $C_{12}Li$, C_8M, $C_{24}M$, $C_{36}M$, M = K, Rb, Cs) in which an alkali metal is introduced between graphite layers (intercalation) undergoes a colour change, as shown in Fig. 3.21 when a voltage is applied with the passing in and out of the alkali metal [64]. With a cell construction using a liquid electrolyte including lithium ions, a fast response of 200 ms is obtained with an applied voltage of 3 to 5 V. It is also possible to use a solid ion conductor such as Li_3N.

A colour change has also been reported [65] in an InN thin film combined with a liquid electrolyte to which a voltage is applied. Experiments have also been reported [66] on the use of iron complex materials such as Prussian blue ($Fe_4(Fe(CN)_6)_3$) and similar materials.

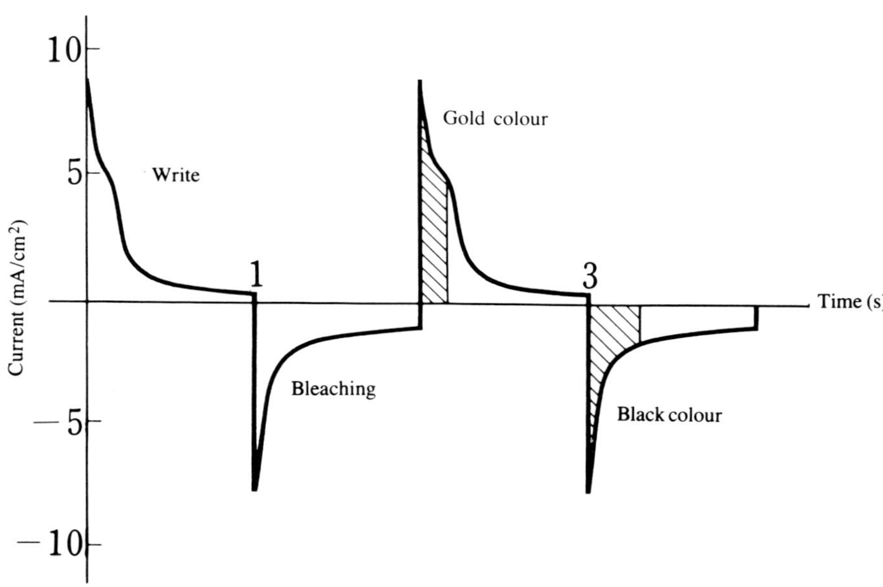

Figure 3.21 *Coloration changes in a graphite laminar compound [64]*

3.4 ORGANIC ELECTROCHEMICAL DISPLAY TECHNIQUES

3.4.1 Viologen-based ECDs

(1) Viologen derivative materials

Organic electrochemical displays using viologen (4,4-dipyridium compound) were some of the first to be investigated, and their development is fairly advanced. The viologen derivatives shown in Fig. 3.1(2) have been found [5] to combine the advantages of a plating reaction with a dye reaction, have relatively good response speed, and, as shown in Fig. 3.22, have attracted attention as materials that can be adapted to different colours by changing the substitution groups. By using a solution of more than one viologen derivative in a single cell, a

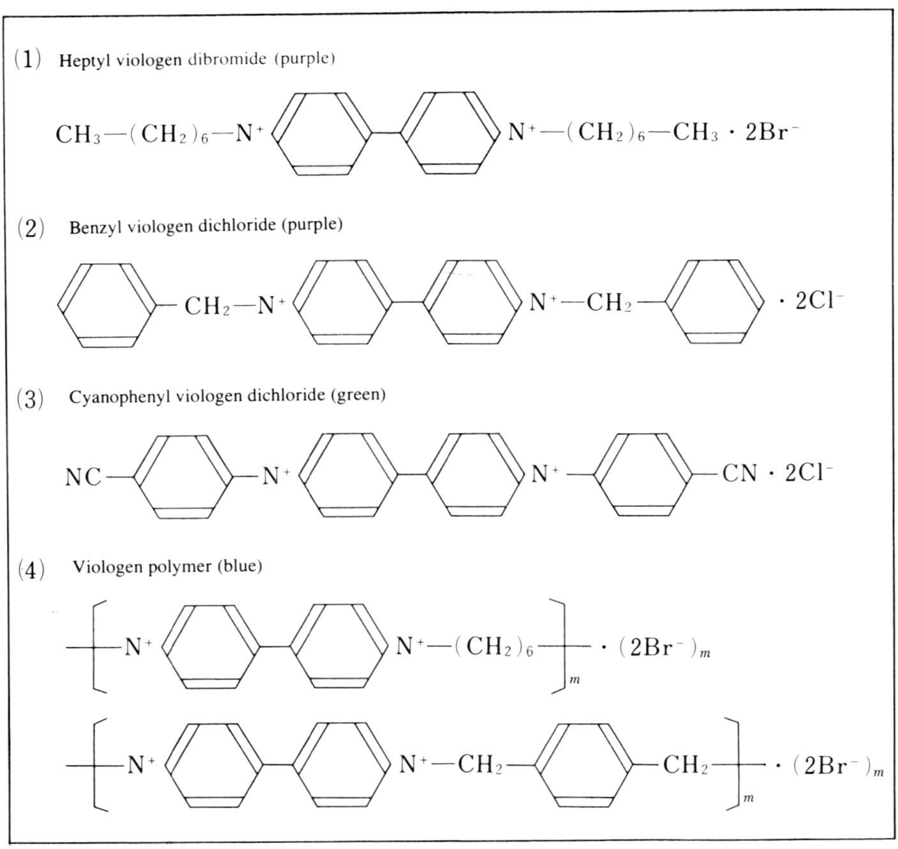

***Figure 3.22** Various colorations of viologen derivatives*

multicolour display can be produced because of the difference in threshold values for coloration [67].

As has already been mentioned, the degradation of viologen-based electrochemical displays was reported almost immediately. In order to overcome this problem proposals have been made to make the substitution groups at opposite ends of the viologen structural formula extremely asymmetric [68], or to make the bromine ion a phosphate radical [69]; polymerization has also been suggested as a method of preventing recrystallization.

(2) Memory type materials

In a viologen-based ECD, insoluble radical cations adhere to the electrode, and in an aqueous solution, even if the voltage is turned off, the coloured film on the electrode is preserved; the viologen derivative is termed an organic memory type material. Other display materials similar to viologen have also been proposed, one example being the 4-benzoyl pyridium derivative shown in Fig. 3.23, and the possibility of a display element with memory capability using an aqueous electrolyte has been discussed [70].

(3) Diffusion type cells

The term diffusion type [20] is applied to an electrochemical display where the base state is reduction with the materials coloured and the display is effected by oxidizing the display materials. In order to form the display enclosure with the viologen maintained in the reduced coloured state, amines are added and titanium oxide or similar is added as a background.

The anode is used as the display electrode and the oxidized bleached portion is displayed as white (background colour) within the coloured reduced portion.

If the applied voltage is removed, then diffusion causes a return to the base state of the reduction colour, and this is the basis of bleaching, but in practice the process is made complete by reversing the polarity of the display electrode. The memory capability disappears, but the instability of the reduction display state in a viologen system is avoided and a lifetime of 10^8 cycles has been reported.

The biggest defect of viologen systems is the degradation of erasability, and other methods of overcoming this from the beginning include the use of a non-

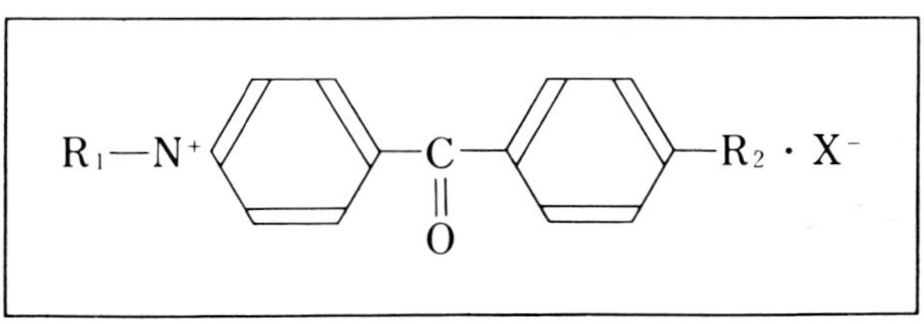

Figure 3.23 *Benzoyl pyridinium derivative [70]*

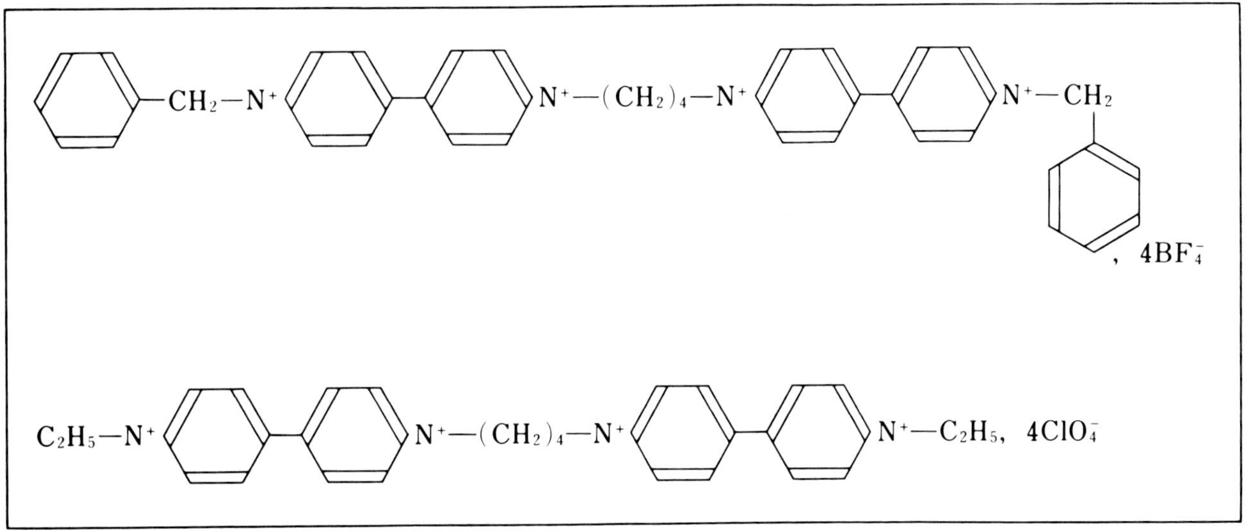

Figure 3.24 *Viologen dimers[14]*

aqueous solvent and the method whereby a salt is put directly into solution [71]. In the former case methanol or acetone is used as the solvent, and thus insoluble salts are not formed. The applications are restricted by the sacrifice of the memory capability, but an improved lifetime can be anticipated.

(4) Polymer materials

To date, with conventional viologen derivatives there have been the problems of pollutant appearing on the surface of the electrode and traces of the erased coloration film remaining. Two reasons can be given for this: (a) there is only a narrow voltage range in which the oxidation–reduction reaction can be carried out stably and (b) in the memory state the thin film properties formed can easily change. In order to solve these problems a polymerized viologen material has been synthesized. For example, the dimers shown in Fig. 3.24 only change to the crystalline form with difficulty, and the oxidation–reduction reaction has good reversibility, so that they offer the possibility of a long-life viologen ECD [14].

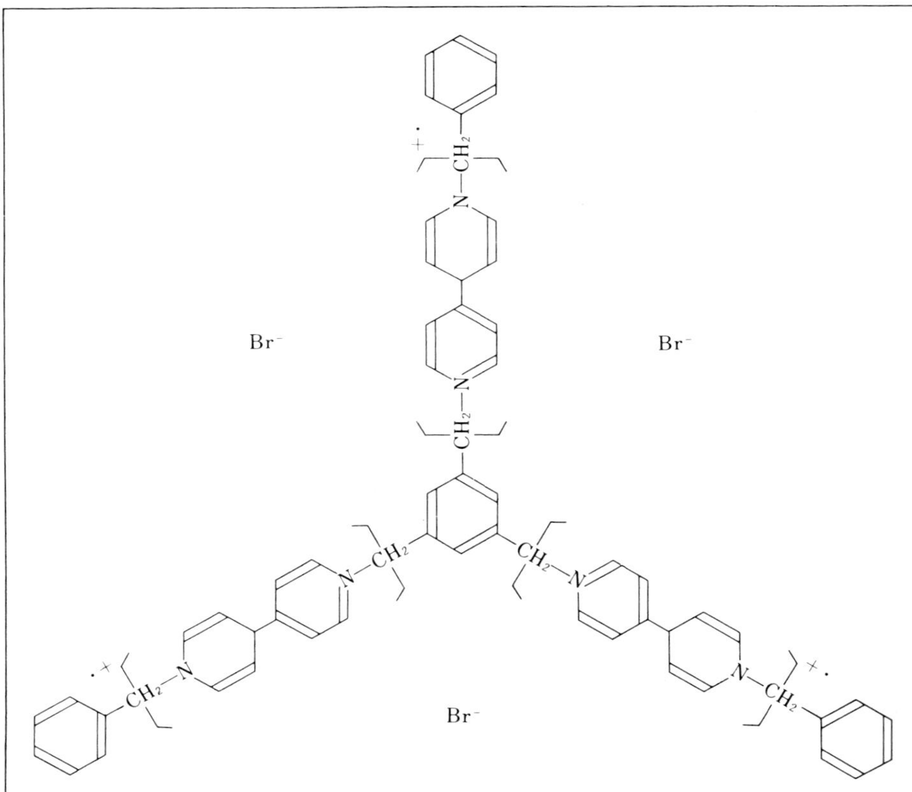

Figure 3.25 *Viologen trimer [72]*

The trimer shown in Fig. 3.25 and higher polymer materials have also been proposed [72]. In a viologen polymer, the oxidation–reduction voltage range is widened compared with the monomer, and the stability of the memory state is improved, but these features do not vary greatly according to the degree of polymerization.

3.4.2 *Organic metal complex type ECDs*

The change in absorption spectrum when a voltage is applied to the diphthalocyanin rare earth compound thin film shown in Fig. 3.26 has been known for a long time, and recently a report of an electrochemical display using a combination of this thin film with a KCl or NaCl aqueous electrolyte has been published [73]. A film of Lu-diphthalocyanin formed by the vacuum evaporation method on a transparent electrode exhibits a colour change when a potential of ± 1.5 V is applied, from red–orange ($+V$) to green ($V=0$) to blue ($-V$). The response time is 50 ms or less, and because of the speed even at low temperatures it has a wide operating temperature range. The power consumption is small, 0.5 to 1.5 mJ/cm^2, and the lifetime is approximately 5×10^4 cycles. There have also been reports of a colour change under an applied voltage in an Er-diphthalocyanin vacuum evaporation film [74].

In an ECD cell in which a cobalt bipyridine composite (Co(2,2′-bipyridine)$_3 \cdot 2NO_3$) is dissolved in a KNO_3 solution, then as shown in the following equation an oxidation–reduction reaction occurs with a change between monovalent and divalent at the display electrode and between divalent and trivalent on the counterelectrode; the monovalent dark blue deposit is utilized for the display [75]:

$$[Co(bipy)_3]^{2+} \underset{-e}{\overset{+e}{\rightleftarrows}} [Co(bipy)_3]^{1+}\downarrow$$

$$[Co(bipy)_3]^{2+} \underset{+e}{\overset{-e}{\rightleftarrows}} [Co(bipy)_3]^{3+}$$

3.4.3 *Polymer type ECDs*

An ECD using a modified polymer as shown in Table 3.9(a) has been described [76] in which tetrathiafulvalene (TTF) is combined with polychloromethyl styrene. Because of the production of radical cations by the electrochemical oxidation, the thin film of TTF (pale yellow) formed on the electrode changes colour to TTF^+ (green) or TTF^{2+} (purple). Since the coloration is a conductive macromolecular thin film fixed to the surface of the electrode, there are none of the defects of the viologen system or of display loss through diffusion, and like the diphthalocyanin complex, the characteristics are good: good response, multicolour possibilities, and memory capability.

A multicolour display is also possible with a TTF derivative (Table 3.9(b)) or

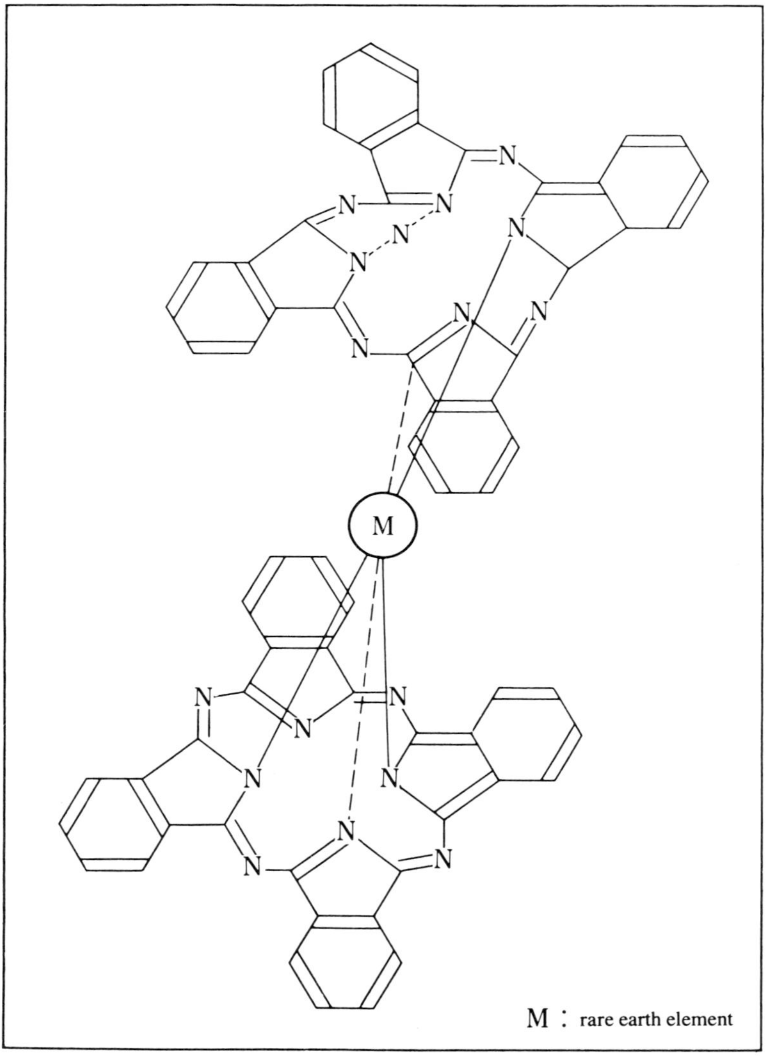

Figure 3.26 Diphthalocyanin rare earth compound structure [73]

pyrazoline pigment (Table 3.9(c)) in pendant form. Other materials considered include polymers with a viologen structure on a side chain or the main chain and polymers having a pigment radical such as a phthalocyanin ring, porphyrin ring, or pyrazoline ring.

A fully solid state organic ECD has also been described recently [77] in which pyrazoline or TTF is dissolved in a macromolecular solvent such as polymethacronitrile together with $LiClO_4$, and this is coated and dried to a thin film. An

ECD cell (Fig. 3.27) using a film formed by casting a viologen polymer complex and a conductive powder on the electrode is stable with respect to an electrolyte, and with an operating voltage of 1 V or less has a fast response of 30 ms and a long lifetime of at least 10^7 cycles [78].

3.4.4 Optically activated ECDs

The first type uses an optical semiconductor. If a photoconductor layer is laminated on a coloration layer then information can be written by an incident light beam.

For example, an ECD cell having a tungsten oxide film and a CdS film laminated between a pair of electrodes was described in 1969 [4]. A construction using amorphous silicon as the photoconductive layer has also been reported [79].

In the second type, a material that is itself photosensitive is used. The pyrazoline compound shown in Fig. 3.28 shows a very large increase in optical absorption due to a combination of a photochemical reaction and an electrochemical reaction. In a cell filled with a chloroform solution of this substance between the electrodes, if a voltage is applied while irradiating with 400 nm light, radical cations are produced and a strong blue–green coloration is induced [80].

3.4.5 Organic solution type ECDs

In comparison with a viologen type display in which coloration seeds are deposited, an organic solution type ECD is a display which uses both coloration seeds and bleaching seeds dissolved within an electrolyte solution. The construction is normally simple: transparent electrode/solution (porous background)/electrode; this allows the vivid colours that are a characteristic of organic pigments to be utilized. An example is a solution in which a supporting salt is added to an anthraquinone type reduction coloration pigment. As a further development of this type, a cell with both an anthraquinone type pigment and an oxidation coloration material such as TTF or styryl pyrazoline sealed in solution in the cell offers the possibility of a three-colour display (blue/red/background) [81].

3.5 APPLICATIONS OF ELECTROCHEMICAL DISPLAYS

3.5.1 Drive techniques for ECDs

In an electrochemical display the coloration density is controlled by the amount of electric charge, and the basic principle of the driving method is therefore a constant charge drive. In practice, however, a constant voltage drive or constant

Table 3.9 *TTF and pyrazoline polymer display characteristics [76]*

Polymer	Absorption peak wavelength (nm) (colour change)	Coloration efficiency $\Delta OD/mC \cdot cm^{-2}$	Time required for 1 mC to flow τ/mC (ms)
(a) 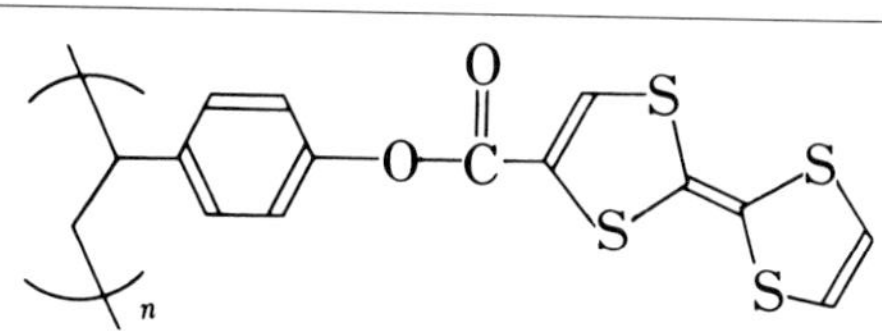	515 (orange ⟶ brown)	0.044	200
(b) 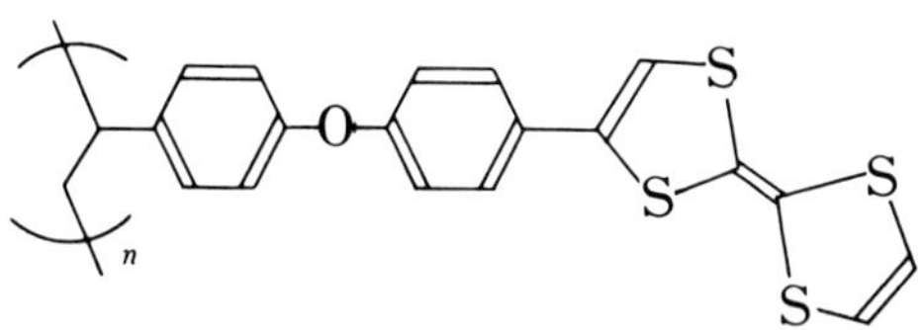	650 (yellow ⟶ green)	0.057	150

(c)	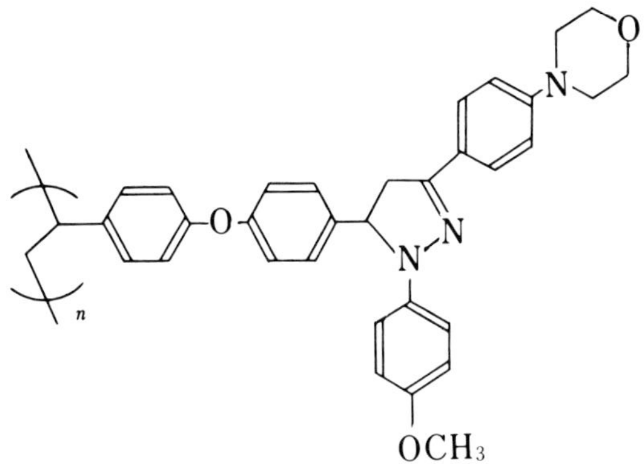	510 (yellow ⟶ green)	0.11	50
(d)	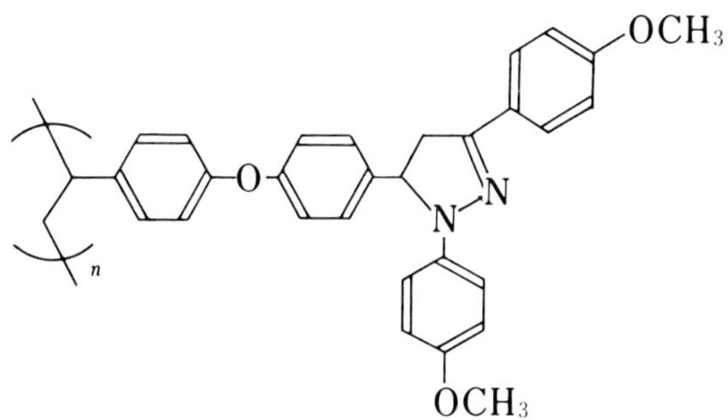	554 (yellow ⟶ red)	0.25	100

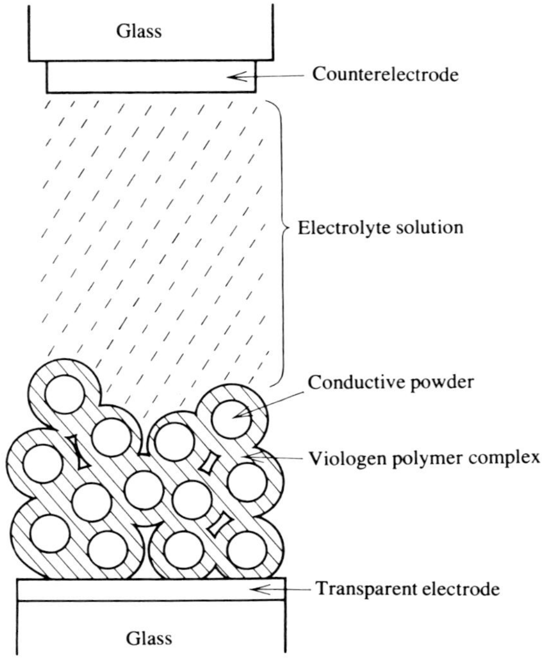

Figure 3.27 *Cell using viologen polymer complex [78]*

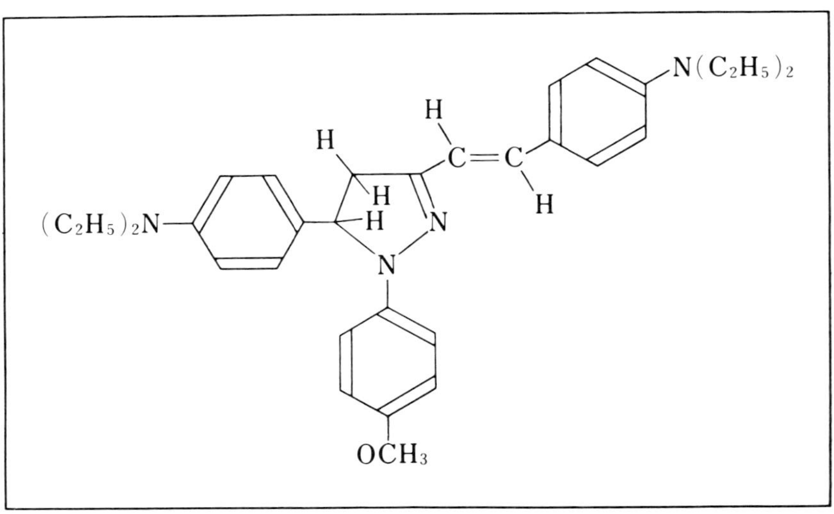

Figure 3.28 *Pyrazoline compound used in optically activated ECD [81]*

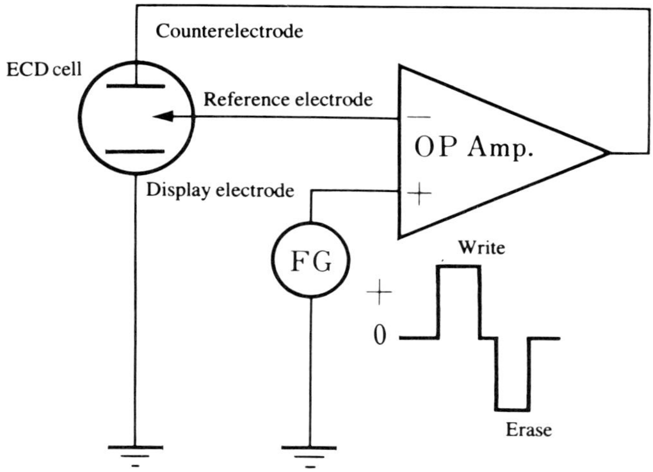

Figure 3.29 Voltage-limiting drive method

current drive is used. In this case it is necessary to control the charge amount in the display operation and preserve the balance of charge in coloration and bleaching. Since an electrochemical display normally has a memory function, once the display is effected, it is maintained even if the voltage is cut off, so that in order to remove the display it is necessary to apply a voltage of the opposite polarity.

As has already been pointed out, the length of time for which the memory function is effective depends on the materials used and the cell construction, and, for erasing, the circuit impedance is also related to the memory time [82]. Then, as well as fluctuations of the electrode potential and counterelectrode reactions, care must be taken over the particular characteristics of electrochemical displays such as the temperature dependency of the electrochemical reaction.

In a constant current drive, the coloration density is controlled by: current × time = charge amount. However, the voltage to maintain the current constant is restricted by the maximum voltage that can be applied to the element. The constant voltage drive is a very simple drive method but requires extra circuitry to solve the problems of temperature compensation* and display area variation.† In the potential regulation drive method which uses a third reference electrode as well as the display electrode and the counterelectrode (Fig. 3.29), the potential of the reference electrode is used as a datum and the charge balance between coloration and bleaching is controlled using the display electrode potential.

In a practical electrochemical display device with a large number of segments a three-electrode drive is somewhat complicated, and for that reason a two-

*Since the response increases with an increase in temperature, a circuit to correct for this is required.
†In order to obtain the same degree of coloration, the charge amount must be varied with the display area.

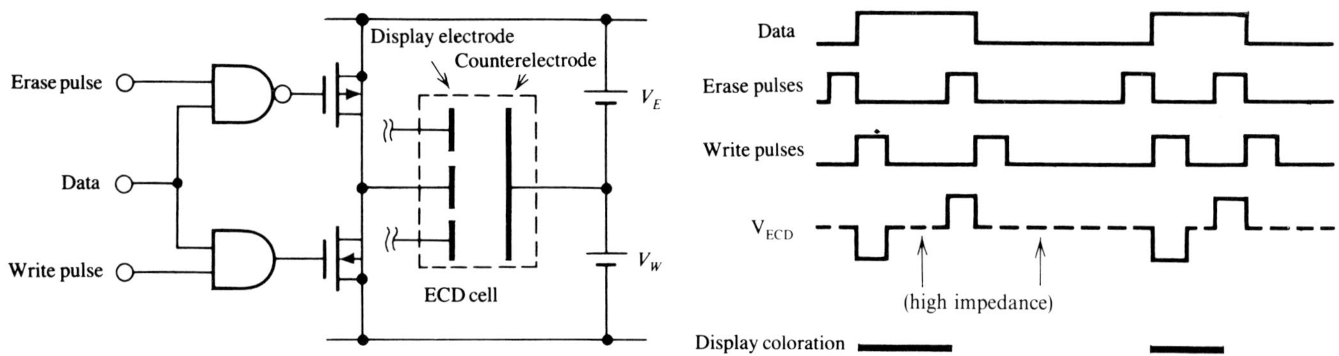

Figure 3.30 *Timing chart for example basic ECD circuit*

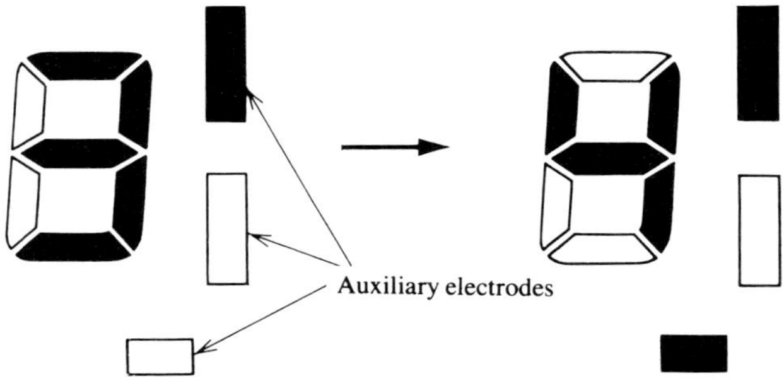

Figure 3.31 *Example display change using the charge transfer method [30]*

electrode drive is adopted, and the charge amount is controlled by regulating the pulse width, waveform, and applied voltage. The basic circuit for the case when a reduction coloration material such as WO_3 is used is shown in Fig. 3.30 together with the drive pulse waveforms. A switching element is turned on and off according to the write and erase pulse signals, and when the potential on the display electrode is a negative voltage corresponding to the write voltage V_W the coloration occurs, whereas when a positive voltage is applied the display is erased. The switching element is maintained in the off state while there is no input signal. In order to maintain the coloured state no applied voltage is necessary, but if the display electrode and counterelectrode are simply left at the same potential, a current will flow between the electrodes because of the electromotive force produced in the coloured state, and the coloration state will change. Therefore, except while switching, the circuit must be maintained in a high impedance state.

A method that has been developed for controlling the balance of charge injection and release is the charge transfer method [30]. In addition to the display electrodes, auxiliary electrodes are provided to store charge; when a segment is erased, the released charge that is surplus, which is not being used for the coloration of a different segment, is stored on these auxiliary electrodes. The coloration density of an ECD is determined by the surface charge density maintained by the electrode, so that if charge transfer is carried out between electrodes of the same surface area the coloration density can be maintained at a constant value. Figure 3.31 shows an example display using this technique.

3.5.2 Clock and watch displays

Possible applications for electrochemical displays include: clocks, watches, calculators, displays for gauges, terminal displays, announcements, signboards, advertisements, vehicle displays, optical valves, and optical shutters. At present, however, not only is the reliability of ECD elements inadequate but, by comparison with liquid crystal displays, there are the disadvantages of higher power consumption, slower response, and the difficulty of dynamic drive

Figure 3.32 *Watch displays (Ebauche)*

operation; inevitably, therefore, the range of applications is limited. Therefore, exploiting the memory capability, which is a characteristic of electrochemical displays, it is reasonable to target displays that require a low frequency of update.

For this reason, the principle applications of electrochemical displays which have been marketed are clocks and watches. The circuit for a clock comprises an oscillator, a frequency divider circuit, a time counter, a control circuit, a pulse signal circuit, a decoder, and an ECD driver. When the coloration/bleaching pulse signal from the pulse signal generator circuit and the display output signal from the segment decoder are synchronized, the segments are switched and a coloration or bleaching operation is carried out. Examples of watches with minute displays using a WO_3 electrochemical display are shown in Fig. 3.32.

3.5.3 Large-scale displays

There are a number of reasons why electrochemical display devices are appropriate for large-scale displays: the spacing between electrodes is not critical, there is no viewing angle dependency, and there are no difficulties on a large scale with forming the materials (solid layers) or sealing (liquid). However, since a charge amount depending on the display area is required, the voltage drop arising from transparent electrodes which have high resistance means that a fast display is difficult, and the display will tend to be mottled. Even for a fixed pattern display some ingenuity is therefore required over the electrode construction and the drive method. If these points are overcome, then large-scale display emphasizing decorativeness and requiring few changes will be an ideal application for electrochemical displays. A transmission type of cell, furthermore, lends itself to application in a projection display system or as an optical modulator. An optically activated cell could also be used for image recording using an optical writing technique.

3.5.4 Matrix displays

Because electrochemical displays do not exhibit a clear threshold value and because of the memory capability, multiplexing drive is difficult. Segment displays are therefore the general rule and there are few examples of matrix displays. In principle, if an active matrix method is adopted in which there is a switching element incorporated into each pixel then multiplexing drive is possible, but in view of the limitations of large-area implementation and cost in this technique, an early implementation is unlikely. However, an experimental ECD has been made [83] using the TFT drive technique [84]. This has a 64×64 array of switching FETs on a 2.5 inch silicon wafer, which are used for multiplexing drive of a viologen derivative. The electrolyte used is a mixture of orthophosphate anions and hypophosphite anions, and the electrodes used are silver electrodes formed by the pulse plating method; with short switching times satisfactory display contrast was obtained, and with an operating voltage of ± 2 V a response of between 2 and 15 ms was obtained.

For organic-based electrochemical displays, it would be possible to implement a multiplexing drive by restricting the memory capability. Normally in an ECD coloration seeds are produced on one electrode to produce the display, but the bleaching speed can be controlled by the addition of oxidation or reduction agents* that do not contribute to the display. In this type of system a dynamic drive is possible. As an example, a matrix addressing display has been described [85] which uses an imidazole material as the electrochemical material and has 20×144 dots. The display is refreshed at a rate of 33 Hz to maintain the display contrast, and bleaching is carried out naturally by the added bleaching agent as described above.

3.5.5 Multicolour displays

Because of the brilliance of colour, which is a characteristic of electrochemical displays, they offer the chance of multicoloured displays in the future. A certain

*If the coloration material is a reduction coloration material then an oxidizing agent will be used for the bleaching.

degree of colour variation is possible in both tungsten oxide systems and iridium oxide systems, but they do not offer multicoloured facilities to the same extent as organic materials. The technique whereby coloration seeds of different colours are deposited selectively on the same display element segment by means of a change in electrode potential or polarity is special to the organic liquid materials. Recently a macromolecular electrochromic material modified by coloration and bleaching materials has been developed, making a new type of multicolour display possible. Electrochemical displays using diphthalocyanin rare earth composites or TTF polymers make it possible to display different colours on the same portion by means of stepwise variation in the applied voltage.

3.6 OUTLOOK

There are a number of requirements for the practical implementation of electrochemical displays, including: increased lifetime (at least 10^8 cycles), improved response (100 ms or less), and matrix drive techniques. The most serious of these is the question of lifetime. Research into developing electrochemical display devices really needs to concentrate on finding a way to maintain the reversibility of the electrochemical reaction. To date there have been no examples of devices where the reversible reaction accompanying the movement of charge carriers can be carried out 10^7 or 10^8 times. Moreover, in order to obtain high display contrast the amount of movement of the charge is increased and correspondingly the lifetime is shortened. In order to overcome these problems, a high-quality electrochromic material must be discovered and a device construction and drive method must be found to exploit it effectively. As this chapter has shown, a number of materials have been considered so far, but apart from the tungsten oxide type which have had some success in practical applications, fully reliable devices have not been achieved and there still seems to be some way to go until their implementation. Research and development is going on, however, over the whole range from the fundamental mechanism to application, and the electrochemical display looks likely to take its place in the range of passive display devices on the market in the not too distant future.

BIBLIOGRAPHY

Kmetz, A.R., and F. K. von Willisen (eds.): *Nonemissive Electrooptic Displays*, Plenum, New York (1976).

Pankove, J. (ed.): *Display Devices, Topics in Applied Physics*, Vol.40, Springer, New York (1980).

REFERENCES

[1] G. C. Sziklai: United States Patent, No. 2632045 (1953).
[2] S. Zaromb: *J. Electrochem. Soc.*, **109**, 903 (1962).

[3] R. W. Pohl: *Naturwissenschaften*, **20**, 932 (1932).
[4] S. K. Deb: *Appl. Optics Suppl.*, **3**, 192 (1969).
[5] C. J. Schoot, J. J. Ponjee, H. T. van Dam, R. A. van Doorn, and P. T. Bolwijn: *Appl. Phys. Lett.*, **23**, 64 (1973).
[6] R. S. Crandall and B. W. Faughnan: *Appl. Phys. Lett.*, **28**, 95 (1976).
[7] B. W. Faughnan, R. S. Crandall, and M. A. Lampert: *Appl. Phys. Lett.*, **27**, 275 (1975).
[8] M. L. Hitchman: *Thin Solid Films*, **61**, 341 (1979).
[9] B. Reichman, A. J. Bard, and D. Laser: *J. Electrochem. Soc.*, **127**, 647 (1980).
[10] S. K. Mohapatra: J. Electrochem. Soc., **125**, 284 (1978).
[11] J. P. Randin and R. Viennet: *J. Electrochem. Soc.*, **129**, 2349 (1982).
[12] J. Bruinink and P. van Zanten: *J. Electrochem. Soc.*, **124**, 1232 (1977).
[13] K. Belinko: *Appl. Phys. Lett.*, **29**, 363 (1976).
[14] J. Bruinink, C. G. A. Kregting, and J. J. Ponjee: *J. Electrochem. Soc.*, **124**, 1854 (1977)
[15] J. P. Randin: *J. Electron. Mater.*, **7**, 47 (1978).
[16] B. W. Faughnan and R. S. Crandall: Ch. 5 in J. I. Pankove (ed.), *Topics in Applied Physics*, Vol. 39, Springer Verlag, New York (1980).
[17] H. Morita and H. Washida: *Appl. Phys.*, **51**, 488 (1982).
[18] K. Yamanaka: *J. Appl. Phys.*, **54**, 1128 (1983).
[19] W. C. Dautremont-Smith: *Displays*, **3**, 3 (1982).
[20] M. Yaguchi: *Denshi Tsushin Gakkai-shi (Journal of the Electronic Communications Institute)*, **62**, 807 (1979).
[21] M. Green, W. C. Smith, and J. A. Weiner: *Thin Solid Films*, **38**, 89 (1976).
[22] R. Hurditch: *Electron. Lett.*, **11**, 142 (1975).
[23] A. DiPaola, F. DiQuarto, and C. Sunseri: *J. Electrochem. Soc.*, **125**, 1344 (1978).
[24] O. F. Schirmer, V. Wittwer, G. Baur, and G. Brandt: *J. Electrochem. Soc.*, **124**, 749 (1977).
[25] B. W. Faughnan, R. S. Crandall, and P. M. Heyman: *RCA Rev.*, **36**, 177 (1975).
[26] H. Morita and H. Washida: *Jpn. J. Appl. Phys.* **19**, L228 (1980).
[27] H. Morita and H. Washida: Denshi Tsushin Gakkai Gijutsu Kenkyu Hokoku (Technical Research Report of the Electronic Communications Institute), ED-79-24 (1979).
[28] S. Kondo and N. Yoshiike: Television Gakkai Gijutsu Hokoku, (Technical Report of the Television Institute), IPD-44-2 (1979).
[29] K. Yamanaka: *Jpn. J. Appl. Phys.*, **21**, 926 (1982).
[30] J. Tabata, S. Miyoshi, and N. Kaneko: Denshi Tsushin Gakkai Gijutsu Kenkyu Hokoku (Technical Research Report of the Electronic Communications Institute), ED-80-85 (1980).
[31] K. Matsuhiro and Y. Masuda: *Proc. SID*, **21**, 101 (1980).
[32] R.D. Giglia and G. Haacke: *Proc. SID*, **23**, 41 (1982).
[33] S. K. Deb: *Phil. Mag.*, **27**, 801 (1973).
[34] S. K. Deb and R. F. Shaw: British Patent, No. 1 186 541 (1970).
[35] Y. Hajimoto, M. Matushima, and S. Ogura: *J. Electron. Mater.*, **8**, 301 (1979).
[36] I. Shimizu, M. Shizukuishi, and E. Inoue: *J. Appl. Phys.*, **50**, 4027 (1979).
[37] H. J. Stocker, S. Singh, L. G. Van Uitert, and G. Zydzik: *J. Appl. Phys.*, **50**, 2993 (1979).
[38] M. Shizukuishi, I. Shimizu, and E. Inoue: *Jpn. J. Appl. Phys.*, **20**, 2121 (1980).
[39] Y. Takahashi: *Optics*, **11**, 365 (1982).
[40] R. Fujiwara, S. Kawate, and I. Shimizu: *Proceedings of Japan Display'83*, Kobe, p. 50 (1983).
[41] M. Watanabe, Y. Koike, T. Yoshimura, and K. Kiyota: *Proceedings of Japan Display'83*, Kobe,. 372 (1983).
[42] M. Green and K. S. Kang: *Thin Solid Films*, **40**, L19 (1977).
[43] G. G. Barna: *J. Electron. Mater.*, **8**, 153 (1979).

[44] M. Miyamura, S. Tomura, A. Imai, and S. Inomata: *Solid State Ionics.*, **314**, 149 (1981).
[45] K. Taira, F. Kanai, and S. Kurita: *Appl. Phys.*, **49**, 155 (1980).
[46] T. Yoshimura, M. Watanabe, Y. Koike, K. Kiyota, and M. Tanaka: *Jpn J. Appl. Phys.*, **22**, 152 (1983).
[47] B. Tell and S. Wagner: *Appl. Phys. Lett.*, **33**, 537 (1978).
[48] B. Tell and F. Wudl: *J. Appl. Phys.*, **50**, 5944 (1979).
[49] B. Tell: *J. Electrochem. Soc.*, **127**, 2451 (1980).
[50] E. Inoue, H. Kokado, and A. Izawa: *Appl. Phys.*, **43**, 54 (1974).
[51] T. C. Arnoldussen: *J. Electrochem. Soc.*, **123**, 527 (1976).
[52] B. W. Faughnan and R. S. Crandall: *Appl. Phys. Lett.*, **31**, 834 (1978).
[53] E. K. Sichel and J. I. Gittleman: *J. Electron. Mater.*, **8**, 1 (1979).
[54] S. Sato and Y. Kiyono: *Vacuum*, **24**, 568 (1981).
[55] B. Reichman and A. J. Bard: *J. Electrochem. Soc.*, **127**, 241 (1980).
[56] S. Gottesfeld, J. D. E. McIntyre, G. Beni, and J. L. Shay: *Appl. Phys. Lett.*, **33**, 208 (1978).
[57] G. Beni and J. L. Shay: *Appl. Phys. Lett.*, **33**, 567 (1978).
[58] K. S. Kang and J. L. Shay: J. Electrochem. Soc., **130**, 766 (1983).
[59] W. C. Dautremont-Smith, L. M. Schiavone, G. Beni, and J. L. Shay: *SID 80 Digest*, p.122 (1980).
[60] C. E. Rice and P. M. Bridenbaugh: *Appl. Phys. Lett.*, **38**, 59 (1981).
[61] S. Gottesfeld: *J. Electrochem. Soc.*, **127**, 272 (1980).
[62] J. Duchene, R. Meyer, and G. Delapierre: *IEEE Trans. Electron Devices*, **ED-26**, 1243 (1979).
[63] L. G. Van Uitert, I Camlibel, R. M. DeLaRue, T. R. Kyle, R. Pawelek, S. Singh, H. J. Stocker, and G. J. Zydzik: *Appl. Phys. Lett.*, **34**, 232 (1979).
[64] P. Pfluger, H. U. Kunzi, and H. J. Guntherodt: *Appl. Phys. Lett.*, **35**, 771 (1979).
[65] O. Takai: *Proceedings of Japan Display'83*, Kobe, p. 58 (1983).
[66] T. Ataka, T. Sakuhara, M. Shigeno, and K. Iwasa: *Proceedings of Japan Display'83*, Kobe, p.384 (1983).
[67] M. Yamana: *Jpn. J. Appl. Phys.*, **15**, 2469 (1976).
[68] G. G. Barna and J. G. Fish: *J. Electrochem. Soc.*, **128**, 1290 (1981).
[69] R. J. Jasinski: *J. Electrochem. Soc.*, **124**, 17 (1977).
[70] N. Yoshiike, S. Kondo, and M. Fukai: *J. Electrochem. Soc.*, **127,** 1496 (1980).
[71] H. T. van Dam: *J. Electrochem. Soc.*, **123**, 1181 (1976).
[72] H. Sato and T. Tamamura: *J. Appl. Polymer Sci.*, **24**, 2075 (1979).
[73] M. M. Nicholson and R. V. Galiardi: *SID 78 Digest*, p.24 (1978).
[74] M. Yamana: *Appl. Phys.* **48**, 441 (1979).
[75] K. Murao, M. Sato, and N. Shito: *Digest of Technical Papers*, The 8th Conference on Solid State Devices, Tokyo, p.113 (1976).
[76] F. B. Kaufman, A. H. Schroeder, E. M. Engler, S. R. Kramer, and J. Q. Chambers: *Appl. Phys. Lett.*, **36**, 442 (1980).
[77] Y. Hirai and C. Tani: *Appl. Phys. Lett.*, **43**, 704 (1983).
[78] T. Masumi, K. Nomura, K. Nishioka, H. Deguchi, and H. Ono: *Proc. SID*, **23**, 245 (1982).
[79] M. Shizukuishi, I. Shimizu, and E. Inoue: *Jpn. J. Appl. Phys.*, **20**, 2359 (1981).
[80] R. V. Pole, G. T. Sincerbox, and M. D. Shattuck: *Appl. Phys. Lett.*, **28**, 494 (1976).
[81] T. Ueno, Y. Hirai, and C. Tani: *Proceedings of Japan Display '83*, Kobe, p.66 (1983).
[82] H. Morita: *Jpn. J. Appl. Phys.*, **21**, 655 (1982).
[83] D. J. Barclay, C. L. Bird and D. H. Martin: *SID 80 Digest*, p.124 (1980).
[84] D. J. Barclay, C. L. Bird, and D. H. Martin: *J. Electron. Mater.*, **8**, 311 (1979).
[85] A. G. Arellano, G. S. Keller, P. J. Melz, M. D. Stattuck, and C. V. Wilbur: *SID 78 Digest*, p.22 (1978)

Chapter 4

Plasma display panels (PDPs)

4.1 INTRODUCTION

Digital displays as in panel type displays began with NIXIE display indicators. These display characters by light emission in the form of the required character, achieved by superimposed cathodes shaped as numerals or symbols etc. They have played a considerable role in the development of electronic displays. In a plasma display panel (PDP), instead of superimposed electrodes, segment electrodes are used. These may be arranged in a matrix of columns and rows.

From the point of view of their operation, PDPs may be classified into two types: the alternating current discharge type (a.c. type) and direct current discharge type (d.c. type). These may be further subdivided into the refresh type and memory type, based on the different methods of drive. They have many advantages: the construction is of small depth, they are not subject to flickering, a large display contrast ratio can be used, the screen can be made larger than in the case of other panel type displays, the response speed is rapid, they are of the active type, and multicolour displays can be produced by utilizing phosphors. A PDP was the first type of panel display to be made commercially available. Technological and market development has essentially been targeted on two fields.

The first of these fields is the field of display devices linked chiefly to computers. Another field is that of colour image display devices for wall-hanging televisions. Both of these have been supported by rapid progress in semiconductor technology, and advances in compactness and weight reduction. However, from the point of view of display quality and cost, etc., they have not yet reached the stage where they can compete with CRT displays. For this reason, research and development is proceeding in many places aimed at enabling them to catch up with CRTs.

This chapter gives an outline of the development of PDPs. After this, a description is given of the physical phenomena on which PDPs are based, their construction, their principles of operation, drive systems, and applications. There then follows a discussion of present problems and future trends.

4.1.1 *What is a plasma display?*

A 'plasma display' is the general term for a flat display utilizing the light emission that is produced by gas discharge. In a panel in which the discharge light emission is directly observed, a rare gas (of which neon is usually the main constituent) is sealed into the enclosure. In the case of colour display panels, display is effected by stimulating a phosphor applied, for example, to the front face of the display cell to emit light in response to ultraviolet rays from, for example, xenon. Such displays may be classified in accordance with their electrode structure as 'indirect discharge' types (AC-PDPs), wherein the electrodes are covered with a dielectric where they are exposed to the discharge space, and 'direct discharge' types (DC-PDPs), in which the electrodes are directly exposed to the discharge space.

4.1.2 *Development and history of plasma displays*

The first plasma displays were digital displays based on NIXIE display indicators. From this, flat multidigit display tubes were developed. Column and row matrix type plasma displays such as are used at present were developed in 1954 [1] (direct current discharge type (d.c. type)) and in 1966 [2] (alternating current discharge type (a.c. type)). In both cases, discharge light emission from a rare gas occurs at small dots arranged in the form of a matrix, and letters or diagrams etc. are displayed by combinations of dots.

Whereas the a.c. type operates by indirect discharge, the electrodes being coated with a dielectric, the d.c. type operates by direct discharge, the electrodes being directly exposed to the discharge space. Thus, these two types have contrasting discharge arrangements. Development and improvement have therefore proceeded with these two types independent of and in competition with each other.

(1) Alternating current discharge type (AC-PDP)

In the AC-PDP type, the electrodes are coated with a dielectric, and the display is operated by an a.c. discharge.

In the original type of panel invented by Bitzer and Slottow [2] of Illinois University in the USA, as shown in Fig. 4.1, the discharge path is restricted to 'discharge gaps' by means of a glass sheet in which are formed small holes arranged in matrix fashion in the regions where the row and column electrodes cross each other. Subsequently, improvements were made in respect of the electrodes and the pressure of the enclosed gas etc., for instance by Owens Illinois Inc. in the USA, and a construction was thereby obtained in which the glass sheet formed with small holes was dispensed with [3–9]. Panels have now been developed wherein the surface of the dielectric layer is coated with a protective film of MgO etc., preventing deterioration of the discharge and lowering the voltage [10–12].

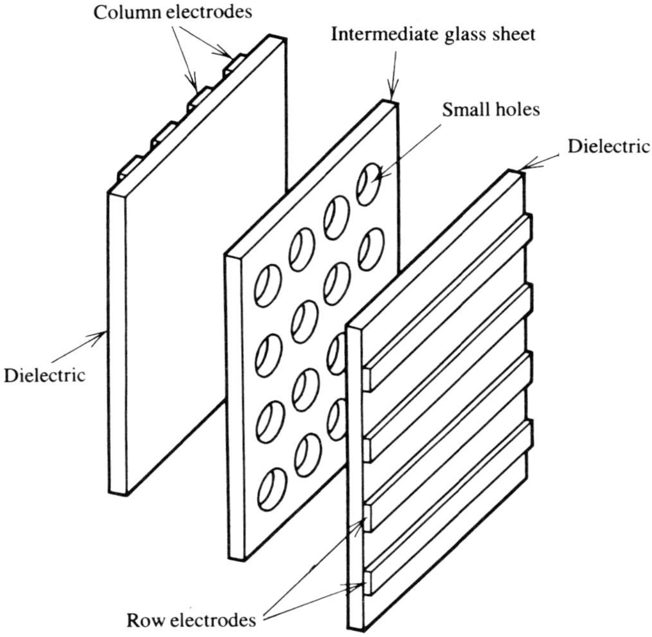

Figure 4.1 *Construction of AC-PDP as initially developed*

Improvements in panel construction have been made hand in hand with increased resolution and size. Ar first, the pitch of the discharge cells was about 0.6 mm. However, this has recently been reduced to about 0.3 mm. IBM in the United States has made commercially available [13] a large panel giving a 160 characters × 62 rows (9920 characters) display, with cell pitch of 0.36 mm. Photonics in the United States have developed an ultralarge panel of 80 cm × 60 cm, provided with 1596 × 1212 cells of pitch 0.5 mm [14].

Research into colour displays is also proceeding. At first, a method was investigated wherein phosphor was applied to the front face glass in the regions where the electrodes cross over each other. However, the phosphor deteriorates in a short time due to collision with positive ions. For this reason, panels have been announced wherein a colour display is produced by photoluminescence of a phosphor produced by ultraviolet excitation of xenon and which operates under conditions of surface discharge by applying a phosphor to the inside surface of a glass sheet on the front face side, a glass plate at the rear face side being arranged with column and row electrodes separated by a dielectric layer.

In all panels that are practically used, a gas is enclosed of which neon is the main constituent, and the red–orange colour light emission produced by neon gas discharge is utilized for the display. The operating voltage is 90 to 150 V, and the luminance is 150 to 250 cd/m^2.

In general, AC-PDPs are driven by applying an a.c. pulse. Both refresh operation and memory operation are used. Recently, to simplify the drive circuitry, a method has been developed wherein the discharge dot is shifted by applying a 'shift pulse'. Advances have been made, both in Japan and abroad, in embodying the drive circuitry in IC form.

Fujitsu employ a monolithic IC drive in which eight drive lines including gate circuitry are packaged [15]. Nippon Denki employ an IC drive that uses a package of 40 lines and consisting of a hybrid between a high-voltage drive circuit chip and low-voltage shift register chip [16]. Texas Instruments in the United States have commercialized a monolithic IC drive of 32 drive lines in which DMOS high-voltage drive circuitry and CMOS shift registers are packaged on a single chip [17].

(2) Direct current discharge type (DC-PDP)

A display wherein the electrodes are operated in the direct current discharge state and exposed to the discharge space is called a DC-PDP display, and was first proposed by R. L. Skellet [1]. A panel in which 100 rows and 20 columns of fine wire were arranged and enclosing neon gas was manufactured on a trial basis in 1954. However, with this panel, it was impossible to fire a plurality of cells simultaneously. In 1963, D. W. Moore [1] developed a prototype panel with neon enclosed in its interior, in which a plurality of cells can be fired simultaneously, a glass sheet formed with a matrix of small holes being arranged between two parallel conducting films coated on the inside face of two glass sheets in the region of the points of intersection of the electrodes, as shown in Fig. 4.2. However, this device is liable to crosstalk. In an effort to improve this, J. E. Thompson *et al.* [1] showed that firing of cells could be selectively achieved by connecting resistances in series with the electrodes.

Subsequently, De Boer [18] announced a television display wherein 100 row anodes of platinum wire and 40 column cathodes are arranged aligned with the positions of holes in an insulating sheet of aluminium oxide formed with small holes, these being sandwiched by two sheets of glass with neon sealed into the enclosure. This panel was manufactured on a trial basis.

These panels are of simple construction, but discharge does not occur immediately on application of voltage. Rather, there is a statistical lag in firing, and statistical scatter appears in failure of firing and in the luminance of the discharge cell. To prevent this, all DC-PDPs manufactured at present are provided with a priming discharge arrangement*. The effect of the priming discharge is to stabilize operation by supplying charged particles or excited atoms beforehand into the discharge space of the cell where firing is to take place. A typical example is the self-scanning panel display (SSPD) [19] developed by Burroughs in the United States. In this display, charged particles are supplied to the display cells by constant scanning of the priming discharge provided at the rear of these cells. In general, DC-PDPs are of fairly complicated construction,

*Also called pilot light.

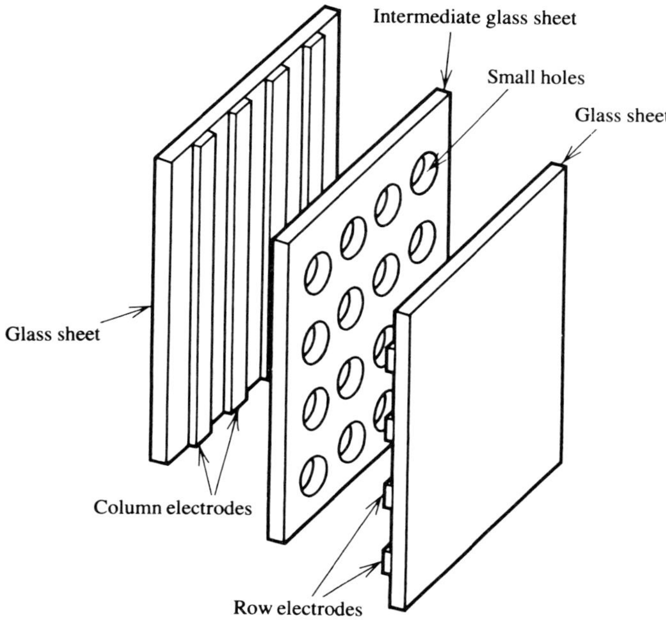

Figure 4.2 *Construction of DC-PDP as initially developed*

but their shape can be modified as desired, and various types of panels are proposed [16, 18, 20–31].

As well as improvements in panel construction, advances are also being made in improving the definition and increasing the size of such panels. The pitch of the first discharge cells was about 1.0 to 1.25 mm, but panels of 0.4 mm pitch have recently been made available as commercial products by Oki Denki for the display of Chinese of Japanese characters. Sony has also announced [28] a large panel provided with 1024 × 512 cells of 0.2 mm pitch.

Additionally, since DC-PDPs can easily introduce phosphors, colour display panels are being developed in many places [23–26, 30, 31]. The method of achieving a colour display is that a phosphor is applied to the inner surfaces of small holes formed in a cell sheet, these constituting a discharge path, or to the inside face of the glass sheet at the front of the display cell. The phosphor is stimulated to emit light by ultraviolet rays from xenon.

Normally, DC-PDPs are operated by a refresh system. However, in the case of colour display panels or large panels, this method produces insufficient luminance. The use of memory panels is therefore being investigated in various places. Methods of constructing memory panels are as follows: (a) connecting a series resistance to each discharge cell [33]; (b) operating the cells with a high-frequency, narrow-width pulse [32, 33]; (c) use of material of small secondary electron emission coefficient for the cathode; (d) combining an AC-PDP with the DC-PDP [34].

In the panels that are practically used at present, neon is the main enclosed gas and the display is usually effected by the red–orange colour produced by neon discharge. However, Oki Denki and Okaya Denki have developed a green display panel by application of a phosphor. The operating voltage of this DC-PDP is 180 to 250 V [35, 36], and its luminance is 120 to 200 cd/m^2.

Previously, hybrid ICs were chiefly used for the drive circuitry. However, in recent years, there has been a shift to high threshold voltage MOSICs and monolithic LSI packages incorporating shift registers and latch memories etc. have also been developed.

4.1.3 Physical phenomena of plasma display discharge

A PDP is a device utilizing the discharge phenomenon. The discharge phenomenon as utilized in PDPs is a functional extension and development of the existing phenomenon, and also constitutes an unknown field whose characteristics are different from those of devices using the existing modes of discharge light emission. For example, in a PDP, the cells constituting the pixels are operated by a minute gap discharge, and the pressure of the enclosed gas is an intermediate gas pressure of several tens to 100 times that of the conventional devices. The discharge phenomenon itself is therefore a topic for research. We shall outline the physical phenomena of PDP discharge contrasted with the basic discharge phenomena that are already known.

A PDP discharge cell is equipped with a cold cathode, and is operated under glow discharge conditions. The basis of its operation is that a large number of discharge cells can be turned on and off reliably at high speed. Consequently, the firing conditions are vital.

When voltage is applied to the discharge cells, the electrons emitted from the cathode are accelerated by the electrical field and collide with the enclosed gas molecules, exciting them and producing ionization, resulting in the production of excited atoms or electrons and positive ions, which migrate towards the anode. Some of the electrons produced by ionization migrate towards the anode, while some of the positive ions return to the cathode. The initiation of insulation breakdown, i.e. discharge within the gas, occurs by a synergetic effect between secondary electron emission (α effect) due to collision of positive ions with the cathode and ionization of the gas by electrons (α effect). The condition for its occurrence is expressed by the following equation [37–42], where l is the interelectrode distance:

$$\int_0^e \alpha \mathrm{d}x = \log_e\left(1 + \frac{l}{\gamma}\right) = F(\gamma) \tag{4.1}$$

In the above equation, α is the primary Townsend ionization coefficient and γ is the secondary electron emission coefficient. The former depends on the applied voltage. In the case of the rare gas enclosed in a PDP, taking the gas pressure as P,

Table 4.1 Rare gas discharge constants

	He	Ne	Ar	Kr	Xe
$A(\mathrm{cm}^{-1}\cdot\mathrm{Torr}^{-1})$	4.4	8.2	29.2	35.7	65.3
$B(\mathrm{V}^{1/2}\cdot\mathrm{cm}^{1/2}\cdot\mathrm{Torr}^{1/2})$	14.0	17.0	26.6	28.2	36.1
$k_+(\mathrm{cm}^{3/2}\cdot\mathrm{Torr}^{1/2}/\mathrm{V}^{1/2}\cdot\mathrm{s})$	4.1×10^4	1.7×10^4	8.3×10^3	5.4×10^3	$40. \times 10^3$

it can be expressed by the following known equation [37–41]:

$$\frac{\alpha}{P} = A \exp\left[- B\left(\frac{P}{E}\right)^{1/2}\right] \tag{4.2}$$

Assuming parallel plate electrodes, on solving equations (4.1) and (4.2), the firing voltage V_f in the uniform electrical field is given by the following equation:

$$V_\mathrm{f} = \frac{B^2(Pl)}{\{\log_e [A(Pl)/E(\gamma)]\}^2} \tag{4.3}$$

where $E = V_\mathrm{f}/l$ and A and B are constants. The values of A and B in rare gas are shown in Table 4.1 [38, 39].

Equation (4.3) expresses V_f as a function of Pl, and is known as Paschen's law. Since in an actual discharge cell the electrodes are not normally parallel plate electrodes, a correction must be applied to the electric field. Figure 4.3 shows an example of a Paschen curve expressing the relationship between V_f and Pl. The value of Pl in the discharge cell is selected in a range greater than the minimum value of this voltage curve.

The atoms that are excited by collision with electrons or gas molecules release

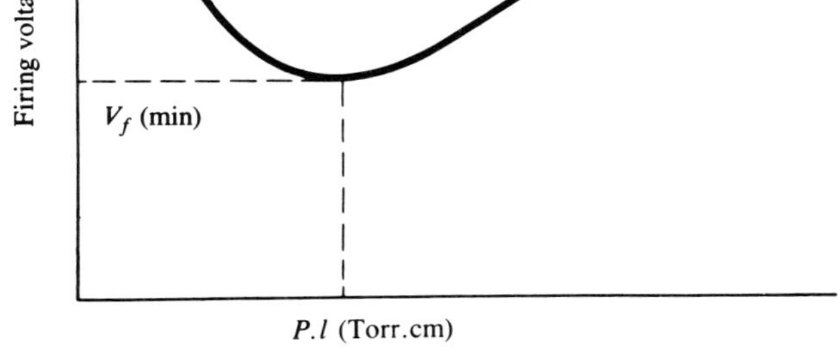

Figure 4.3 *Relationship between uniform field firing voltage and Pl*

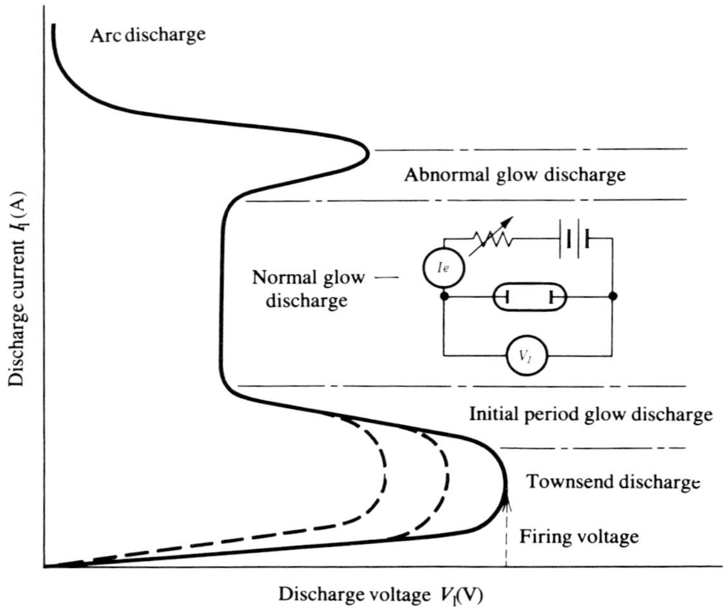

Figure 4.4 *Diagram of discharge voltage–current characteristic of cold cathode discharge tube*

some or all of their energy as light within a period of about 10 to 8 seconds, and then return to their normal state or to a state of lower excitation. The wavelength λ that is emitted when a transition occurs from excited state E_1 (eV) to state E_2(eV) is expressed by the following equation [37, 42]:

$$\lambda = \frac{12\,400}{E_1 - E_2} \quad \text{Å} \; (1\text{Å} = 10^{-8}\,\text{cm}) \tag{4.4}$$

Some excited states are not capable of emitting light, and are called metastable states. Such metastable states last about 10 to 4 seconds. When a metastable atom receives energy from another electron or photon, it shifts to a higher excited state, and then either emits light or returns to the normal state, releasing its energy. The 'Penning' effect in a mixed gas discharge, to be described later, utilizes metastable atoms.

A typical cold cathode discharge voltage current characteristic is shown in Fig. 4.4. The initial weak current range is the Townsend discharge region. When the voltage rises, the first glow discharge region, having a negative characteristic, is reached. This region is highly unstable. When the current is further increased, stable discharge occurs in the regular glow discharge region and the discharge voltage is maintained constant. When the current is still further increased, the region of abnormal glow discharge is entered. Increased discharge voltage results in heating of the cathode and a shift to arc discharge. The discharge cells of a DC-

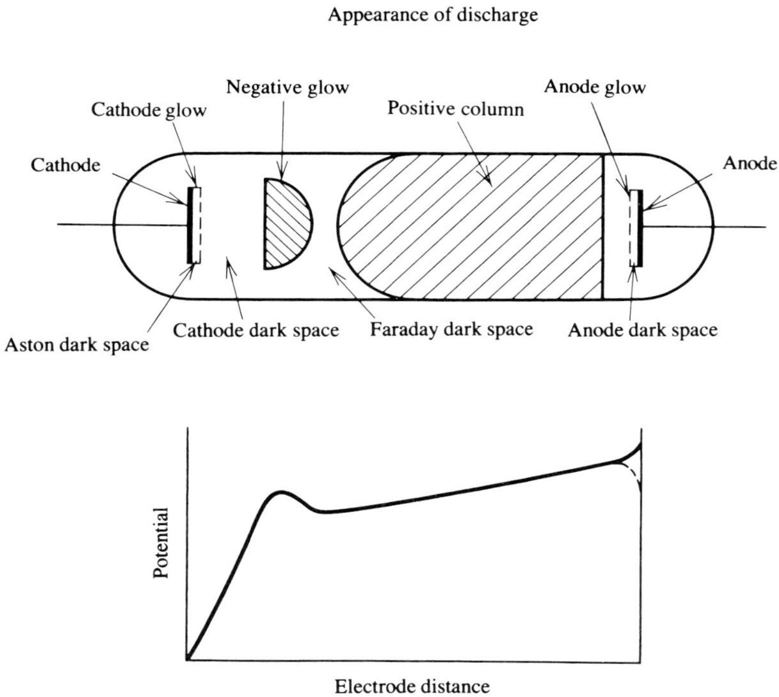

Figure 4.5 *Appearance of discharge and corresponding potential distribution*

PDP operate in the region from normal glow discharge to abnormal glow discharge.

Figure 4.5 shows the potential distribution and external appearance of the region between the cathode and anode during glow discharge. This external appearance is not always the same. For example, the positive column does not appear with short gaps, as in normal discharge cells. The cathode glow and anode glow are often not seen. The negative glow gives the essential character of the glow discharge, and always appears.

The most important part in a glow discharge is the cathode drop region. Most of the discharge voltage depends on this. When the relationship between the cathode drop voltage V_C, cathode current density j, and thickness d of the cathode drop region is obtained for a normal glow discharge and abnormal glow discharge, the cathode drop characteristic expressed by the following equation is obtained [37–41]:

$$\frac{j}{P^2} = \varepsilon_0 k_+ \frac{V_C^{3/2}}{(Pd)^{3/2}} \tag{4.5}$$

where P is the gas pressure, k_+ is the mobility coefficient of the ions (see Table 4.1), and ε_0 is the permittivity of vacuum.

The following equations are obtained when the normal cathode drop voltage V_{cn} and normal converted current density j_n are derived for parallel plate electrodes. These may be used as standards for the operating voltage and operating current of a DC-PDP discharge cell.

$$V_{cn} = 1.92\left(\frac{B^2}{A}\right)[\log_e(1+\gamma^{-1})] \tag{4.6}$$

$$\frac{J_n}{P^2} = (1.794 \times 10^{-12})\varepsilon_0(k_+ AB^2)\left[\frac{1+\gamma}{\log_e(1+\gamma^{-1})}\right] \tag{4.7}$$

where A and B are constants related to the primary Townsend ionization coefficient α, their values being shown in Table 4.1. The value of γ is in general larger for substances of smaller work function. V_{cn} falls with an increase in γ, while j_n/P^2 increases. Consequently, to supply the desired current at low voltage, a cathode material should be selected which has a γ that is as large as possible.

4.1.4 Characteristics of plasma displays

Since in PDPs the gas that constitutes the discharge medium is uniformly sealed into the interior of the panel, the operating characteristic and luminance of the discharge cells are kept uniform, and the PDP has the following characteristic advantages:

1. It is suitable for large screen displays.
2. Since the elements are self-luminous, they are bright and easy to see.
3. Since the display is constructed in the form of a flat sheet, diagrams are not distorted.
4. Luminance control can be easily achieved.
5. The response of the display is rapid.
6. Colour displays can be obtained.
7. Good resistance to vibration and high mechanical strength.
8. Long life.
9. Suited to mass production.

Since the construction and operating principles of AC-PDPs and DC-PDPs are different, a brief account of their technical features is given below.

(1) Alternating current discharge type (AC-PDP)

The construction of a typical AC-PDP is shown in Fig. 4.6. Its characteristic features are that column and row matrix electrodes are coated with a dielectric layer, and the display is operated by indirect discharge. Two glass sheets held with a separation of about 0.1 mm enclose in the space between them a mixed gas of which the principle constituent is neon. The characteristics are determined by a protective film of MgO. Research has also been carried out into use of an initial

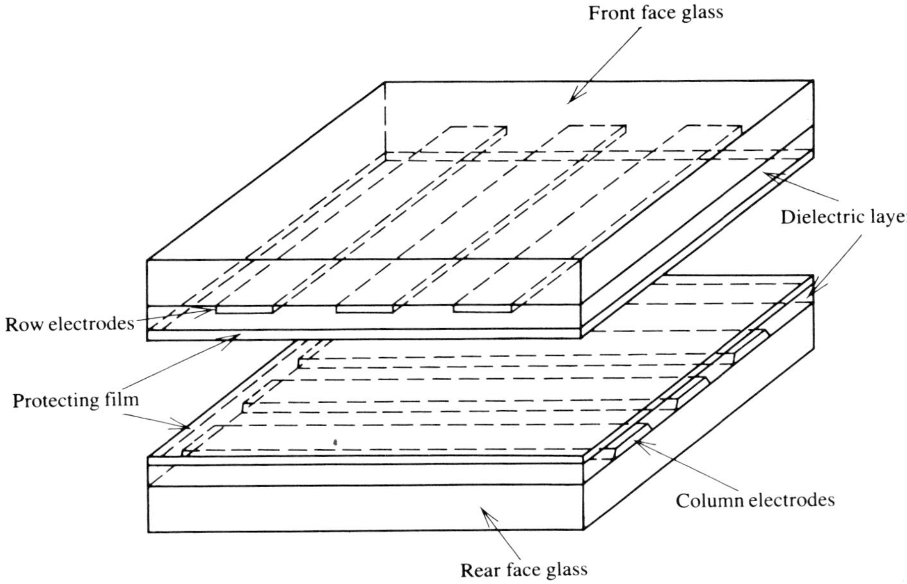

Figure 4.6 *Basic structure of AC-PDP*

protective layer of SiO_2 or Al_2O_3 to prevent deterioration of the dielectric layer caused by ionic impacts. MgO has a deterioration-preventing effect, and has favourable characteristics in that its discharge voltage is low, thanks to a small work function. These characteristics have enabled the development and practical application of long-life AC-PDPs.

Since discharge cells are capacitatively loaded, they are driven by applying an a.c. pulse voltage. Their modes of operation can be classified into refresh operation, which does not utilize memory of the discharge cells, and memory operation, which does utilize it. The luminance is proportional to the number of times of discharge, i.e. the number of times of repetition of the pulse voltage. In refresh operation, large display capacity results in reduced luminance, so this method is mostly used in panels of small display capacity [43, 44].

(2) Direct current discharge type (DC-PDP)

A DC-PDP can be of any desired construction, and various types of DC-PDP panels have been put into practice. Fig. 4.7 shows as a typical example the construction of a self-scanning panel display (SSPD). Its characteristic features are that column and row electrodes are exposed to the discharge space and act directly in discharge, the discharge cells being separated by insulating sheets or barriers provided with small holes. As in the case of an AC-PDP, a mixed gas of which the main constituent is neon is used as the enclosed gas. However, a trace of helium is added to prevent damage to the cathode by ionic impacts. Partitions are

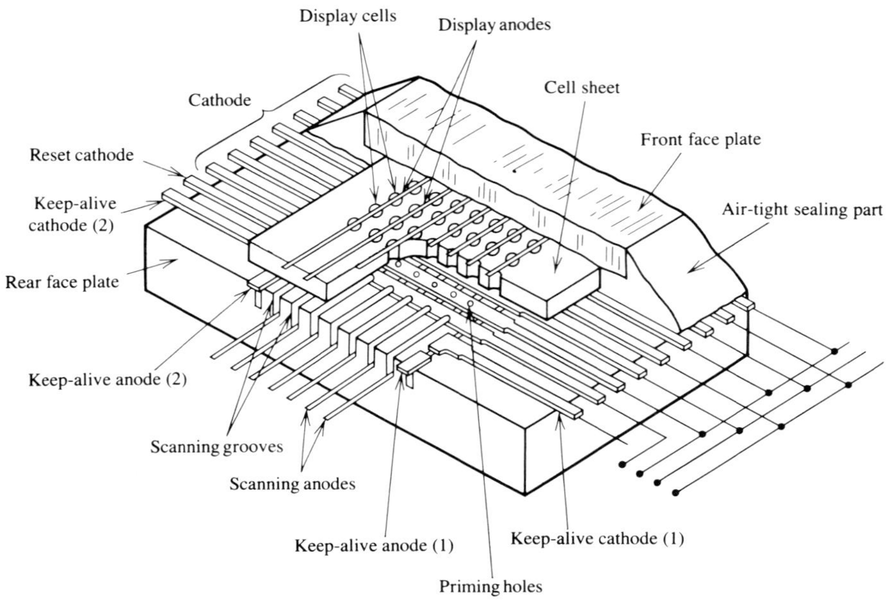

Figure 4.7 *Structure of self-scanning PDP*

effective in preventing colour crosstalk in the case of a colour display or preventing spreading of the discharge. The response speed of the display cell can be increased by providing priming discharge regions that act to prevent statistical variations in the lag time of firing when voltage is applied to the display cell. The priming discharge is made to perform a scanning action by application of a clock pulse to the cathodes, which are connected in multiphase, as in a pulse motor. In scanning, discharge is generated in the display cells in synchronisation with the timing of the discharge of the priming discharge regions. Letters or patterns can be displayed over the whole face of the panel by repetition of this process.

4.2 STRUCTURAL COMPONENTS AND LAYOUT OF A PLASMA DISPLAY

AC-PDP and DC-PDP display panels do have some common points in respect of the materials used to construct the panels, such as glass plate, type of enclosed gas, and phosphor used in the colour displays. However, since their construction is different and the cathode material etc., which play a vital role in operation, are also different, they will be discussed separately.

4.2.1 Alternating current discharge type (AC-PDP)

The construction of an AC-PDP is simple. The column and row matrix type panel that is used in practice at present has a construction as shown in Fig. 4.6,

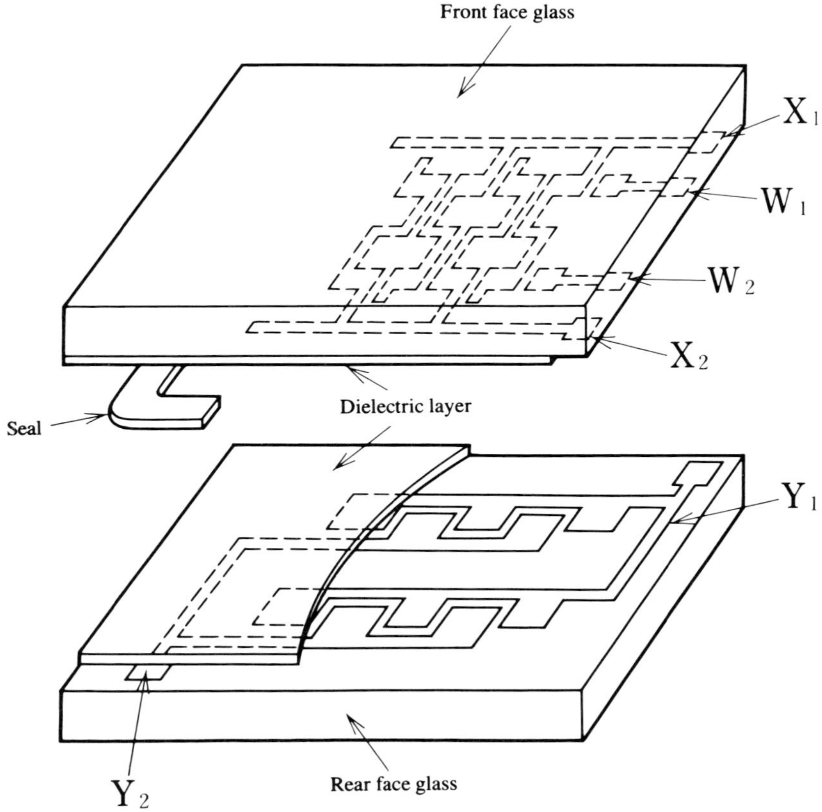

Figure 4.8 *Construction of self-shift AC-PDP (meander electrode type)*

described above. To simplify the drive circuitry, a panel with an electrode construction as shown in Fig. 4.8 and equipped with a shift function has been implemented as a product [45]. Ordinary glass is used as the base material of the panel, in all cases. The electrodes are arranged on the inside face of this glass and are made of a conductive paste consisting of silver etc., a three-layer conductive film (Cr–Cu–Cr), or Cu–Al alloy conductive film etc., deposited by printing or evaporation using a pattern mask [46]. For the dielectric layer with which the electrodes are coated, glass paste or SiO_2 etc. is used, this being deposited by printing or evaporation. As a protective film, a triple layer of metal oxide thin films such as MgO is deposited on the outer surface by printing or evaporation. The glass plates are fixed with column and row electrodes intersecting at right angles and separated by a gap of about 0.1 mm. The periphery of the panel is sealed in a gas-tight manner by frit glass. An exhaust port provided at the back face of the panel is fitted on to a vacuum device, and the gaseous impurities within the interior of the panel are degassed during vacuum evacuation in a constant-temperature tank at high temperature. As a Penning gas to lower the firing voltage, a mixed gas obtained by adding a trace of argon or xenon to the neon gas representing the main constituent is sealed in at a pressure of 300 to 600 Torr.

The operating voltage of an AC-PDP is about 90 to 150 V. Hybrid ICs or high threshold voltage MOSICs are used in the drive circuit. However, there has recently been a change to monolithic LSIs with shift registers, latch memory, counter, and decoder, etc., incorporated in a single package.

4.2.2 Direct current discharge type (DC-PDP)

The construction of DC-PDPs can be modified in any desired way, and various types of panel have been proposed. However, the main types of column and row matrix panels that have been practically utilized are the SSPD of Burroughs shown in Fig. 4.7 and the self-scanning I-type panel [47] obtained by improving the SSPD as shown in Fig. 4.9, the Oki Denki/Okaya Denki type panel [36], wherein the angle of view of the display is increased by provision of cathodes at the front face, as shown in Fig. 4.10, and the Matsushita Electronics type panel [27], which is equipped with a display discharge function and priming discharge in a single discharge cell, as shown in Fig. 4.11 [27]. Panels of other types are scarcely met with.

As the basic material of the panels, conventional sheets of glass are used. Regarding the electrode materials, in the Matsushita Electronics type panel, conductive paste such as nickel paste is used for the cathode, while silver paste is used for the anode. These pastes are applied to the glass plate by printing. In the SSPD and Oki Denki/Okaya Denki type panels, ribbon-shaped electrodes are

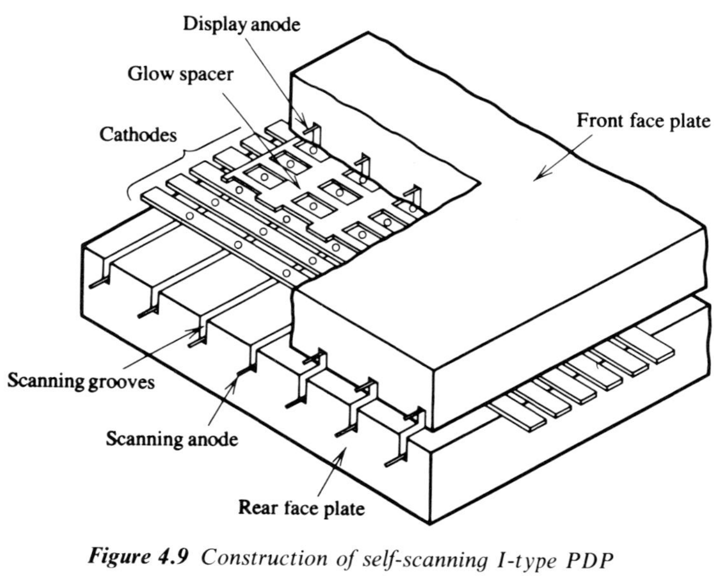

Figure 4.9 *Construction of self-scanning I-type PDP*

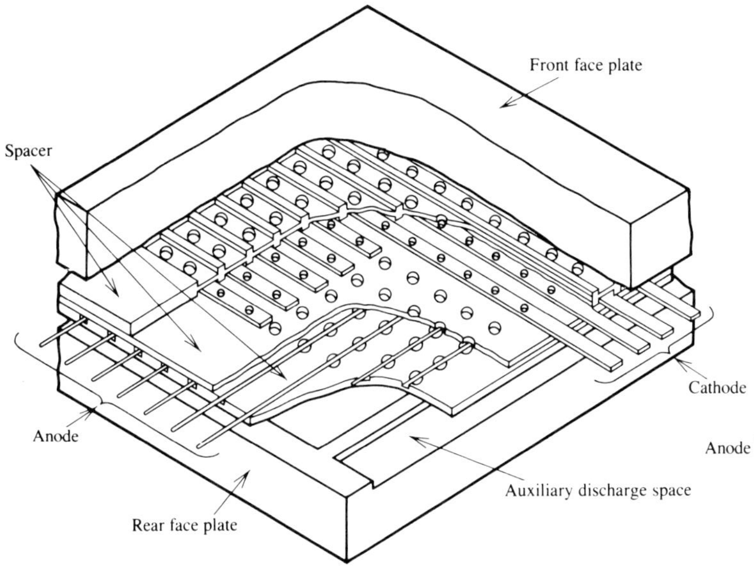

(a) Perspective view of panel

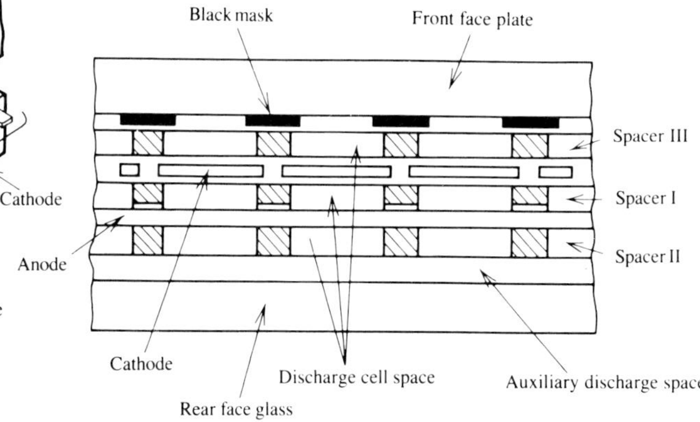

(b) Cross-sectional view of discharge cell

Figure 4.10 *Construction of Oki Denki/Okaya type PDP*

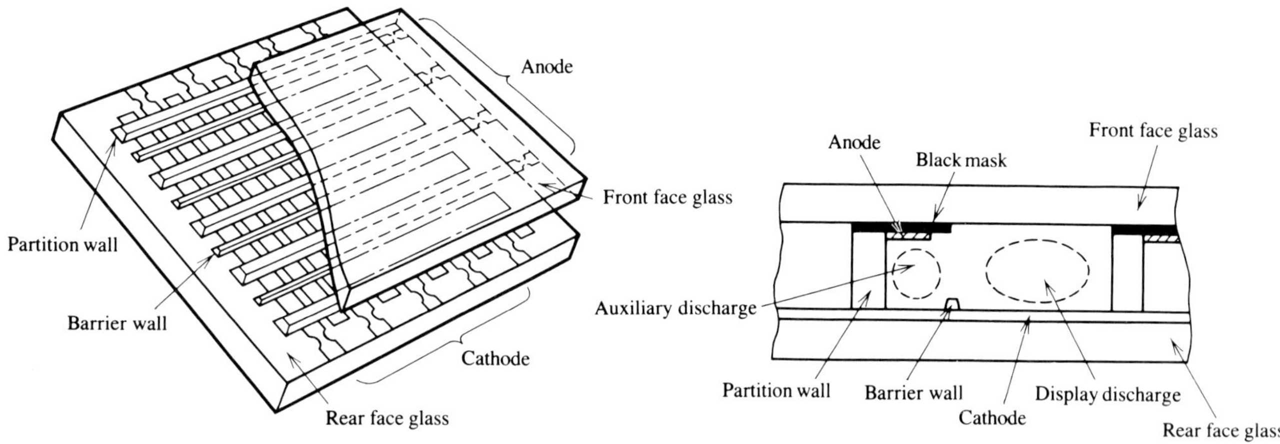

(a) Perspective view of panel

(b) Cross-sectional view of discharge cell

Figure 4.11 *Structure of Matsushita type PDP*

used which are chemically etched using a thin film of Fe–Ni alloy or Fe–Ni–Cr alloy as a mask pattern. As the spacers used to separate the discharge cells, photosensitive glass formed with holes is used in SSPDs, while mica formed with holes is used in self-scanning I-type panels. In the case of the Oki Denki/Okaya Denki type panel, glass sheet formed with holes by chemical etching is used. The scanning anodes are in all cases embedded in grooves formed by chemical etching in the glass sheet. In the case of the Matsushita Electronics type pattern, a partition is provided on which a dielectric paste is printed in multiple layers by thick-film printing. For assembly of the panels, a jig is used, the periphery of the glass sheet being sealed in a gas-tight manner using frit glass. For degassing the interior of the panel and sealing in the gas, practically the same steps as in the case of an AC-PDP are carried out. In the case of a DC-PDP, mixed gas is sealed under a pressure of 100 to 300 Torr, and mercury vapour is then sealed into the interior of the panel by heating mercury or enclosing mercury in a small capsule.

The operating voltage of a DC-PDP is 180 to 250 V. A hybrid IC or high threshold voltage MOSIC is used for the drive circuit. As in the case of AC-PDP drive circuits, development of high threshold voltage monolithic LSIs is actively proceeding in many places. It is therefore anticipated that further increases in compactness will be obtained in the future.

4.2.3 Colour display PDPs

As methods of achieving colour display of PDPs, the method of controlling the light-emission level of a phosphor by stimulating it by means of ultraviolet ray emission and by means of visible light emitted by a discharge [48] and the method of altering the light-emission colour in accordance with equilibrium and non-equilibrium energy states in discharge of a mixed gas [49] have been proposed. However, there are problems in applying these methods directly to PDPs. In practice, the method of achieving a colour display by stimulated light emission of a phosphor in response to ultraviolet ray emission of rare gas atoms is more practical. The reason why ultraviolet rays from mercury cannot be utilized, as in a fluorescent lamp, is that the interelectrode gap of the discharge cells is short, and the mercury vapour pressure is low under the operating conditions of a PDP, so sufficient ultraviolet ray intensity cannot be obtained.

(1) Alternating current discharge type (AC-PDP)

The first method that was tried in order to achieve colour display of an AC-PDP was applying phosphor to the neighbourhood of the electrodes of the discharge cells. However, the phosphor was damaged by ionic impacts, so this method could not be practically employed. Research has been carried out both in Japan and overseas aimed at solving this problem. This has led to the development of the colour display AC-PDP shown in Fig. 4.12 by Fujitsu and Nippon Denki [50, 51].

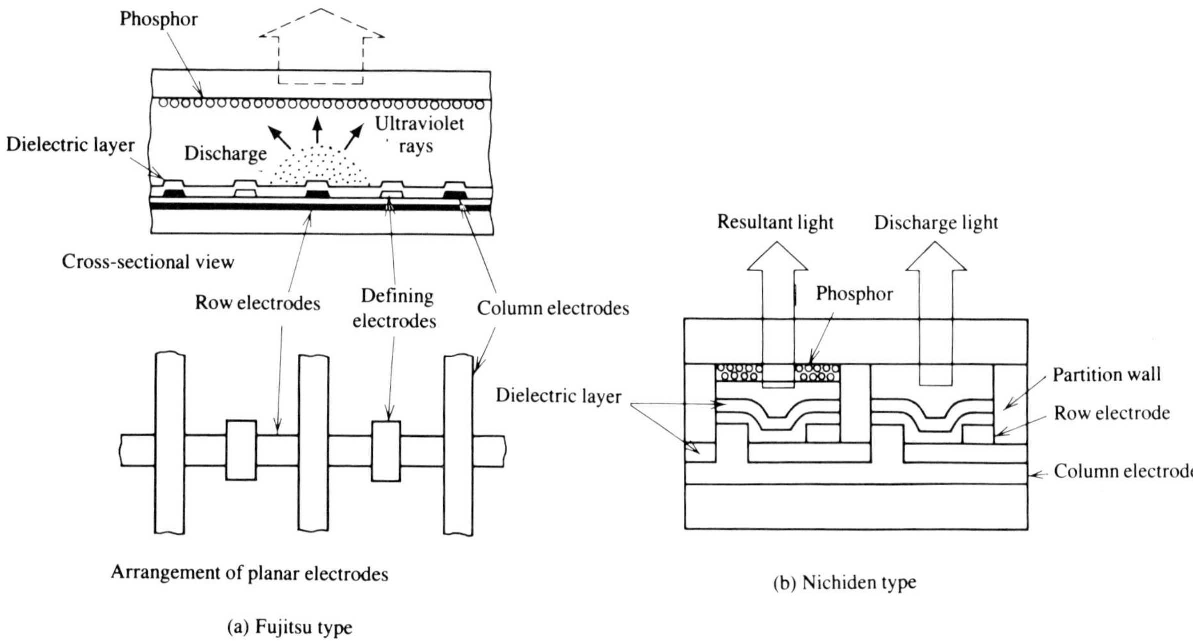

Figure 4.12 *AC-PDP colour display*

In both cases the operation is in the surface discharge mode, and in order to prevent crosstalk the former is provided with an electrode referred to as a defining electrode and the latter is provided with a partition. The enclosed gas used in the former is an He–Xe gas mixture and in the latter is a gas mixture obtained by adding xenon to neon. In the case of the former, a method of colour display is being investigated in which a phosphor of the three primary colours red, green, and blue is applied, but the latter is a two-colour display, produced by the red–orange colour obtained from neon luminescence and a different colour luminescence provided by a phosphor. Both are at present in the development stage. The critical points in successful product manufacture appear to be the prevention of contamination of the phosphor and prevention of optical crosstalk caused by ultraviolet emission from adjacent cells.

(2) Direct current discharge type (DC-PDP)

In a DC-PDP type colour display, phosphor is applied to the insides of the holes in the glass sheet which separates the discharge cells in an Oki Denki/Okaya Denki type panel, and the luminescence in the transmission mode is observed. Companies such as NHK Giken [33] are developing a colour panel wherein phosphor is applied to the inner walls of small holes in the photosensitive glass sheet which separates the discharge cells, as shown in Fig. 4.13, and the luminescence of the phosphor in the reflection mode is utilized. In the former case it is simple to apply the phosphor but the luminance is low. Methods adopted in the application of the phosphor to make it adhere include spraying and printing using a mask pattern. The gas composition and pressure of the enclosed gas are selected in accordance with, for example, the operating voltage, the luminance,

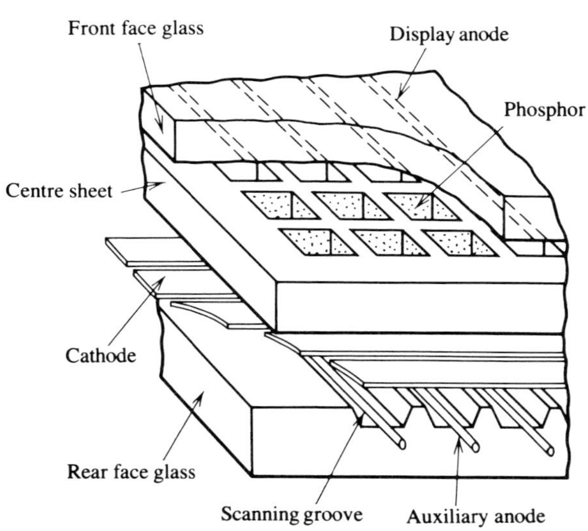

Figure 4.13 *Structure of a self-scanning PDP (NHK Giken type)*

and the scanning characteristics of the secondary discharge. The main compositions used for the enclosed gases are Xe–He, Xe–He–Ar, Xe–Ne–Ar, and Xe–Kr–Ar–He.

The phosphors used for photoluminescence are mainly those that respond to ultraviolet excitation of mercury (wavelength of resonant line 253.7 nm). However, the PDP depends on ultraviolet excitation of xenon (wavelength of resonant line 147 nm) so it has a problem of low luminous efficiency. To improve this, NHK Giken and Dai Nippon Toryo are jointly developing high luminance phosphors such as $BaMgAl_{14}O_{23}$: Eu (blue), $BaAl_{12}O_{19}$: Mn (green), and (Y, Gd)BO_3: Eu (red) [52].

4.3 OPERATING PRINCIPLES AND STANDARD CHARACTERISTICS OF PLASMA DISPLAYS

Unlike the AC-PDP in which the electrodes are covered with a dielectric layer and which operates in the indirect discharge mode the DC-PDP has electrodes that are exposed to the discharge space and it operates in the direct discharge mode. Their operating principles and characteristics are fundamentally different, so they are described below under separate headings.

4.3.1 Operating principles of AC-PDP

The discharge spaces of an AC-PDP are coupled capacitatively to the electrodes through the medium of glass sheets [2]. One discharge cell and its idealized equivalent circuit are shown in Fig. 4.14. In diagram (b) C_W is the electrostatic capacitance of the dielectric layers and protective layers taken together, C_C is the electrostatic capacitance of the discharge space, V_t is the terminal voltage applied to the external electrodes, and V_C is the voltage applied to the discharge cell. If the

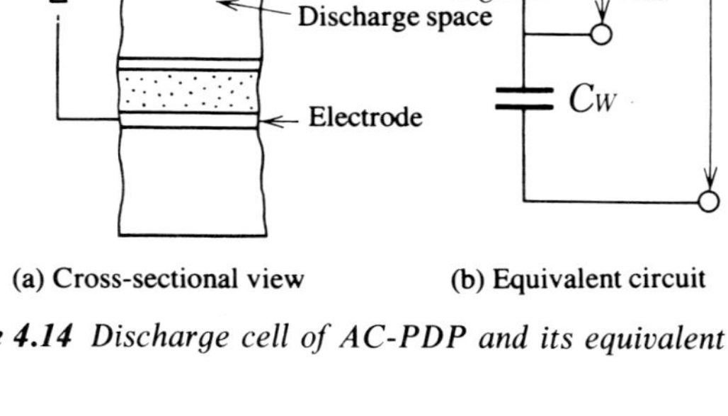

(a) Cross-sectional view (b) Equivalent circuit

Figure 4.14 *Discharge cell of AC-PDP and its equivalent circuit*

value of V_C is taken to be V_{CO} when there are no charged particles in the discharge cell then

$$V_{CO} = \frac{C_0 V_t}{2C_C} + C_0 \tag{4.8}$$

where C_0 is the electrostatic capacitance measured between the anode and cathode. The voltage which appears at the discharge cell is found from equation (4.8) by distributing V_t between the three capacitors. When V_{CO} exceeds the firing potential V_f, discharge is initiated and electrons, positively charged ions, and excited atoms that are not charged are formed within the cell. Of these, the capacitative cell loading causes the electrons and ions to diffuse to the inner walls of the dielectric layers on both sides, giving them an electrical charge. This creates a wall potential V_W of inverse polarity to the external voltage, thereby decreasing the electric field within the cell until the discharge ceases. Because the dielectric layer has a high resistance the charge on the inner walls remains without leaking away.

If now a voltage of reverse polarity is applied, it superimposes a polarity which is the same as the charge within the cell, so discharge again occurs when V_f is exceeded; thus the above-mentioned process is repeated. Once discharge has been initiated, it can be maintained by a voltage lower than V_f. Thus the charge on the inner walls has a memory function. This property can be utilized to perform write, erase, and memory functions. The operating waveform in this situation is shown in Fig. 4.15

The memory characteristic is determined by the wall charge but this is difficult to ascertain. For this reason, in order to express the actual memory characteristic, the following equation is used, showing the memory coefficient M as a function of the firing potential V_f and the minimum potential V_{em} at which discharge is maintained:

$$M = \frac{V_f - V_{em}}{(1/2)V_f} \tag{4.9}$$

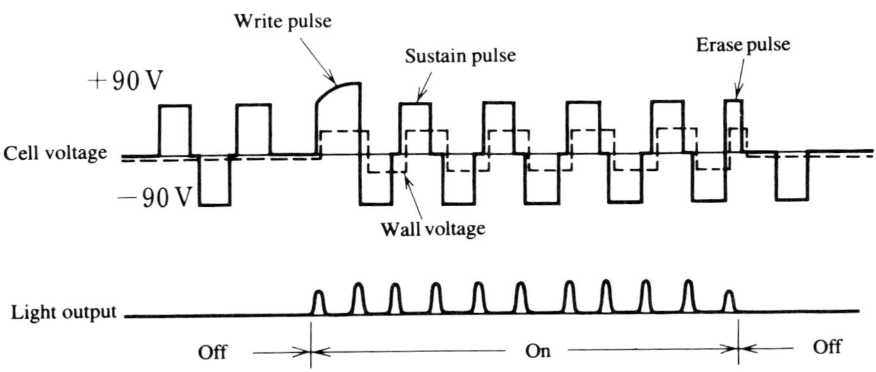

Figure 4.15 *Operating waveform of AC-PDP (memory function)*

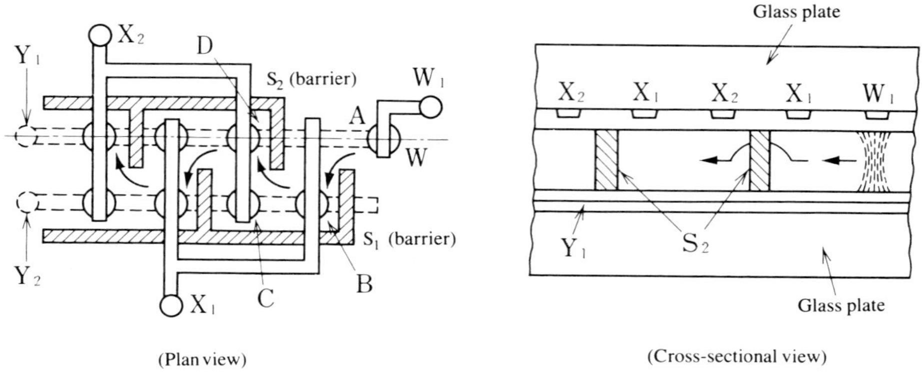

Figure 4.16 Principle of operation of self-shift AC-PDP

Actual panels contain large numbers of cells and there is statistical scatter of the values of V_f and V_{em} for each cell, so, for convenience, the median value of M is obtained and used as a representative value. Under practical conditions V_f is 90 to 150 V at 10 to 200 kHz, the discharge maintenance voltage V_e is 70 to 100 V and the luminance is 120 to 200 cd/m^2. Various methods are adopted for applying the drive waveform and voltage. If the memory function is not utilized, a high-frequency (10 to 500 kHz), non-polarized discharge drive method can be employed.

Examples will now be given of the operating principles of panels with a self-shift function. Figure 4.16 shows a diagram of the operating principle of the M-PDP [45, 53] developed by Fujitsu, which is provided with 'meander' electrodes. A shift unit is constituted by four discharge cells, the electrodes being arranged alternatively on the inner surfaces of two facing glass sheets. The electrodes are all covered with a dielectric layer and the electrode gap, sealing gas, etc., are the same as for the AC-PDP mentioned previously.

The first discharge point is produced in cell W by a write pulse which is applied between the electrode Y_1 and the write electrode W_1, which is arranged independently of it. Next, the display is formed by applying a set of four transfer pulses in succession to electrodes X_1–Y_1, X_1–Y_2, X_2– Y_2, Y_2–Y_1, causing the discharge point to shift in turn to A, B, C, and D. After the discharge has shifted to the end the cell W is reset and the same operation is repeated.

4.3.2 *Operating principles of DC-PDP*

In a simple panel that is driven by increasing the direct current voltage at the discharge cell formed by the points of intersection of the columns and rows of the matrix electrodes that are adjacent to the discharge space, when voltage is applied discharge does not take place immediately and an unevenness of the luminance shows up, which is caused by the statistical time lag of the firing. In order to

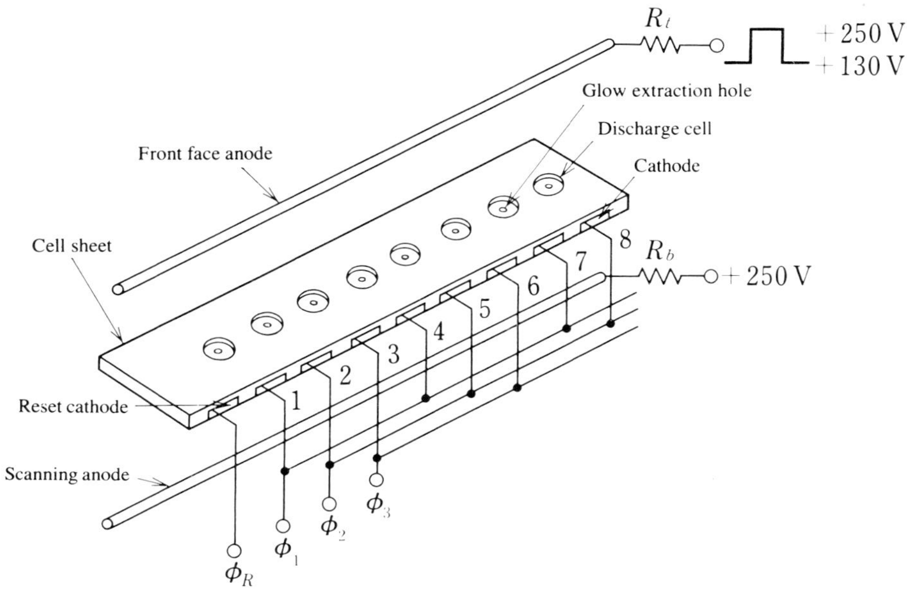

Figure 4.17 *Principle of operation of self-scan panel display*

correct this, DC-PDPs which are used in practice make use of priming discharge. An example of this type is the self-scan panel display (SSPD) [19]. The principle of its operation is shown in Fig. 4.17. The positive display electrodes run in the row direction in front of the small holes in the cell sheet, which separates the discharge cells, and on the reverse side ribbon-shaped cathodes in which there are minute holes run in the column direction, large numbers of them being arranged parallel to one another. Thus a discharge cell is formed at each point of

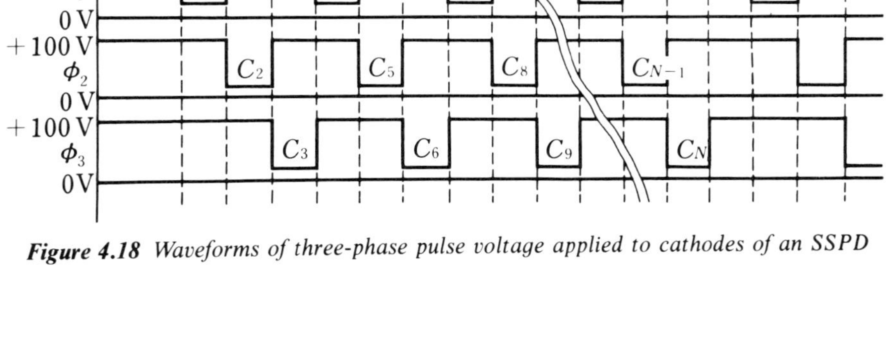

Figure 4.18 *Waveforms of three-phase pulse voltage applied to cathodes of an SSPD*

intersection. On the reverse side of the discharge cell large numbers of reverse side anodes (referred to as scanning anodes) are also arranged parallel to one another in the row direction, and each one is connected to the power source through a resistance.

The way in which the cathodes are connected is as follows. A reset cathode R at one end is separate, while the other cathodes are connected up by dividing them, for example, into three groups. Pulse voltages as shown in Fig. 4.18 are applied to each of these groups. The glow discharge at the reverse side of the discharge cells is first produced between the reset cathode and the reverse side anode. This glow discharge is then shifted by applying a three-phase clock. After the glow discharge has been transferred to the end of the panel, discharge is again produced at the reset cathode and transferred as before. If a positive voltage with the same timing with which the glow discharge is generated at the cathodes on the reverse side is applied to the display anode, the glow discharge passes through the minute holes in the ribbon-shaped cathode and forms a display dot which appears at the front of the cell. When there is no glow at the reverse side a display dot does not appear, even if a voltage is applied. In this method of driving, addressing in the column direction is performed by the display anode, but there is no specific address in the row direction so the drive circuit can be simplified.

As a different method of transferring the glow discharge at the reverse side cathode the Oki Denki/Okaya Denki panel uses what is known as a parallel simultaneous scanning system [36], in which transfer is carried out by applying transfer pulse of the same phase to each of the cathodes in the column direction. In this method the number of circuit elements used is increased but there is a considerable reduction of spurious operation in the transfer of the glow discharge.

In the Matsushita panel [27] each discharge cell space is divided into two regions: a priming discharge region and a display discharge region. The method used is that first of all a weakly luminescent glow discharge is generated in the priming discharge region, and then, when display is to be effected, this priming discharge glow is shifted into the display discharge region and increased in extent. Furthermore (although this has not yet been implemented practically), in the Sony panel a priming discharge can be generated in the non-polarizing discharge mode in the same way as in the AC-PDP, and a system is being tested [28] in which the display cell is operated by utilizing the charged particles produced.

4.3.3 Characteristics of AC-PDP

The AC-PDP discharge cell is capacitively loaded, so when it is in the state in which no discharge is taking place, a charging current flows at the same time as the build-up of the applied pulse voltage. The discharge current flows with a 0.1 to 0.5 μs delay after the voltage build-up, charge transfer takes place over 1 to 5 μs, and the current then stops. The operating voltage is determined by the distance l between the cell electrodes, the type of enclosed gas, the gas pressure P, and the

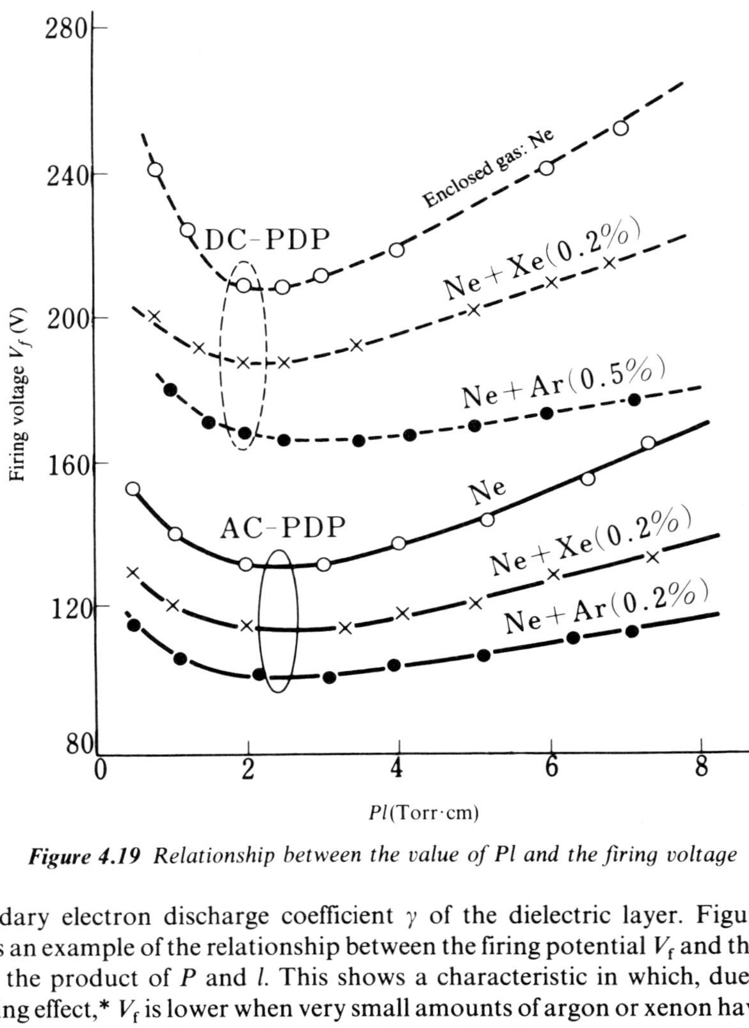

Figure 4.19 *Relationship between the value of Pl and the firing voltage*

secondary electron discharge coefficient γ of the dielectric layer. Figure 4.19 shows an example of the relationship between the firing potential V_f and the value of *Pl*, the product of *P* and *l*. This shows a characteristic in which, due to the Penning effect,* V_f is lower when very small amounts of argon or xenon have been added than when only neon is used as the enclosed gas.

Because there is a delay of 0.1 to 0.5 μs in firing after the voltage build-up, if the

*When two gases A and B are mixed, if A is in a metastable state and has an excitation voltage ${}_A V_m$ which is higher than the ionization potential ${}_B V_i$ of gas B, then the ionization

$$A^* + B \rightarrow A + B^+ + e$$

can take place. An example of this is

$$Ne^* + Ar \rightarrow Ne + Ar^+ + e$$

$$(V_m = 16.6\ V)\ (V_i = 15.75\ V)$$

This type of ionization is known as the Penning effect.

A* and Ne* are metastable states, B^+ and Ar^+ are ionized states, and e is an electron; V_m is the excitation voltage and V_i is the ionization potential.

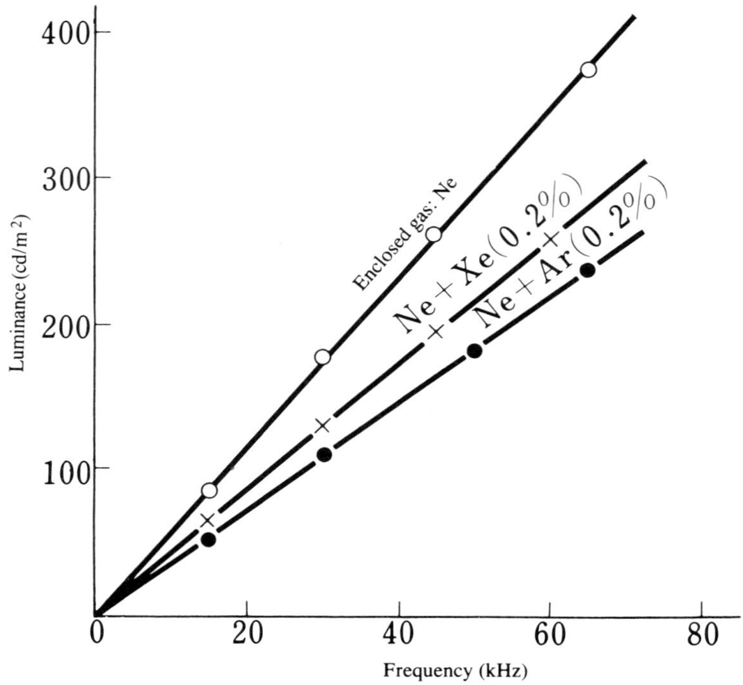

Figure 4.20 *Relationship between frequency and luminance*

rise time of the voltage pulse is faster than this, the luminance increases in proportion to the frequency, as shown in Fig. 4.20. If the rise time of the voltage pulse is slower, then the voltage does not rise to the full extent, discharge becomes difficult due to the development of wall charges, and the luminance is reduced. To correct this, a memory cell which utilizes the wall charge is adopted. When the operating voltage increases the luminance increases, but the extent to which the voltage can be increased is limited by the drive circuit, so in general the luminance is governed by the frequency.

4.3.4 Characteristics of DC-PDP

DC-PDP discharge cells are generally equipped with a priming discharge section on, for example, the reverse side of the cell and operate under conditions in which charged particles are supplied. When the applied voltage is low a statistical lag time appears in the discharge but when a voltage is applied that exceeds the firing potential the cell immediately strikes and changes over to normal glow ignition. Consequently, the firing potential of a DC-PDP discharge cell has properties similar to the restriking potential in, for example, a rectifier discharge tube, and glow discharge is initiated at a value lower than the spark breakdown voltage.

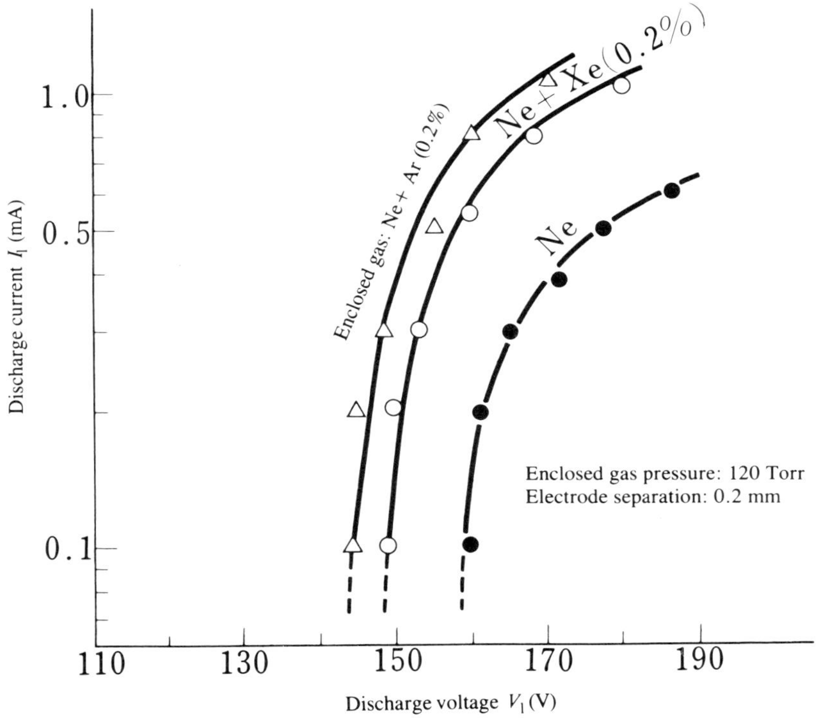

Figure 4.21 *Discharge voltage–current characteristic of DC-PDP*

The operating characteristics are influenced by the quantity of charged particles supplied from the priming discharge section as well as by the various conditions in the AC-PDP cell.

The dotted curves in Fig. 4.19 are sample characteristics of the values of *Pl* and the firing potential V_f. The trends of the curves are as mentioned above. V_f shows a lower characteristic when traces of argon and xenon are added than when only neon has been enclosed.

Figure 4.21 shows an example of the relationship between the discharge maintenance voltage V_l and the discharge current I_f. The value of V_l varies according to the type of enclosed gas; in the low current range the cell operates under conditions of normal glow discharge and V_l is maintained at an approximately constant level. In the higher current range, there is a transition to a state of abnormal glow discharge and V_l gradually increases.

The luminance characteristic increases with increasing discharge current, as shown in Fig. 4.22. In panels in practical use there are a large number of discharge cells, so the luminance changes according to the burning time allotted to the cells within a given period.

In other words, the luminance of the DC-PDP discharge cells varies in

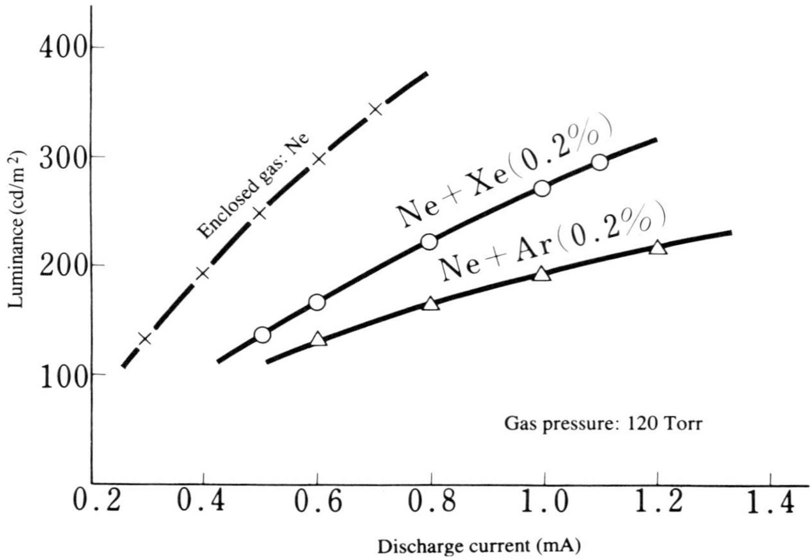

Figure 4.22 *Example of discharge current–luminance charateristic of DC-PDP*

proportion to the value of the discharge current and the pulse width of the drive voltage.

4.4 Drive systems for plasma displays

In a cathode ray tube (CRT), in which dots are sequentially driven, there are only a few electrodes so the circuit connections are simple. In contrast to this, in a matrix display of columns and rows there are a large number of electrodes. Especially in PDPs where the drive voltage is high, this necessitates a large drive circuit and many connections between the panel and the circuitry. Recently, however, new terminal connection components have been developed and the drive circuitry has been converted to IC and LSI form. There are also moves towards cost reduction and improved reliability, achieved by compact mounting. Drive circuit systems can be classified into refresh systems and memory systems. Their basic circuitry is described below.

4.4.1 Drive systems for a.c. discharge type

(1) Refresh system

The column and row matrix type panel is operated by line sequential scanning, so a line sequential scanning circuit and data input signal circuits equal in number to the number of column and row electrodes must be provided. Figure 4.23 shows

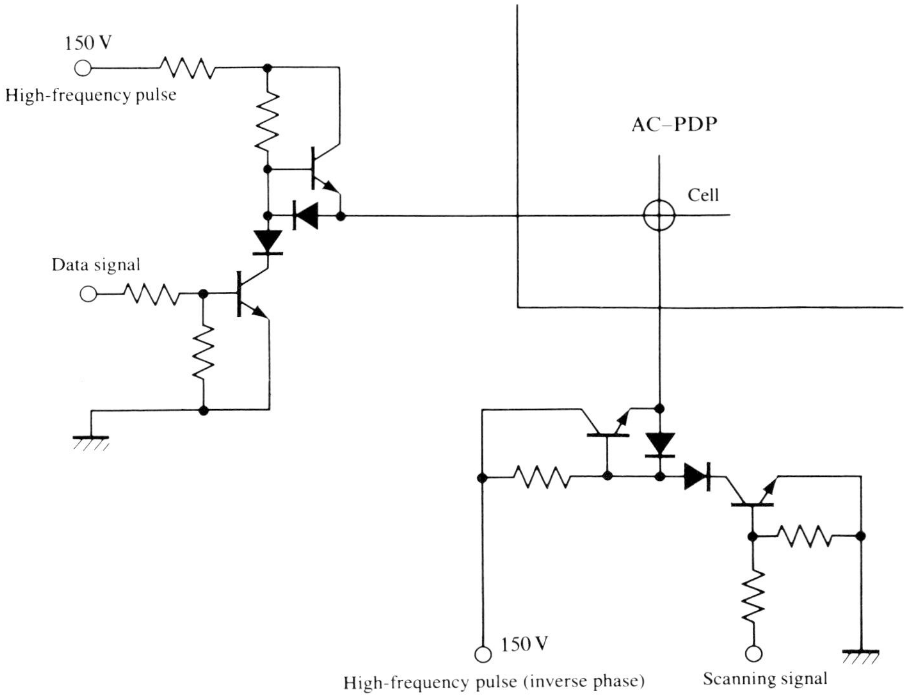

Figure 4.23 *Basic refresh type drive circuit of AC-PDP*

the basic drive circuit in the refresh system [54]. High-frequency pulses in inverted phase are applied to the column and row electrodes respectively. The selected discharge cell is made to emit light by a switching driver operated by the data input signal and scanning signal. The luminance is increased by increasing the frequency during the time assigned by scanning.

In the case of a segment type panel, on the other hand, drive is effected by replacing the circuit corresponding to the data input signal circuit for row drive by a segment input signal circuit, and replacing the circuit corresponding to the line sequential scanning circuit of row drive by a digit sequential scanning circuit.

(2) Memory system

(a) Matrix system In the case of a memory type panel, sustaining voltage to maintain the cell discharge, writing voltage to effect discharge of the cell when it is addressed, and blanking voltage to extinguish the discharge are required.

In the conventional system of writing to cells, the method of writing by superimposing voltages was adopted. However, the method that is now adopted is a phase method, involving modification of the timing or waveform of writing. In

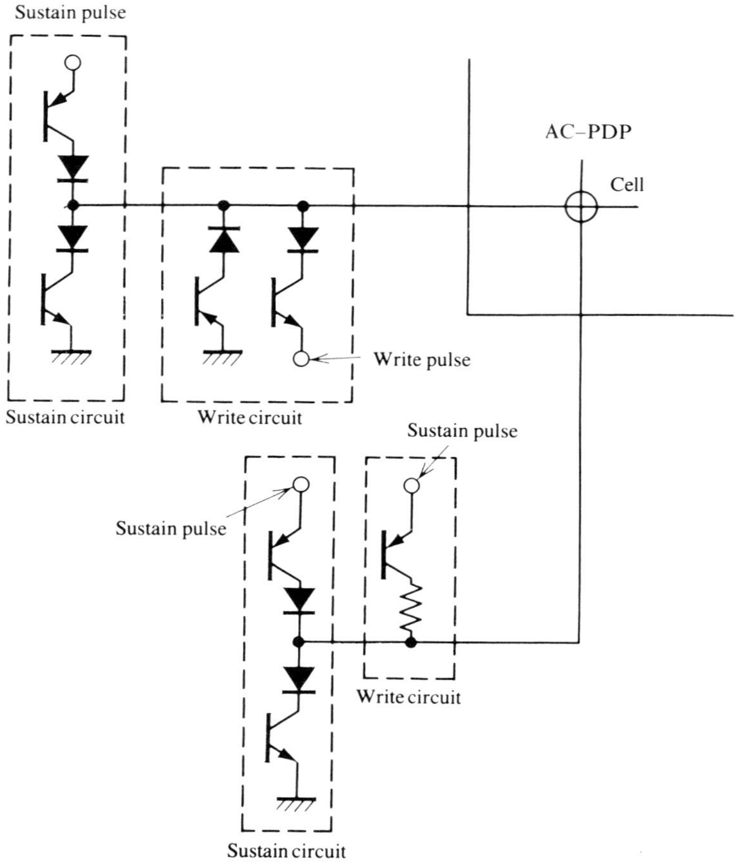

Figure 4.24 *Basic memory type drive circuit of AC-PDP*

this method, writing of data is effected by a resultant pulse obtained by applying positive and negative write pulses with the same phase to the column electrode and row electrode of the selected cell. For blanking, the inverse blanking method, in which a blanking voltage is applied in a narrow-width pulse of the same value as the sustain voltage, but with inverted phase [55, 56], is adopted. The basic drive circuitry is shown in Fig. 4.24.

Alternating current sustain voltage pulses are applied from a sustain circuit to each electrode in the column and row, and the write circuit is operated by a clamp transistor and a pull-up transistor for the row electrode and a pull-down transistor for the column electrode. This method of drive has the advantages that the circuitry is simple and there is a large operating margin.

(b) Self-shift system This is a system that was developed in order to simplify the drive circuitry. Display was effected by shifting the input data in the manner of an

'electric news tape' display by applying four sets of transfer pulses [45, 57]. In this system, once a display pattern was written, it could not be partially rewritten. Recently, however, a meander electrode type panel (MM-PDP) with a compilation function [58] and equipped with a vertical shift panel [59] has been developed.

In the case of an MM-PDP, the four electrodes discussed earlier in the case of an M-PDP are utilized exclusively for scanning, and partial rewriting of a display pattern is achieved by providing dedicated display electrodes that write the input data afresh.

4.4.2 Drive systems for d.c. discharge type

(1) Refresh system

In this system, light emission and display are obtained by sequentially performing switching scanning by application of a line sequential scanning voltage to the cathodes, while high voltage corresponding to the input data is applied to the anode, so that the cathode–anode voltage of the selected cell exceeds the firing voltage. Taking the number of scanned cathodes as n and the refresh period as t_f, the width t_p of the scanning voltage pulse is given by $t_p = (nt_f)^{-1}$. The limiting frequency of visual flicker is selected at a value greater than 60 Hz. From the value of the luminance and the statistical lag time for cell firing, the lower limit of t_p is about 30 μs.

For a panel of large display capacity incorporating a large number of discharge cells, due to the requirements of luminance and compactness of the drive circuitry, a scanning system wherein two columns are scanned simultaneously or a row scanning system, etc., may be adopted.

(a) Self-scanning system To give the priming discharge of the panel a self-scanning function, the cathodes are connected in multiphase to a line sequential drive circuit. The scanning discharge switching circuit is fairly complicated, but use of a self-scanning drive enables the number of components to be very considerably reduced [19, 34]. The basic drive circuit is shown in Fig. 4.25.

PNP type transistors are used for the data input circuit, so when there is no display the switching transistor is cut off, enabling power consumption to be reduced.

(b) Parallel simultaneous scanning system Line sequential scanning drive is achieved in the parallel simultaneous scanning system by connecting all the cathodes to the switching driver. As shown in Fig. 4.26, the system constitutes a resistance bleeder system [36] wherein the display anode side is earthed through a resistance and switching transistor. The selected cell is discharged to produce the display when display data is input in synchronisation with the timing of scanning of the priming discharge region. When the switching transistor of the

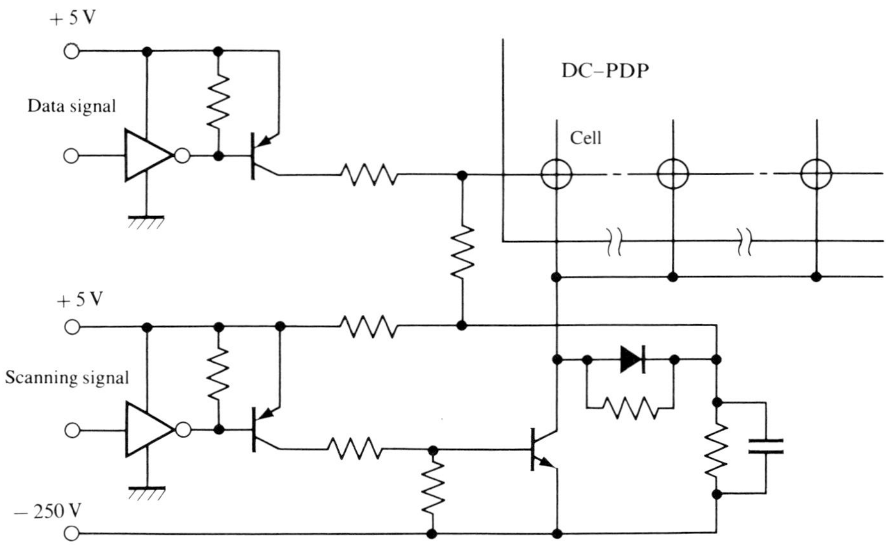

Figure 4.25 Basic self-scan drive circuit of DC-PDP

display anode is turned on, the bleeder voltage obtained by voltage division at the resistor is applied to the display anode, causing it to have a lower voltage than the discharge sustain voltage of the cell, and thus stopping the discharge and display.

This system has the advantage that there is less spurious scanning discharge of the priming discharge regions than in the case of the self-scanning system, but it

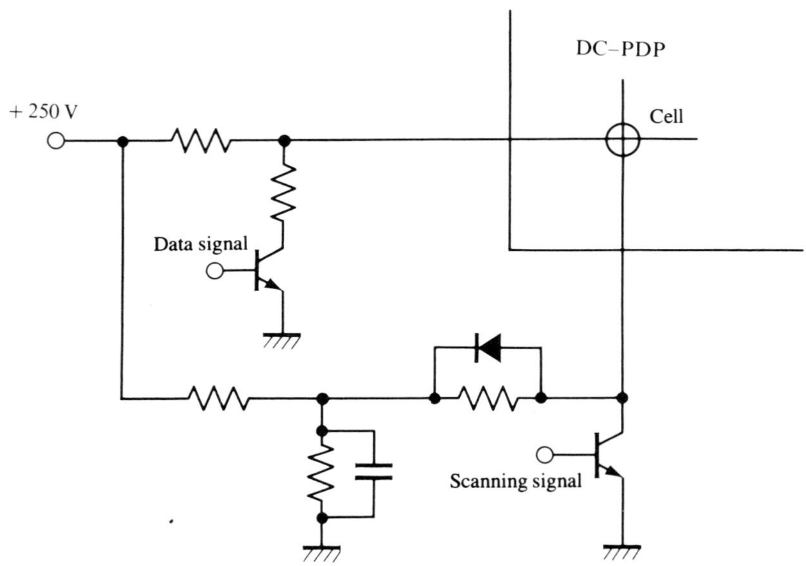

Figure 4.26 Basic parallel simultaneous scanning drive circuit of DC-PDP

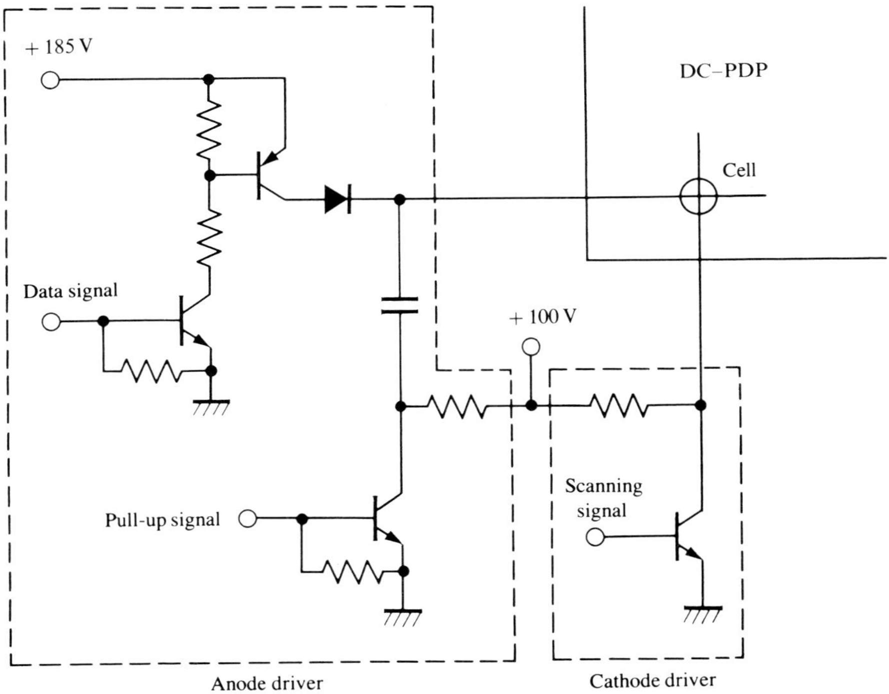

Figure 4.27 *Basic self-scanning pull-up type drive circuit of DC-PDP*

involves an increase in the number of circuit components and increased power consumption when the cells are not lit.

(c) Self-scanning pull-up system In this system, the charging voltage of the floating capacitance between the cathode and anode is used to generate the priming discharge (pilot discharge). A line sequential scanning voltage of 0 V is applied to each cathode during discharge and a line sequential scanning voltage of 100 V is applied on cessation of discharge [27, 53]. The basic drive circuitry is shown in Fig. 4.27.

A small capacitor is connected between each anode that inputs display data, and cell discharge is achieved by using the charging voltage of this capacitor to pull up the applied voltage.

(2) Memory system

Various types of memory drive system have been proposed as described above. However, they suffer from problems such as the panel construction becoming complicated or the discharge voltage increasing. None of them have been implemented in practice. However, one of these systems, namely the pulse memory drive system, is being researched in various places [32, 33]. In this

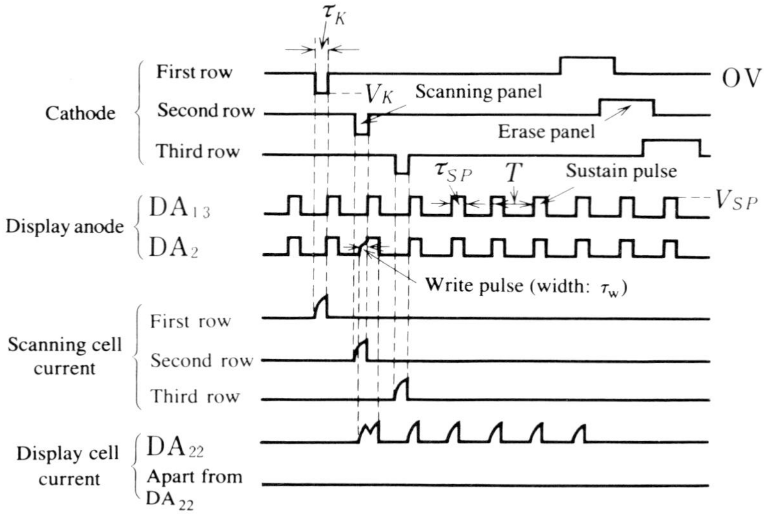

Figure 4.28 Time chart of pulse memory panel drive operation

system, the panel construction is not too complicated and drive is effected by a pulse of narrow width.

When a discharge cell is operated by a narrow pulse in this way, since the distribution of charged particles within the cell space does not reach a steady state, the discharge voltage–current characteristic shows a positive resistive characteristic. Cells to which data are not written therefore continue to emit light intermittently, driven by the pulse discharge, unless blanking is applied.

As an example of this drive system, a memory system of the NHK Giken type will now be described in outline.

In a pulse memory panel, a three-layer electrode structure (Burroughs type panel) is used to decrease the write time when scanning discharge is employed [1]. The time chart of writing and blanking the display cells is shown in Fig. 4.28. A discharge sustain pulse of narrow pulse width τ_{sp} (0.5 μs) is constantly applied to the display anode. A scanning pulse of pulse width τ_k(2μs) is applied to the cathode to create a scanning discharge between it and the scanning anode. A narrow pulse of pulse width τ_W (0.7 μs) is applied to the display anode in synchronism with the timing of the scanning discharge. The firing voltage of the display cell falls due to diffusion of charged particles in the scanning discharge, so that data are written only to the selected cell. For cell blanking, a blanking pulse is applied to the cathode and display anode, thereby lowering the sustain pulse voltage.

4.5 Various types of plasma display panels

Plasma displays are classified in accordance with the shape of their electrodes, as in Table 4.2. In the segment type, a large number of electrodes in a 'rectilinear

***Table 4.2** Types of PDP electrode*

PDP electrode type	Segment type	Custom pattern display
		Numeric/alphanumeric display
		Bar graph display
	Matrix type	Character display
		Graphic display
		Colour image display

figure-eight' arrangement, or a 'four adjacent squares' arrangement, or a ribbon are arranged parallel to each other. In the matrix type, column and row electrodes are regularly arranged.

Table 4.3 shows a classification of plasma displays in accordance with the drive system. AC-PDPs and DC-PDPs have contrasting electrode constructions and drive systems, but development is proceeding aimed at improving performance and reducing costs by combining AC-PDPs and DC-PDPs so that they complement each others' defects.

In this section, we shall describe various types of plasma display, including some that have only been manufactured on a trial basis.

4.5.1 Dot character display panel

This type of panel can display numeric and alphanumeric data. Discharge cells are arranged in rows and columns in a format of 5×7 dots or 7×9 dots. Both AC-PDP and DC-PDP panels of various display capacities have been implemented as products. The latest tendency is to panels of large display capacity, with increased luminance and improved efficiency. Interest is also being shown in high-resolution displays obtained by high-density cell arrangements. AC-PDPs have an advantage in terms of luminance because they have a memory function, due to the wall charge. However, multicolour displays have problems of optical crosstalk, which have so far prevented their implementation as a product. In the case of DC-PDPs, since the panel itself does not have a memory function, research is going on into ways of improving luminance and providing them with memories of various types.

The Burroughs company in the United States have developed a self-scanning memory panel (SSMP), which combines the self-scanning function of an SSPD (see (2) (b) in Section 4.4.1), as shown in Fig. 4.29, and a memory function utilizing the wall charge of an AC-PDP. They have announced [60] an increase in luminance to about twice that of an SSPD. As a way of improving efficiency, Okaya Denki have developed a technique wherein the cathode base is coated with an emitter substance by a plasma coating technique. Some of these panels are being implemented as products [34] to obtain low-voltage, long-life DC-PDPs.

Table 4.3 *PDP drive system*

- PDP drive system
  - A.c. discharge type
    - Refresh type
      - Simple type
      - Self-shift type
    - Memory type
      - Simple type
      - Self-shift type
  - D.c. discharge type
    - Refresh type
      - Simple type
      - (Auxiliary discharge type (self-scanning, parallel simultaneous scanning, etc.)
    - Memory type
      - Simple type
      - Auxiliary discharge type (self-scanning, parallel simultaneous scanning, etc.)
    - Electron beam excited type
  - A.c./d.c. discharge combination type → Auxiliary discharge type (→ Electron beam excited type)

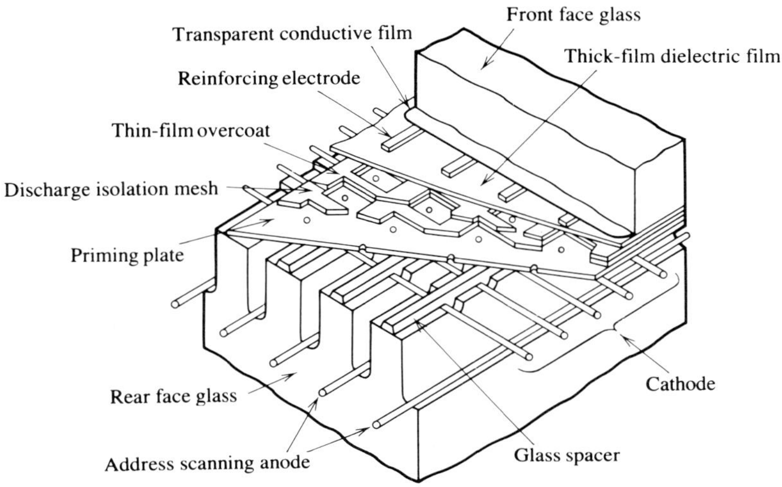

Figure 4.29 *Construction of self-scanning memory panel*

4.5.2 Full dot display panel

In this type of display panel, the discharge cells are arranged uniformly in columns and rows. Such full-dot display panels are utilized for graphics displays or displays of diagrams, etc., as well as for displays of Chinese or Japanese characters. Colour television displays are also being developed. As in the case of character display panels, the most important tasks in further development of these panels are in obtaining high-resolution displays by high density cell arrangements and increasing the size, luminance, and efficiency of the panel.

One example of this is the increase in display resolution and panel size that is being achieved both in Japan and abroad. Regarding AC-PDPs, Photonics in the United States implemented as a product a panel having 1024 × 1024 cells of pitch 0.3 mm, and a panel of improved resolution (0.25 mm pitch) is also under development. Panels of 80 × 60 cm having 1596 × 1212 cells of pitch 0.5 mm have also been announced [14]. IBM in the United States have produced in product form a large panel capable of displaying 160 characters × 62 rows (9920 characters), with cells of 0.36 mm pitch, the screen size being 34.1 × 27.3 cm [13].

Regarding DC-PDPs, Sony have announced a high-resolution panel equipped with trigger electrodes for priming discharge, as shown in Fig. 4.30 [28]. For panel operation, the following method is adopted: charged particles are generated in the form of an indirect discharge by trigger electrodes coated with a dielectric layer. These charged particles are utilized for direct current discharge of the display cells. Two types of panel, of 1024 × 512 cells and 512 × 256 cells, have been manufactured on a trial basis. The cell pitch of the former is 0.2 mm and that of the latter is 0.3 mm. These panels give the highest display resolution achieved

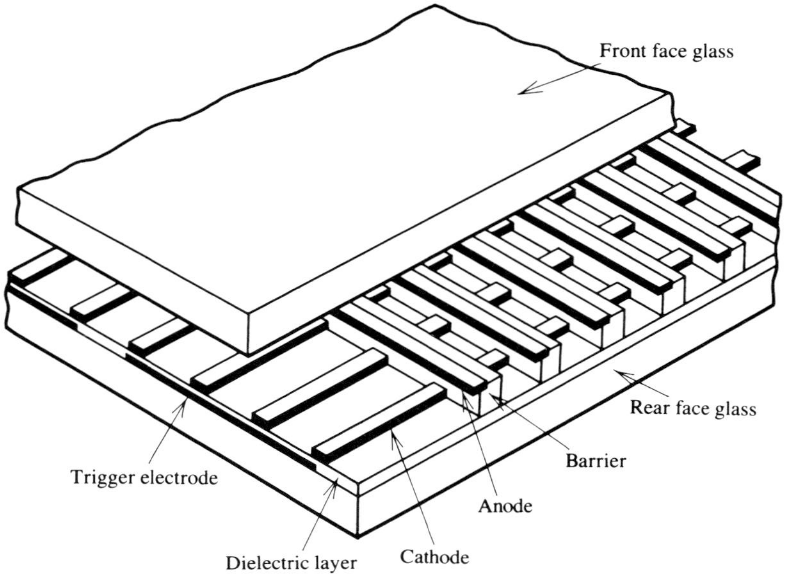

Figure 4.30 *Structure of DC-PDP of indirect discharge type, with trigger electrode*

up to the present. The luminance of the former is 35 cd/m^2, while that of the latter is 85 cd/m^2.

The refresh drive system is not suited to obtaining high luminance. Methods of adding a memory function have therefore been investigated. In particular, various methods have been announced aimed at colour displays. One of these is the memory drive method using pulse discharge developed by NHK Giken mentioned above [33]. Hitachi Central Research Laboratories have developed a green phosphor excitation pulse memory panel utilizing ultraviolet emission from a Townsend discharge. They announce that luminance of 510 cd/m^2 and light emission efficiency of 3.3 lm/W have been obtained [61]. The West German firm Telefunken have developed a memory panel with four electrode layers and three cell layers. They announce that this is of high luminance and gives a high efficiency colour display panel [62]. In a prototype panel, the cell pitch is 1 mm, and a white display obtained from the three primary colours red, green, and blue has a luminance of about 275 cd/m^2 and luminous efficiency of 2.1 lm/W.

In parallel with this research and development, basic research into gas compositions aimed at increasing the radiated intensity of ultraviolet light and analysis of the physical properties of the plasma, as well as materials research aimed at lowering the voltage, is proceeding.

4.5.3 Segment display panel

These are display panels with segment cathodes, normally in a rectilinear figure-eight arrangement, arranged in multiple rows. They are utilized for numerical display. The type of enclosed gas etc. are the same as in the case of dot matrix display panels. The display is produced by light emission from glow discharge. There are two discharge systems: the alternating current discharge system and the direct current discharge system. The alternating current discharge type is driven by a square wave alternating current voltage of 10 to several tens of kilohertz. The approximate characteristic is: drive voltage 130 to 180 V, 0.2 to 2 mA per digit.

Direct current discharge type display panels are operated by a time division pulse drive. Digit skipping drive is used instead of sequential drive. Spurious discharge is prevented by making the separation between adjacent digits larger than the deionizing time. The characteristic is: drive voltage 170 to 230 V, about 1.5 mA per digit.

4.5.4 Bar graph display panel

In this type of display panel, a large number of short-segment cathodes are arranged in parallel rows. The glow discharge formed between these and the

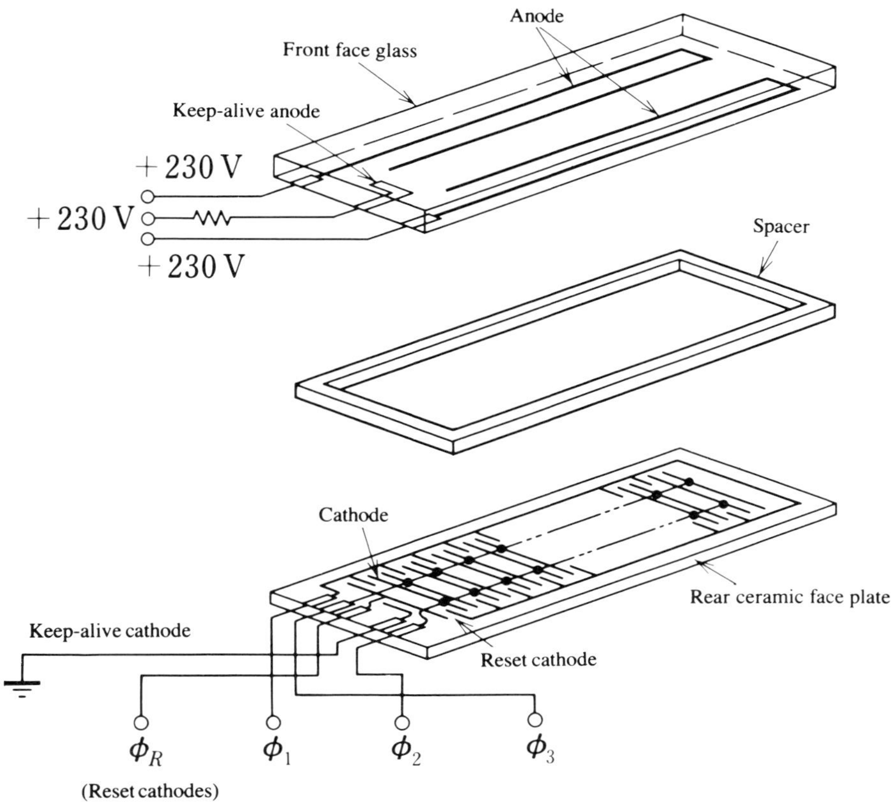

Figure 4.31 *Structure of bar graph display panel*

anode is sequentially shifted in response to the signal, so as to display a level by forming a continuous luminous bar [63]. The circuit layout is simple, since even though there are a large number of segment cathodes only a few terminal leads need be provided. Multicolour displays can be achieved by utilizing phosphors [64].

As shown in Fig. 4.31, rows are formed on the surface of a rear plate consisting of sheet glass, which is coloured black, for example, by printing a large number of short parallel segment cathodes using conductive paste. The front face consists of a transparent sheet of glass, coated on its inside with a transparent electrically conductive film constituting the anode. The periphery is sealed in a gas-tight manner, and a mixed gas whose principal constituent is neon is enclosed. The end one of the segment cathodes is independent and constitutes a reset cathode. The other cathodes are connected in a three-phase connection at every third cathode, so that a continuous bar light emission is formed by sequentially shifting the glow discharge by the same principle as in an SSPD. The reset cathode performs the function of returning the glow to the starting point after it has shifted to the extreme end.

In a practical bar graph display, usually the number of bars provides either one channel or two channels, and the number of cathode segments is either 201 or 101. In each case, one segment is a reset cathode. Note that, although the drawing shows a three-phase connection, five-phase and six-phase connections are also employed. The pulse width of the signal to the segment cathodes is 50 to 100 μs, and the clock frequency is set at a value sufficiently larger than 60 Hz to avoid flicker.

4.5.5 Capital letter display panel

Each of the various plasma displays discussed up to this point are mainly employed in personal displays. In contrast to these, in recent years there has been an increasing demand for displays for public information purposes. Capital letter display panels can meet the requirements of many types of public information display. The discharge type may be either the alternating current discharge type or direct current discharge type. The form of display may be classified into the segment type and dot matrix type.

In the case of the segment type, numerical displays are produced using large-segment electrodes arranged in rectilinear figure-eight configuration to form one digit or several digits. The enclosed gas and the method of drive, etc. are the same as in the case of the segment display panels described above. There are various types of panel, providing figure heights of about 40 to 150 mm.

In the dot matrix type, large dot electrodes are arranged in 5×7 dot, 7×9 dot, and 16×16 dot format. The first two are used for numeric, alphanumeric, and Japanese katakana symbol displays, while the last is utilized in displays of Japanese or Chinese characters. The enclosed gas etc. are the same as in the case of the segment type, and various types of panel of figure height 40 to 150 mm are available.

There are also large display panels using 7×30 to 100, 9×30 to 100 and 16×70 to 150 large dot electrodes arranged to provide figure heights of 40 to 150 mm. The first two of these are used for fixed-letter displays of numeric, alphanumeric, and Japanese katakana symbols, and as scrolling displays such as 'electric news tape' displays, while the latter is used for fixed-letter displays or scrolling displays of Chinese or Japanese characters.

4.6 APPLICATIONS OF PLASMA DISPLAYS

CRTs are still mainly used for displaying large quantities of information. However, with the development in recent years of information processing technology, there has been a demand for display devices having functions and features that cannot be obtained with CRTs. Of these, PDPs make possible large screen displays that are not obtained with other panel displays, and in fact meter-sized panels have been implemented as products. Colour displays can easily also be obtained, and a degree of colour purity not inferior to CRTs can be obtained. They also have the characteristic advantages that screen size and display mode can be altered as required. Because of this, PDPs are employed in a wide range of applications. The main fields of application will be discussed separately below.

4.6.1 Character displays

This type of display panel has a format of 5×7 dots or 7×9 dots, and is used to display numerals, letters of the alphabet, and Japanese katakana symbols. In the field of small to medium capacity displays, providing about 10 to 500 characters, PDPs are superior to CRTs, which give too small a display capacity for their volume.

There is also a market demand for compactness in the field of office automation, mainly represented by microcomputers, which have become common in recent years. Their self-luminosity and compactness have enabled PDPs to make steady advances in these fields, and further progress is anticipated.

Small panels providing 10 to 50 characters are employed in a wide range of applications, such as simple cash dispensers, POS (point of sale) terminals, and displays for measuring instruments.

Medium-size panels capable of displaying 50 to 500 characters are employed in displays such as in ticket machines, minicomputer systems, computer peripheral terminals, terminals of medical equipment, banking terminals, warehouse management terminals, numerically controlled machine tools, vehicle monitors of trains, and plant measuring devices, etc. As an example, Fig. 4.32 shows a bank terminal display device using PDPs.

Recently, use of PDPs in device monitors etc. at the pilot's seat of an aircraft have expanded, and demand for these displays continues to diversify.

Large display panels of 500 to 2000 characters are in competition with CRTs.

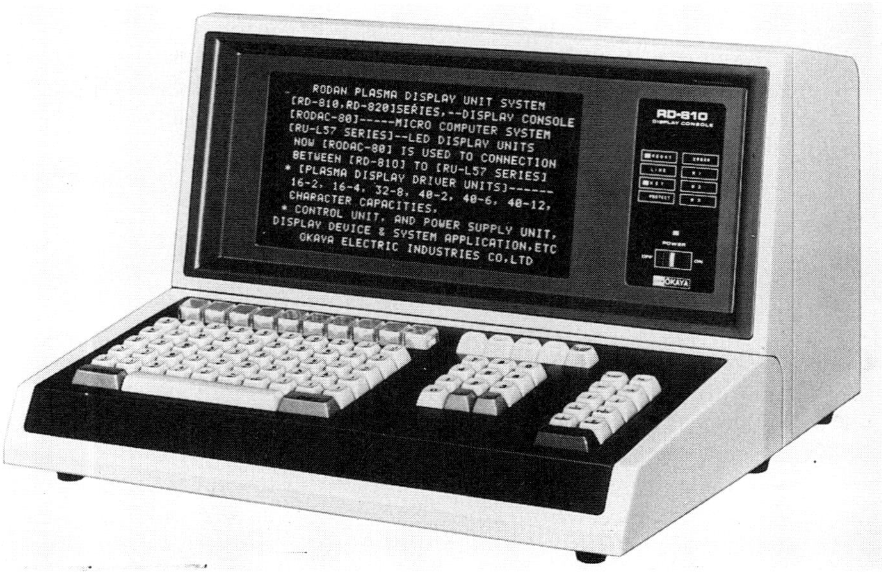

Figure 4.32 *Example of computer terminal display device using PDP*

However, apart from not being subject to the fluctuation and distortion that occur with CRTs, PDPs have advantages such as compactness. They are therefore used in displays such as those of computer terminals, traffic system terminals, news system terminals, aircraft control systems, control panel terminals, etc. Further development to meet future needs is expected.

Ultra-large panels of 2000 characters and up are used in displays such as computer terminals and control consoles. Further expansion of their range of application is anticipated.

4.6.2 Graphics displays

Graphics display panels show wide divergence in their fields of application. They are versatile in that they can display both graphs and diagrams. Also, since they can display Chinese or Japanese characters, they are employed in the display terminals of various types of devices in combination with word processors.

Ultrasmall panels providing a number of dots less than 3000 provide convenient displays for symbols, graphs, and Chinese or Japanese characters. They are therefore applied to communications monitors for schools and companies, surveillance monitors for power sources and transformers, etc., received signal displays such as anticriminal devices and warning devices in buildings and hotels, displays for sensing indoor abnormalities, window guides and displays indicating which client will be attended to next in financial organizations and hospitals, fare displays in stations, and terminal displays of OCRs (optical character readers).

Small panels of 3000 to 30 000 dots are used in linguistic educational equipment, terminals of medical equipment, terminals of minicomputer systems, plant measurement devices, guides to side shows in supermarkets and department stores, etc., guides for showrooms and exhibitions, convenient graph and diagram displays such as display terminals of various types of devices in combination with word processors, and Chinese or Japanese character displays.

Medium-scale panels with dot numbers of 30 000 to 300 000 are used in display terminals of various types of devices in combination with word processors, teletype monitors, surveillance monitors such as of numerically controlled machine tools and production equipment, process surveillance monitors for systems or plants, etc., and the monitors of OCRs for hand-written Chinese or Japanese characters.

Large panels of 300 000 dots and up are used in diagram processing terminals, computer terminals, and radar, etc.

4.6.3 Segment displays

These are mostly based on segments arranged in the form of a rectilinear figure-eight. Small panels providing a figure height of 20 to 40 mm are utilized in the displays of various types of measuring device, ECR (electronic cash register), share price, and POS (point of sale) terminals, and displays of bus fares, etc. Large panels providing figure heights of 40 mm and up have found an increasingly wide range of application, and will be discussed below in section 4.6.7.

4.6.4 Bar graph displays

Bar graph displays are a system for continuously displaying and tracking data. They are utilized in analogue displays. Regarding their resolution, a precision of about the same as a 0.5 per cent class* meter is obtained. Also, by utilizing the features of multicolour displays obtained using phosphors, their applications have expanded into a wide field such as displays for measurement devices, meters for medical equipment, sound level displays in sound studios and stereos, and flow meter and pressure meter displays.

4.6.5 Picture displays

Although they have not reached the stage of practical application, much research is going on both in Japan and overseas into colour television displays; research

*This means a meter ranked as being such that the range of error between the true value and the measured value obtained by reading the meter is less than 0.5 per cent.

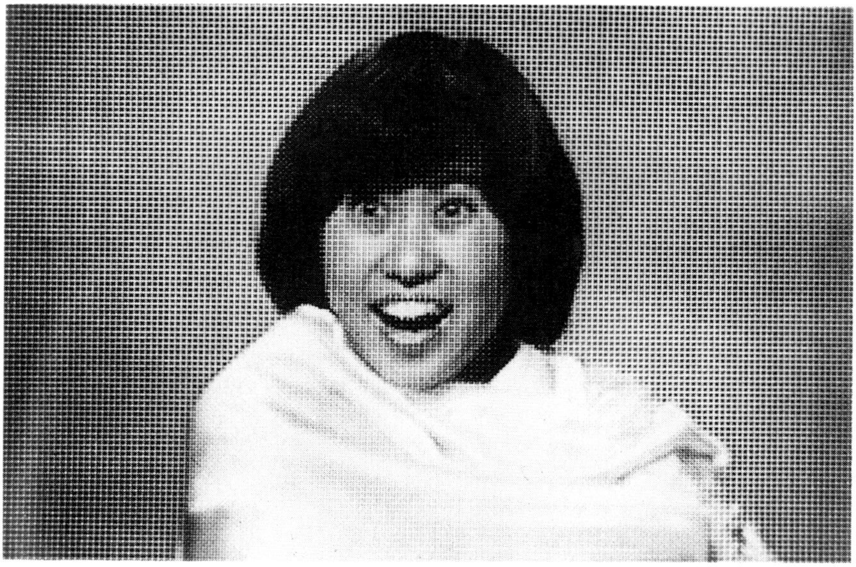

Figure 4.33 Six inch colour television picture using DC-PDPs (NHK Giken)

announcements have reached an enormous number [4, 14, 20–22, 26, 27]. For display panels, usually DC-PDPs are adopted because the intensity of light emission is easy to control. For television display a line sequential scanning system is adopted, wherein the picture signal is first stored and then converted as a simultaneous signal for each horizontal period. To this end, signal hold memories are inserted at each row electrode. An example of the construction is as follows, in outline.

The display device consists of a processing unit for signal processing and a panel unit to drive the display panel. The processing unit uses an A/D converter to encode the television picture input signal to produce a PCM (pulse code modulated) signal. Data corresponding to pixels (discharge cells) in a single row of the display panel is stored in a line memory for each horizontal line, and this stored data is then transferred to a pulse width modulator. The modulated PWM (pulse width modulation) signal is supplied to the panel unit. The panel unit receives the PWM signal, divides it into its constituent signals, and drives the display anodes by a driver, to display the television picture. Figure 4.33 shows a PDP screen used to display a received colour television picture at NHK Giken.

4.6.6 Colour displays

These generally utilize the photoluminescence of phosphor that is produced in response to ultraviolet rays from a rare gas discharge. Colour purity is not inferior to that of CRTs. However, the low luminance is a problem. Green light emitting DC-PDP panels have been practically implemented as products [29]. Since the light emission is in the region of high visual sensitivity, there is said to be little eye fatigue.

Panels utilizing phosphors in bar graph displays have also been implemented, and are employed in, for example, sound level displays of stereos.

4.6.7 Public information displays

Rectilinear figure-eight segment display panels, 5 × 7 dot matrix display panels, and 16 × 16 dot matrix display panels are utilized in shops both in Japan and abroad for public information displays. In each case, the character height is 40 to 150 mm. The first two types of display panels are used for petrol pump displays in petrol stations, large digital clocks in stations and buildings, date displays, process management boards in factories, production line display boards, and various types of sports score displays. The latter type of display is utilized to display station names in, for example, buses, and in guide displays in buildings and hotels.

Matrix displays of letter height about 40 to 150 mm of 7 × 30 to 100 dots and 16 × 70 to 150 dots are utilized in public information displays in shops both in Japan and abroad. The type of display may be divided into a static letter display and a scrolling display. Panels of the first type are used in public information displays such as in supermarkets or stores, while panels of the latter type are used in, for example, destination or area and station name displays of route buses, guide displays in city station constructions, 'electric news tape' displays, and commercial displays. Further expansion of these applications is expected. An example of a public information display is shown in Fig. 4.34.

4.6.8 Panel type light sources

By providing a different electrode construction from that of a PDP, and by altering the shape, a panel light source can be constructed. Panel type light

Figure 4.34 *Public announcement display using AC-PDPs*

sources using phosphors have excited interest regarding backlighting of liquid crystal displays. Other fields of application that are being investigated are as light sources for use in the charging of toners of electrostatic recording type copiers. Further applications in fields other than displays are expected to be developed.

4.7 OUTLOOK

Various types of panel displays have been developed. To meet requirements for displays capable of handling large amounts of information utilizing the features of plasma displays, efforts have been made aimed at increasing panel size, providing colour display, decreasing the discharge cell pitch, and improving display quality. The most important tasks are to increase luminance and efficiency. From this point of view, the most important research topics are development of convenient panel constructions for large panels, and research into materials and manufacturing techniques, including basic research on suitable cathode materials to enable lower voltages to be used, gas compositions to enable high luminance colour displays, and uniform and stable memory function. In particular, research is being carried out in many places into cathode materials and conductive materials using organometal compounds, with the object of opening up new applications for panel displays. Improvements in panel construction are also anticipated.

Colour display PDPs are mainly based on photoluminescence of phosphors. However, recently Siemens in West Germany[64] and Lucitron in the USA[65], have been researching colour display panels utilizing electroluminescence of phosphors produced by plasma electrons generated by discharge. Also, though this will not be described in detail here, in 1984 Siemens announced a new type of colour television display at the Japanese International Display Conference. This was noted because it is expected to start a new trend in colour displays. Regarding drive circuitry, high withstand-voltage PNP type LSIs are being developed both in Japan and abroad. Improvements are also being made in components such as connectors used to connect the panel and circuitry. It is anticipated that these will lead to even more compactness of the display device. Improvements in production techniques are sought, as well as further simplification of panel structure, with the aims of mass production, lower costs, and better reliability. Solution of these problems will be the deciding factor in further development of PDPs.

BIBLIOGRAPHY

Denki Gakkai (Electrical Association): *Hodenkan Benran (Discharge Tube Manual)*, Ohmu-sha (1957).

Denki Gakkai (Electrical Association): *Hoden Handbook (Handbook of Discharge)*, Ohmu-sha (1974).

Oishi, Ogoe, and Nakayama: *Gazo Display (Image Display)*, Korona-sha (1975).
Pankove, J. I.: *Display Devices, Topics in Applied Physics*, Springer-Verlag, Berlin, Heidelberg, N.Y. (1980).
Sherr, S.: *Electronic Display*, a Wiley–Interscience Publication, N.Y. (1979).
Television Gakkai (Television Association): *Saikin no Display Sochi (Modern Display Devices)*, Nippon Hoso Shuppan Kai (1974).

REFERENCES

[1] R. H. Willson: Nov/Dec., p. 39 (1968).
[2] D. L. Bitzer and H. G. Slottow: The plasma display panel—A digitally addressable display with inherent memory, Fall Joint Computer Conference, AFIPS, *Conf. Proc.*, **29**, 541 (1966).
[3] S. Sherr: *Electronic Display*, A Wiley–Interscience Publication, N.Y. (1979).
[4] Television Gakkai (Television Association): *Saikin no Display Sochi (Modern Display Devices)*, Nippon Hoso Shuppan Kai (1974).
[5] J. I. Pankove: *Display Devices, Topics in Applied Physics*, Springer-Verlag, Berlin Heiderberg, N.Y. (1980)
[6] Oishi, Ogoe, Nakayama: *Gazo Display (Image Display)*, Korona-sha (1975).
[7] R. L. Johnson, D. L. Bitzer, and H. G. Slottow: *IEEE Trans. Electron Devices*, **ED-18**, 642 (1971).
[8] N. Nakayama and S. Andoh: *Proc. SID*, **13/1**, First Quarter, 61 (1972).
[9] W. E. Johnson and L. J. Schmersal: *Proc. SID*, **13/1**, First Quarter, 56 (1972).
[10] T. Urade, T. Iemori, M. Osawa, N. Nakayama, and I. Morita: *IEEE Trans. Electron Devices*, **ED-23**, 313 (1976).
[11] M. O. Aboelfotoh: Biennial Display Research Conference, p. 62 (1978).
[12] M. O. Aboelfotoh, K. C. Park, and W. A. Pliskin: *J. Appl. Phys.*, **48**, p. 2910 (1977).
[13] IBM Catalog, NG-518-5126-0 (1983).
[14] T. J. Soper, R. E. Ernsthausen, D. R. Willis, and D. K. Wedding: SID Int. Symp. Dig. Tech. Paper, p. 162 (1982).
[15] Andoh: *Electronics*, **25** (8), 825 (1980).
[16] Y. Takekawa and A. Yano: *NEC Rev. Dev.*, **54**, 27 (1979).
[17] P. Curran, T. Engibous, and J. D. Spencer: *Electronics*, **55** (7), 130 (1982).
[18] Th. J. de Boer: SID Int. Symp. Dig. Tech. Paper, p. 193 (1968).
[19] G. E. Holz: *Proc. SID.*, **13**, 2 (1972).
[20] Y. Amano: *IEEE Trans. Electron Devices*, **ED-22**, 1 (1975).
[21] C. J. Chodil, McDe Joule, and J. Markin: *IEEE Trans. Electron Devices*, **ED-20**, 1098 (1973).
[22] S. Watanabe, A. Watanabe, Hirokami, Kanizawa, and Shiobatake: *Oki Denki Jippo (Oki Denki Bulletin)*, **39**, 46 (1973).
[23] Mikotoride, Shinada, Mishima, Fukushima: Kara terebi hyoji yo 8-inchi yokochu memori paneru, Terebi Gakkai Giho (8-inch positive column memory panel for colour T.V. display), *Technical Report of the Television Association*, **2** (16), 31 (1978).
[24] Kamegaya: *Shingaku Shi (Journal of Communications)*, **61** (6) 637 (1978).
[25] R. Kaneko, T. Kamegaya, M. Yokozawa, H. Matsuzaki, and S. Suzuki: SID Int. Symp. Dig. Tech. Paper, p. 46 (1978).
[26] Sakai: *Shingaku Shi (Journal of Communications)*, **B62** (10), 893 (1979).
[27] Nakagawa, Okamoto, Atsumi, Inakuma, Tagawa, and Akutsu: *National Technical Report*, **25**, 1256 (1979).
[28] Y. Amano, K. Yoshida, and T. Shionoya: SID Int. Symp. Dig. Tech. Paper, p. 160 (1982).

[29] J. Koike, T. Kojima, R. Toyonaga, A. Kagami, T. Hase, and S. Inaho: SID Int. Symp. Dig. Tech. Paper, p. 150 (1980).
[30] Y. Okamoto and M. Mizushima: *IEEE Trans. Electron Devices*, **27** (9), 1778 (1980).
[31] Kojima, Toyonaga, Sakai, Tashima, Sega, Kuriyama, Koike, and Murakami: *Terebi Gakkai Giho (Technical Report of the Television Association)*, **2** (16), 45 (1978).
[32] G. E. Holz: SID. Int. Symp. Dig. Tech. Paper, p. 36 (1972).
[33] Murakami, Kaneko, Toyonaga, and Sega: Television Gakkai Zendai (Television Association National Conference), 5–5 (1983).
[34] Watanabe, Kamegaya: Denshi Tsushin Gakkai Giho (Technical Report of the Electronic Communications Association), IE-81–26, p. 55 (1981).
[35] Kamegaya: *Terebi Shi (Television Journal)*, **27** (5), 363 (1973).
[36] Watanabe: *Denshi Gijutsu (Electronic Technology)*, **20** (11), 16 (1978).
[37] Denki Gakkai (Electrical Association): *Hoden Handbook (Handbook of Discharge)*, Ohmu-sha (1974).
[38] A. L. Word: *IEEE Trans. Electron Devices*, **ED-10** (4) 255 (1963).
[39] Matsuzaki: Denki Gakkai (Electrical Association), Plasma Kenshi (Plasma Research Bulletin) EP-78–32 (1978).
[40] A. von Engel: *Ionized Gases*, Oxford University Press, London (1965).
[41] J. D. Cofine: *Gaseous Conductors*, Dover Publications, Inc., N.Y. (1958).
[42] Denki Gakkai (Electrical Association): *Hodenkan Benran (Manual of Discharge Tubes)*, Ohmu-sha (1957).
[43] Andoh, Umeda: *Shingaku Shi (Journal of Communications)*, **65** (7), 720 (1982).
[44] Akutsu: *Denshi Zairyo (Electronics Materials)*, **19** (2), 52 (1980).
[45] S. Andoh, K. Oki, and K. Hoshikawa: SID. Int. Symp. Dig. Tech. Paper, p. 78 (1977).
[46] V. Brusic, F. M. D. Heurle, R. D. MacInnes, E. I. Alessandrini, J. Angilello, J. J. Dempsey, and M. Sampogna: *IBM J. Rev.*, **22** (6), 647 (1978).
[47] D. E. Miller, J. Ogle, R. Cola, B. Caras, and T. Maloney: SID Int. Symp. Dig. Tech. Paper, p. 146 (1980).
[48] R.A. Cola: *Electronics*, **12** (7), 66 (1971).
[49] Fujino, Tatsuko, Sato, Kubo, and Sakaya: Obutsu Gakkai Zendai (National Conference of the Applied Physics Association), 27P–K4, p. 269 (1978).
[50] Yoshikawa, Shinoda: Terebi Gakkai Giho (Technical Report of the Television Association), ED-545, IPD53-13, p. 69 (1980).
[51] Ishikawa and Kimura: Denshi Tsushin Gakkai Zendai (National Conference of the Electronics Communication Association), p. 1144 (1981).
[52] J. Koike, T. Kojima, R. Toyonaga, A. Kagami, T. Hase, and S. Inaho: *J. Electrochem. Soc.*, **126** (6), 1008 (1979).
[53] Umeda: Denki Gakkai Tokyo Shi Dai Shinpojiumu (Electrical Association, Tokyo Branch, Symposium), S. 2–6, p. 21 (1977).
[54] Yasuda, Iwakawa, Nishida, and Yano: Saikin no AC-gata plasma display device (Modern AC-type plasma display devices), *Nichiden Giho (Nichiden Technical Reports)*, **129**, 78 (1979).
[55] Andoh: *Electronics*, **25** (8), 825 (1980).
[56] Andoh and Umeda: *Shingaku Shi (Journal of Communications)*, **65**, 720 (1982).
[57] Shinoda, Oki, and Andoh: *Terebi Gakkai Giho (Technical Report of the Television Association)*, **2**(3), 99 (1978).
[58] Yoshikawa, Sato, and Andoh: *Terebi Gakkai Biho (Technical Report of the Television Association)*, **2** (16), 25 (1979).
[59] Sato and Kurahashi: *Denshi Tsushin Gakkai Giho (Technical Report od the Electronic Communications Associations)*, ED79–67, p. 25 (1979).
[60] G. Holz, J. Ogle, N. Andreadakis, J. Siegel, and T. Maloney: SID Int. Symp. Dig. Tech. Paper, p. 148 (1981).
[61] Mikotoride, Shinada, and Shirai: *Terebi Gakkai Giho (Technical Report of the Television Association)*, IPD67–7, pp. 37–42 (1982).

[62] M. F. Schiekel and H. Sussenfach: SID Int. Symp. Dig. Tech. Paper, p. 148 (1980).
[63] R. Marshall: *Electronics*, 28 Sept., p. 130 (1978).
[64] A. Schauer: *Electronics*, **55** (25), 128 (1982).
[65] D. Glaser, M. Dejule, C. J. Whelchel, C. S. Stone, A. Sobel, and J. Makin: SID Int. Symp. Dig. Tech. Paper, p.136 (1983).
[66] Burroughs: Application Not. Bulletin, No. S101.
[67] Y. Okamoto, E. Mitani, A. Sumioka, and T. Okabe: *Proc. SID*, **21** (4), 321 (1980).

Chapter 5

Electroluminescent displays (ELDs)

Electroluminescent displays (ELDs) are classified as flat face active displays in which light emission is produced by application of an electric field to a phosphor that is applied onto a transparent organic film or glass sheet. They are completely solid state, and are therefore easy to handle, and have been implemented as planar light sources or graphic displays. Regarding the colour of light emission, the main colours that have been implemented are yellow–orange and green. However, multicolour displays including, for example, red and blue are also possible.

ELDs in practical use at present are of the high-voltage a.c. type. However, great efforts are being made to develop low-voltage types and d.c. drive types. They can provide typical display functions such as memory function, optical image memory, optical erasure, and direct electrical reading. Further utilization and development of these is anticipated.

5.1 INTRODUCTION

5.1.1 What is electroluminescence?

EL is the abbreviation for electroluminescence. This is the phenomenon of emission of light when an electric field is applied to a phosphor, usually a semiconductor. Electroluminescence may be of two types: 'true electroluminescence' and 'charge injection electroluminescence'. In true electroluminescence, emission of light is produced when a voltage is applied directly or indirectly between electrodes. In charge injection electroluminescence, light is emitted when an electric field is applied to a *pn* junction in a single crystal semiconductor, such as GaAs, by injection of carriers from electrodes directly fixed to the junction. The light-emitting diodes described in Chapter 7 are of this latter type. The term 'electroluminescence' on its own means true electroluminescence.

5.1.2 Electroluminescence and displays

ZnS type electroluminescence was discovered by G. Destriau of France in 1936. When transparent electrode material (nesa glass) was discovered in 1947, the Sylvania Company in the United States etc. began a full-scale development programme aimed at utilizing this to provide a planar light source. Electroluminescent displays of this type emit light when an a.c. voltage is applied and are called dispersed type AC-ELDs.

Devices utilizing this type of electroluminescence include photoamplifiers, consisting of an assembly of EL and photoconductive films, and X-ray image intensifiers. In the 1960s television image display devices using matrix electrode electroluminescent panels were announced by the research departments of various companies in Japan, starting with NHK, and including Mitsubishi Electric and Matsushita Electric [1–3].

The dispersed type d.c. electroluminescence, in which a dispersed electroluminescent material is made to emit light by being driven by d.c. voltage, was first developed by a group from Vecht (UK). Considerable development work is being carried out on matrix panels for graphics display, and also on development of multicolour displays. Multipixel (224 × 224) panels for TV image display were developed by Matsushita Electric at the end of the 1960s. Image formation on a flat screen TV was also attempted [4].

In 1968 a new type of evaporated thin-film electroluminescence called LUMOCEN (luminescence from molecular centre) was developed at the Bell Laboratories in the United States. Much higher luminance has been achieved than with dispersed type electroluminescence [5]. Subsequently, in 1974, Sharp developed a three-layer thin-film a.c. electroluminescent display: high luminance, high stability, and long life were achieved [6]. Drive MOSICs capable of withstanding high voltage and establishment of continuous mass production techniques have enabled the implementation of graphic displays [7]. These are based on three-layer electroluminescent displays and can provide many functions, such as internal memory function or optical writing or optical erasure functions [8, 9]. In 1980, a thin a.c. electroluminescent film was developed by Rohiya Company in Finland using atomic beam epitaxy, and has been practically utilized [10]. Also, at Tokyo Engineering University, MIS construction has been adopted for realizing low-voltage drives, and drive voltage has been greatly improved. At Osaka University, light emission of the colours red, yellow, green, and blue has been obtained by controlling the voltage applied to thin laminated electroluminescent films.

D.c. drive thin electroluminescent films were developed in 1978 by Abdalla *et al.* in France [11]. Also, at Tokyo Engineering University, low-voltage panels have been realized by evaporation of ZnSe light-emitting layer onto a GaAs substrate by the molecular beam epitaxy method [12].

As a guide to the direction of future development of electroluminescence, a classification of the four types of electroluminescence discussed above is shown in Table 5.1.

Table 5.1 Types of EL

- EL
  - Inorganic EL
    - Dispersed type
      - ACEL
      - DCEL
    - Thin-film type
      - ACEL
        - Non-memory type EL
        - Memory type EL
      - DCEL
  - Organic EL — Dispersed type — ACEL

5.1.3 Characteristics of electroluminescent displays

Electroluminescent displays (ELD) can be classified into light-emitting displays, flat screen displays, and totally solid state displays. The display is similar in appearance to a light-emitting diode (LED).

However, compared with an LED, an ELD is larger and is easier to view. The cost per pixel is lower. At the present time, areas of dissatisfaction regarding performance are: colour displays, luminance in large multipixel displays, light-emitting efficiency, etc. Also there are still problems of cost when ELDs are compared with CRTs or LCDs.

5.2 MECHANISMS OF LIGHT EMISSION IN VARIOUS TYPES OF ELECTROLUMINESCENCE

5.2.1 Dispersed type a.c. electroluminescence

In a dispersed type ELD, as shown in Fig. 5.1, a light-emitting layer is obtained by embedding light-emitting phosphors in transparent dielectric, each individual phosphor being independent. This light-emitting layer is sandwiched between a transparent electrode and metal electrode. It is therefore appropriate to consider

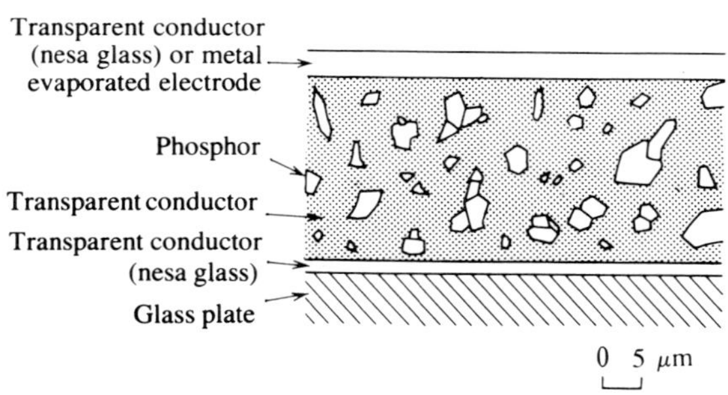

Figure 5.1 Structure of dispersed type EL[3]

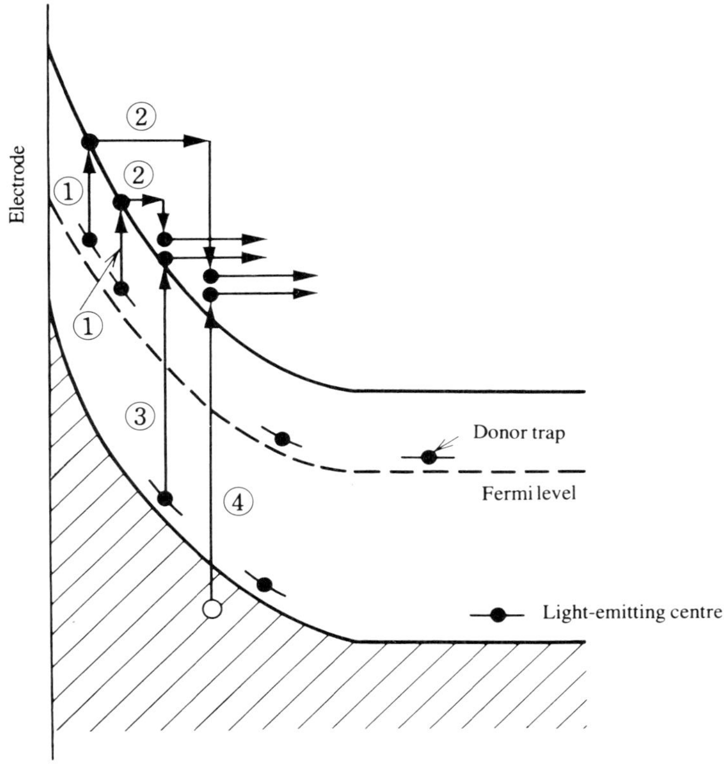

Figure 5.2 *Mechanism of light emission by strong electric field*

the light-emission mechanism in terms of a single phosphor particle. In general, the main type of phosphor used for electroluminescence is ZnS. Cu, Mn, Cl, and Al, etc., are often used as activating agents, providing light-emission centres.

The electroluminescent process may be divided into an excitation step and a light-emission step. First of all, as shown in Fig. 5.2, when a strong electric field is applied to an electroluminescent phosphor, (a) electrons are ejected from deep traps; (b) they are accelerated, so that high energy is obtained; (c) they collide with the light-emitting centres and ionization takes place. In this process, the emitted electrons perform repeated collisions, raising the electrons into the conduction

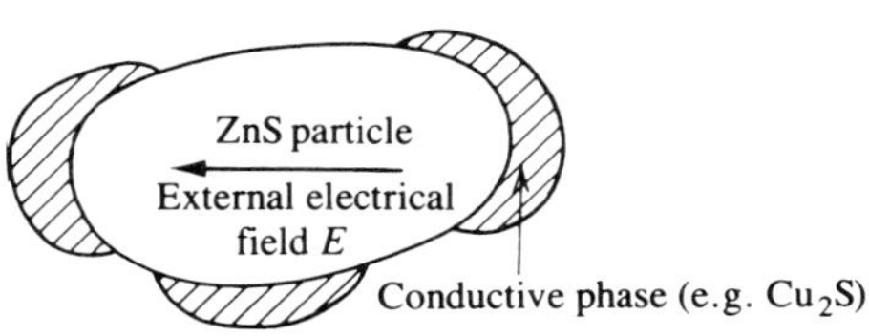

Figure 5.3 *ZnS particle of dispersed type DC-EL*

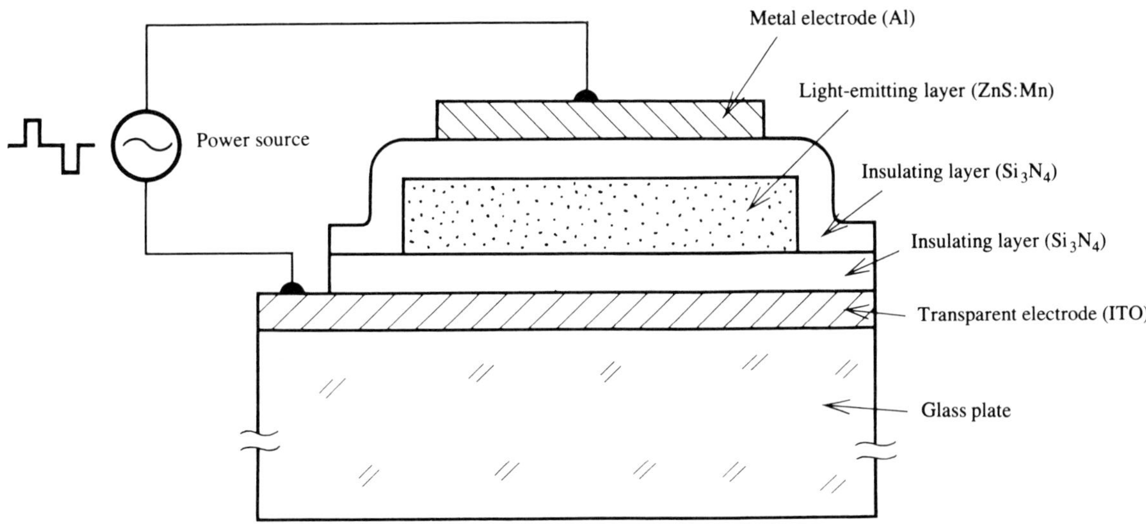

Figure 5.4 *Structure of typical double insulated film EL [6]*

band. On the other hand, electrons that have obtained high energy may (d) directly raise electrons from the valence electron band into the conduction band. The large number of electrons thus obtained in the conduction band show a characteristic light emission when they return to their original state.

5.2.2 Dispersed type d.c. electroluminescence

In dispersed type d.c. electroluminescence, the surface of the phosphor particles is covered with a heterogeneous substance, such as, for example, CuS, and a junction region is obtained by hot-forming. The junction has a diode characteristic. The mechanism of excitation is the same as in the a.c. type, but, due to the rectifying effect, electroluminescence occurs only in response to a voltage in one direction of polarity. Figure 5.3 shows a diagram of a d.c. electroluminescent phosphor particle.

5.2.3 Evaporated thin-film electroluminescence

Figure 5.4 shows a typical three-layer thin electroluminescent film. A light-emitting layer consisting of ZnS doped with Mn is sandwiched between insulating layers. Thus, the light-emitting layer does not directly contact the electrodes. As shown in Fig. 5.5, when voltage is applied, primary electrons are extracted from traps, accelerated, and directed towards the opposite electrode. During this process, emission of light occurs due to recombination. However, some of the electrons are believed to remain in the boundary regions with the insulating layers. Next, when voltage is applied in the reverse polarity, excitation

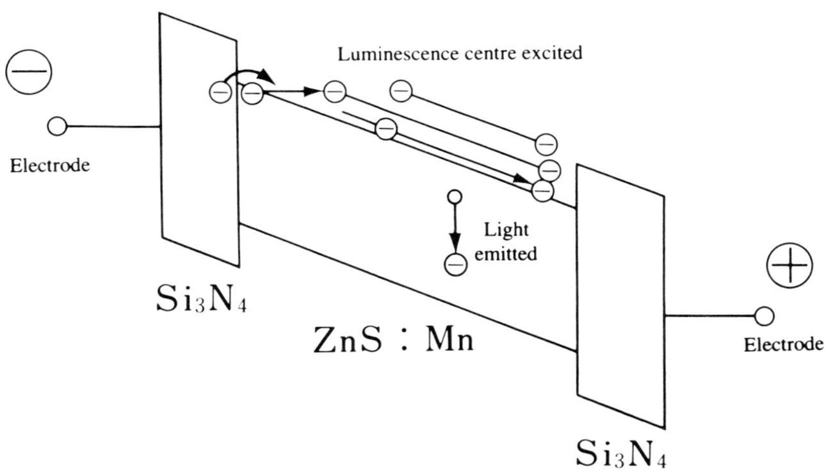

Figure 5.5 *Action when voltage is applied [13]*

ionization takes place in the light-emitting layer. Light emission occurs due to recombination, with the charge remaining within this layer. Some of the electrons then again remain on the boundary surface with the insulating layer on the side of the other electrode.

A thin electroluminescent film only emits light once with a single pulse. In this respect it is different from the dispersed type. Also, a thin electroluminescent film having a memory function has an optical polarization effect [9], as shown in Fig. 5.6; that is when a voltage V_0 of a threshold value is applied to the electroluminescent display, and light is incident from outside, an optical polarization charge is generated on the light-emitting layer, with emission of light. This phenomenon is called the optical polarization effect. Next, if as shown in Fig. 5.7, external light is incident while the electroluminescent display is maintained in a state in which its two electrodes are short-circuited, the

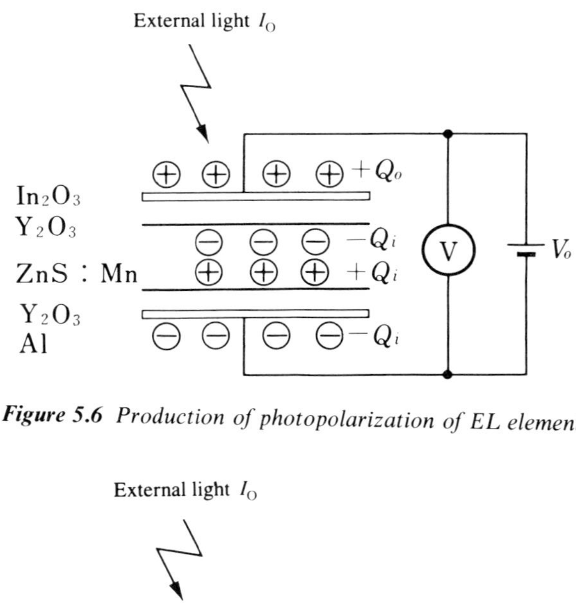

Figure 5.6 *Production of photopolarization of EL element*

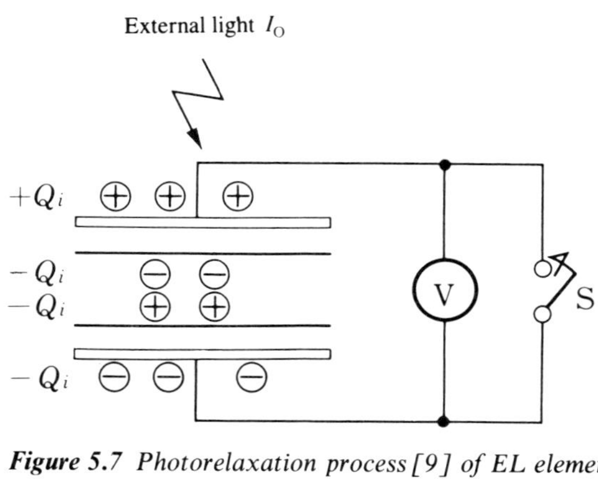

Figure 5.7 *Photorelaxation process [9] of EL element*

Table 5.2 *Main electroluminescent materials[2]*

Type	Activating agent	Sintering conditions		Colour of electroluminescence
		Atmosphere	Temperature (°C)	
ZnS:Cu, Cl	Cu 0.1% NH_4Cl 10%	N_2	1075	Blue
ZnS:Pb, Cu, Cl	Pb 0.003% Cu 0.06% $ZnCl_2$ 5%	Air	950	Blue/green
ZnS:Cu, Al	Al 0.2% Cu 0.2%	$H_2S + H_2O$	1100	Green
ZnS:Mn, Cu	Mn 0.7% Cu 0.3%	Dry H_2S	1100	Orange
Zn(S,Se):Cu, I	ZnSe 100% Cu 2% NH_4I 10%	N_2	1075	Yellow
ZnSiCu	Cu 0.3%	Dry H_2S	1100	Red

polarization charge held in the light-emitting layer is relaxed and disappears. This process is not accompained by any emission of light at all. This phenomenon is called the optical relaxation effect.

In this process, the light that is incident from outside has to have an energy that is larger than the band gap energy of the ZnS. Light of a wavelength shorter than 400 nm is used.

5.3 ELECTROLUMINESCENT DISPLAYS AND VARIOUS TYPES OF MATERIALS

5.3.1 Light-emitting materials

The principal light-emitting material that is used for electroluminescent displays in ZnS. Typical substances that are used as activating agents providing centres of light emission are shown in Table 5.2. The amount of these activating agents that should be added is much greater than in the case of fluorescent lamps etc. This is because the excitation process is different and the conditions during sintering are also different. Figure 5.8 shows various types of light-emitting material and the distribution of light-emission wavelengths. Apart from these, there are reports of uses of Mn (yellow–orange colour), TbF_3 (green colour), PrF_3 (white colour), and DyF_3 (yellow–white colour) as activating agents in thin electroluminescent films. Of these, Mn is most used, on account of its light-emitting efficiency. Figure 5.9 shows an *xy* chromaticity diagram of the light-emission colours produced by these phosphor materials.

The group of Vecht and others has attempted production of multicolour d.c. electroluminescent displays. Apart from the above-described materials, they have used SrS:Ce, Ba_2ZnS:Mn, CaS:Ce, ZnS:Te, Mn, and CaS:Eu, etc. By using

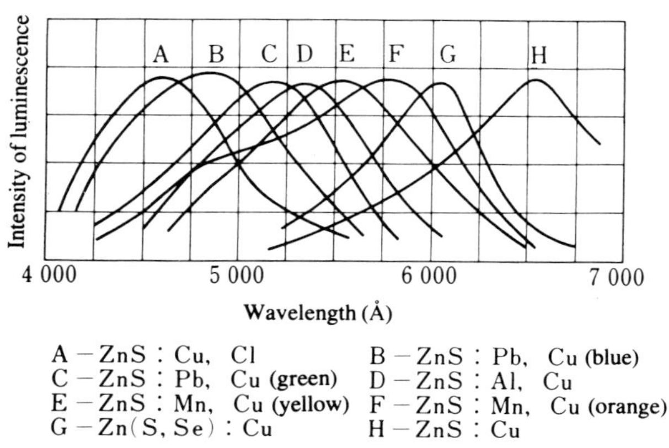

Figure 5.8 *Distribution of luminescence wavelength of various types of EL [2]*

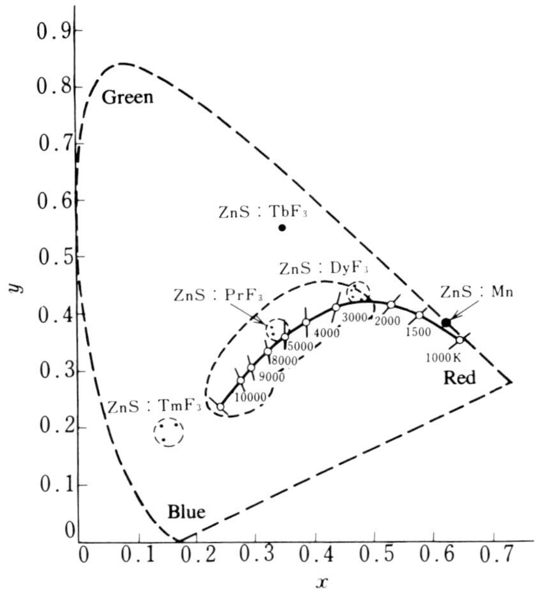

Figure 5.9 *Chromaticity coordinates of luminescence of various EL phosphor materials [14]*

powder electroluminescence in which local light-emitting centres* such as transition metals or rare earth elements like Mn, Ce, or Eu are introduced into a phosphor matrix such as SrS or CaS, light emission is obtained over a narrow spectral width, as shown in Fig. 5.10 [15].

5.3.2 Dielectric materials

In the case of inorganic dispersed type a.c. electroluminescence, a dielectric material is provided on one face of the light-emitting layer in order to obtain high impedance. High dielectric materials such as $BaTiO_3$ etc. may be used as this dielectric material. In organic dispersed type a.c. electroluminescence, high permittivity (ε from 8 to 15) polymer dielectric materials are used. Such materials are strong in respect of bending, and so are applied to the manufacture of electroluminescent films. Typical organic dielectric binder materials are shown in Table 5.3. In a thin a.c. electroluminescent film, dielectric materials such as Y_2O_3, Si_3N_4, Sm_2O_3, Ta_2O_3, $BaTiO_3$, and $PbTiO_3$ are used; ε in such materials is very different (from 10 to about 180), but one cannot necessarily say that a simple

*'Local light-emitting centres' means that, though electron migration in light emission is not affected by surrounding atoms or ions, excitation energy is absorbed at light-emission centres, with the result that the electrons in such light-emitting centres are raised from the base state to an excited state, and emit light when they return.

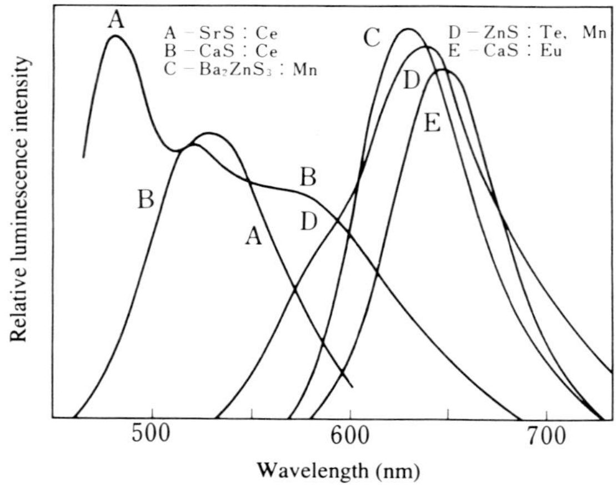

Figure 5.10 *Luminescence spectrum of powder EL produced by local luminescence centres [15]*

Table 5.3 *Organic dielectric binder materials [16]*

NAME OF MATERIAL	PERMITTIVITY (1kHz)	MELTING POINT ETC
Polyvinylidine fluoride	8	160–180°C
Rubber fluoride (three-component copolymer of vinylidene fluorides)	13.8	Pyrolyses (about 250°C)
Vinylidene fluoride-ethylene trifluoride copolymer	15	155°C
Cyanoethyl cellulose	~15	—

increase in ε of the dielectric binder material will result in better electroluminescence. Compatibility with the light-emitting material and/or electrode material etc. and evaporation conditions are important. In particular, problems such as spalling affect device life, so a suitable selection must be made when the device is to be implemented in practical form. For example, if Si_3N_4 is used, there are no problems regarding compatibility with the In_2O_3 of the transparent electrode, but there are problems concerning its compatibility with the back face aluminium electrode, so that a practical device can only be obtained by inserting an intermediate layer of SiO_2.

Furthermore, film-forming conditions are important, since to prevent ingress of moisture into the dielectric insulating layer, a fine thin film is needed containing

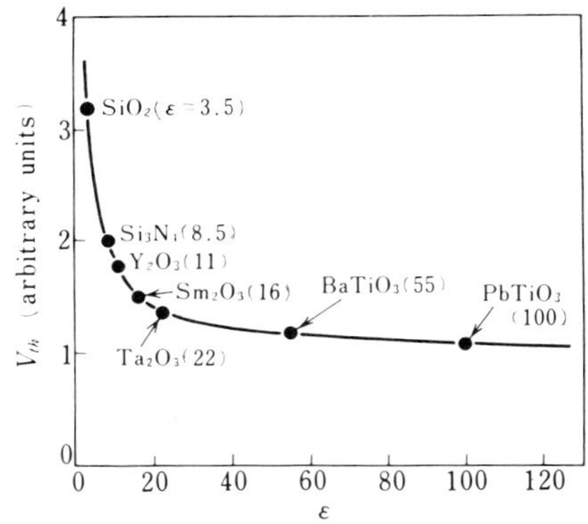

Figure 5.11 *Relationship between luminescence threshold voltage (V_{th}) of double-layer insulated thin-film AC-ELand permittivity of insulation layer*

no pinholes etc. Figure 5.11 shows the relationship between firing voltage and permittivity in a three-layer electroluminescent film with double insulating film. A high-permittivity insulating film may be used in order to lower the applied voltage, but it is necessary to ensure that the electric field intensity in the light-emitting layer is not substantially lowered in order not to lower the luminance. Ta_2O_3 and $PaTiO_3$, etc. have the advantage of not requiring initial ageing [16].

5.3.3 Electrode materials

The main transparent electrode materials used for the front face of the electroluminescent panel are In_2O_3 or SnO_3, or else ITO, which consists of a combination of these. Aluminium is often evaporated onto the back face of the electrode material. However, in the case of a transparent electroluminescent panel, or multicolour panel wherein two ELD layers are superimposed, In_2O_3 may be used for both electrodes.

Apart from transmissivity, resistivity is an important factor. When the device is to be driven at high speed, its speed of response is determined by the product of the electrode resistance and the electrostatic capacity of the pixels, so material of low resistance is required.

5.4 STRUCTURE AND PROPERTIES/FUNCTION OF ELECTROLUMINESCENT DISPLAYS

5.4.1 Dispersed type AC-ELD

As shown in Fig. 5.12, a dispersed type AC-ELD is constructed as follows: a transparent electrode is evaporated onto a glass substrate and a light-emitting

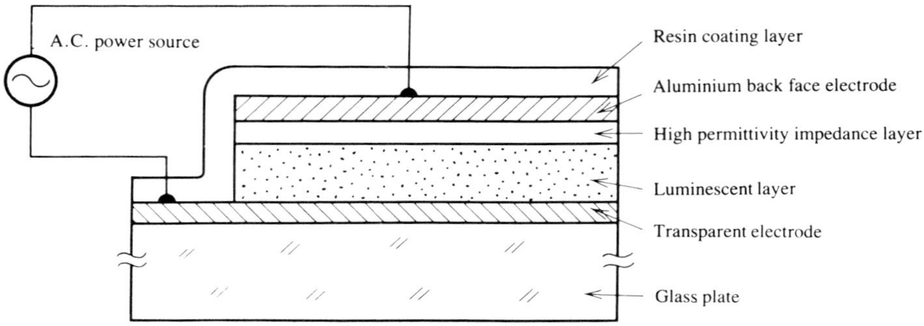

Figure 5.12 *Structure of typical dispersed type AC-ELD*

layer is applied thereon. A high-impedance dielectric layer is provided, and on top of this a back face electrode is evaporated. For mechanical protection, a resin coating is often applied as the final layer. In this construction, when an alternating voltage is applied between the transparent electrode and back face electrode, the luminance increases with an increase in voltage. In the case of a pulse voltage drive, as shown in Fig. 5.13, light emission is observed on both the rise and the fall of the pulse. An applied voltage versus luminance characteristic for a typical panel is shown in Fig. 5.14.

The following relationship exists between the luminance and applied voltage:

$$B = B_o \exp \frac{-C}{\sqrt{V}}$$

where B_o and C are constants determined by the excitation conditions, the structure of the element, and the phosphor material, etc.

In the case of an a.c. type ELD, the luminance also depends on frequency. The relationship is shown in Fig. 5.15. In the low-frequency range, the luminance and

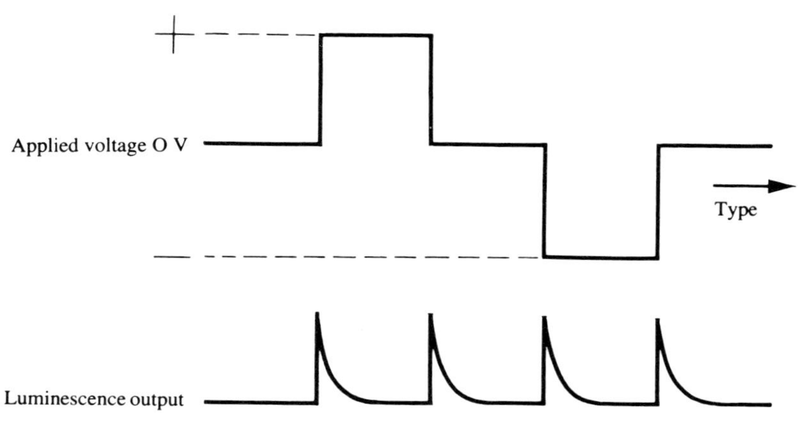

Figure 5.13 *Luminescence output on application of pulse voltage*

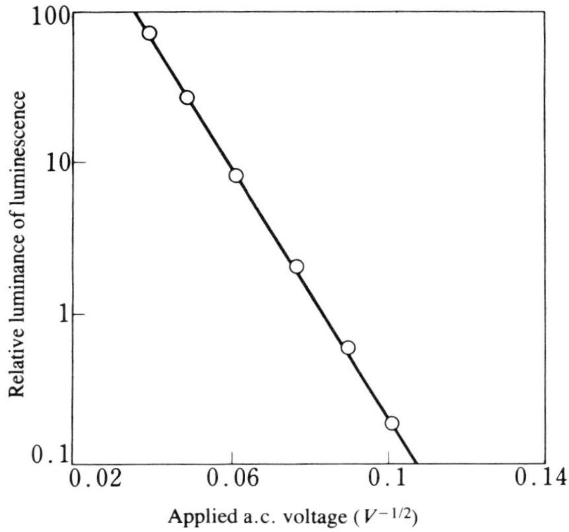

Figure 5.14 *Luminescence luminance characteristic versus applied voltage of typical dispersed type AC-ELD [3]*

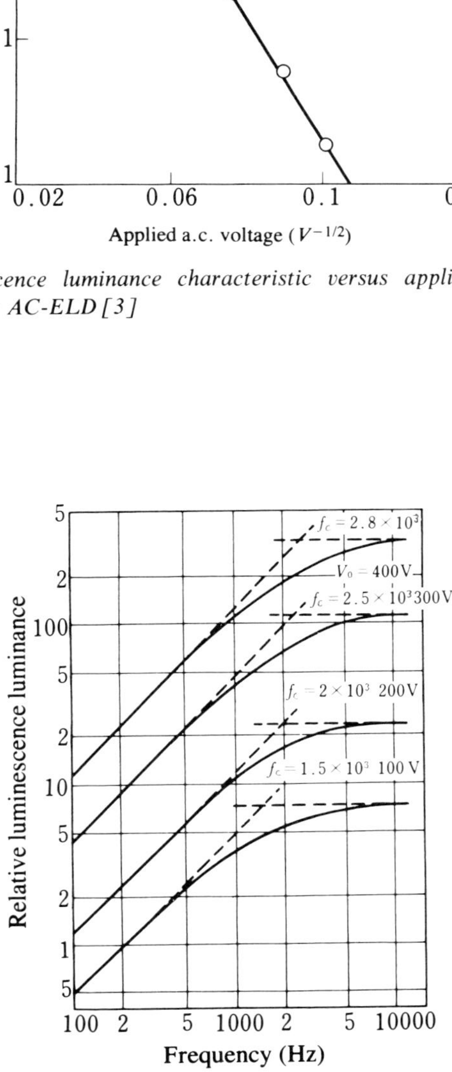

Figure 5.15 *Luminescence luminance–frequency characteristic [3]*

applied frequency are in a proportional relationship, but in the high-frequency region, the luminance gradually becomes saturated. In addition, the luminance shows a temperature dependence, which is marked when the device is driven at high frequency. However, in the range from 0 to 100 °C below 1000 kHz, the luminance does not show much change.

5.4.2 *Dispersed type DC-ELD*

As shown in Fig. 5.16, the construction of a dispersed type direct current ELD is similar to that of an a.c. type ELD, but the electrodes are in direct contact with the light-emitting layer, without interposition of a high dielectric layer. Hot-forming is performed in the manufacturing step, producing a layer of CuS on the surface of the phosphor particles. The arrangement is such that there is a junction surface between this and the interior, so light emission occurs when voltage is applied in one direction only. Consequently, the only drive circuit required is a simple acyclic power source.

5.4.3 *Thin-film AC-ELD*

A thin-film AC-ELD is manufactured by a completely different manufacturing process from that of a dispersed type ELD. Its construction is shown in Fig. 5.17.

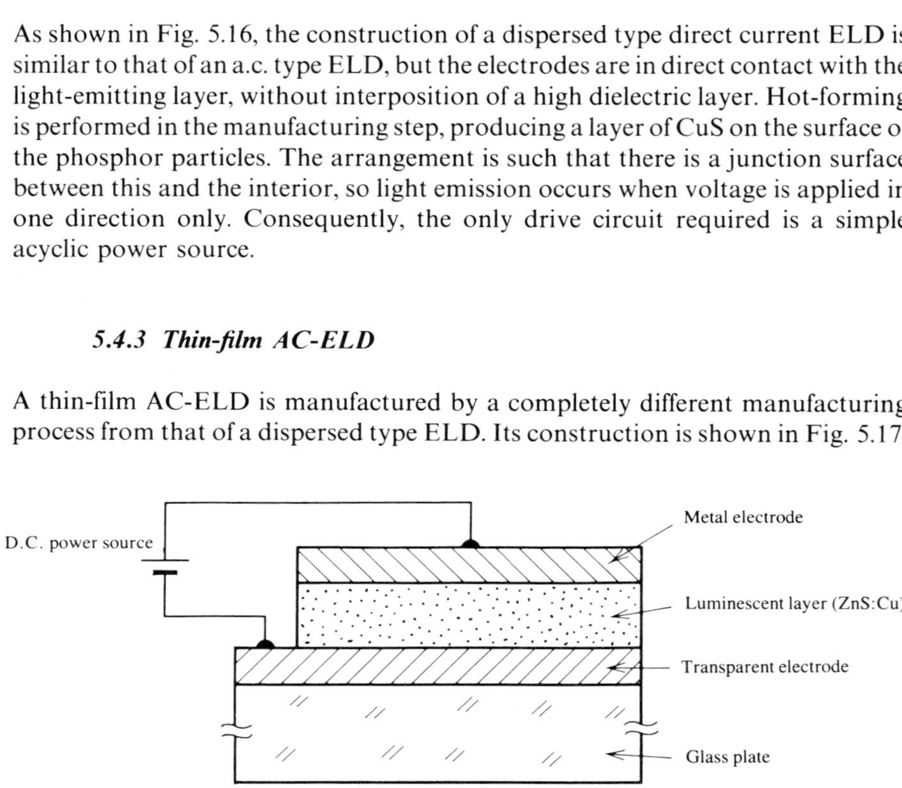

Figure 5.16 *Structure of dispersed type DC-ELD*

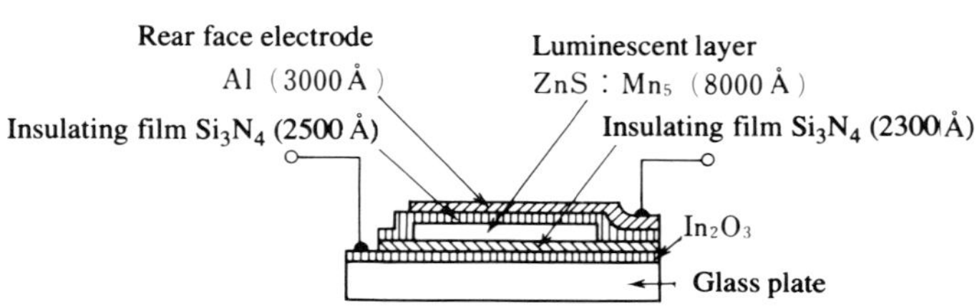

Figure 5.17 *Typical structure of three-layer thin-film ELD*

First of all, a transparent electrode (In_2O_3) is evaporated onto a glass substrate, and an insulating film (Si_3N_4) is evaporated on top of that. A phosphor material (ZnS:Mn) is evaporated onto this insulating film. In this process, if the thickness of the light-emitting layer is increased by increasing the doping amount of, for example, Mn, a memory effect appears, as will be described later. Additionally, an insulating film (Si_3N_4) is evaporated on to this light-emitting layer, and then a back face electrode consisting of aluminium metal is evaporated onto this. In this process, forming treatment may be carried out after evaporating the light-emitting layer, and a step may be added of improving the crystallinity of the phosphor. Also, when the panel is finally assembled, glass is provided to seal in silicone oil to increase the life.

An example of the applied voltage versus luminance characteristic of a thin-film AC-ELD is shown in Fig. 5.18, where *t* indicates the ageing time. Compared with the dispersed type ELD, the threshold voltage is very well defined, with a steep gradient of the characteristic curve. Furthermore, the display is bright because the absolute value of the luminance is large. The reason for this is believed to be that polarization charge is generated in the light-emitting layer when driven by a.c. voltage, and in the light-emitting layer the polarizing voltage due to this charge is superimposed on the external voltage every time the display is driven.

Figure 5.19 shows the applied voltage versus luminance characteristic of a memory effect ELD. When the applied alternating voltage is gradually increased,

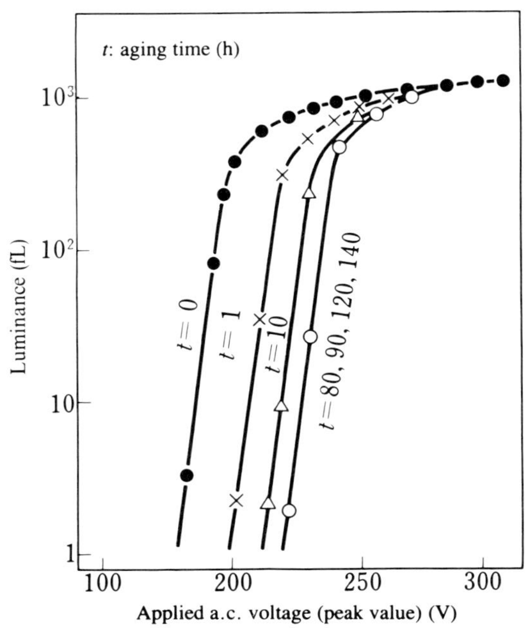

Figure 5.18 *Typical luminance versus voltage characteristic of thin-film AC-ELD*

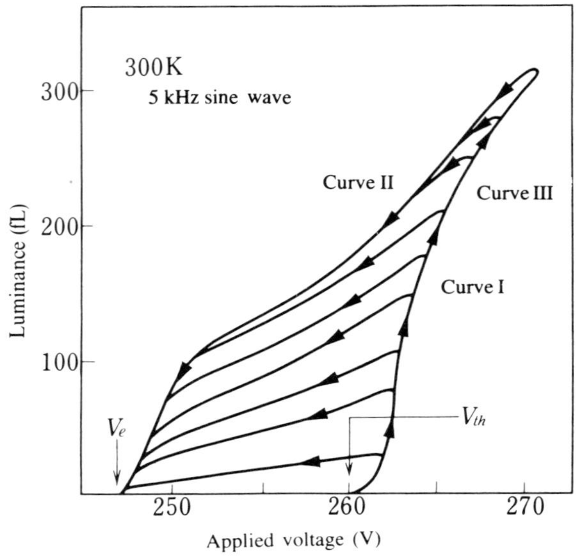

Figure 5.19 Luminance characteristic versus applied voltage of memory effect thin-film ELD

it follows curve I. Next, when the voltage that has now risen is gradually lowered, it follows a curve (II) different from the original curve (I). Since curve II follows a different route, such as curve III, depending on the magnitude of the applied voltage, an infinite number of hysteresis curves can be obtained. This shows that memory of an analogue quantity is possible (see Section 5.5).

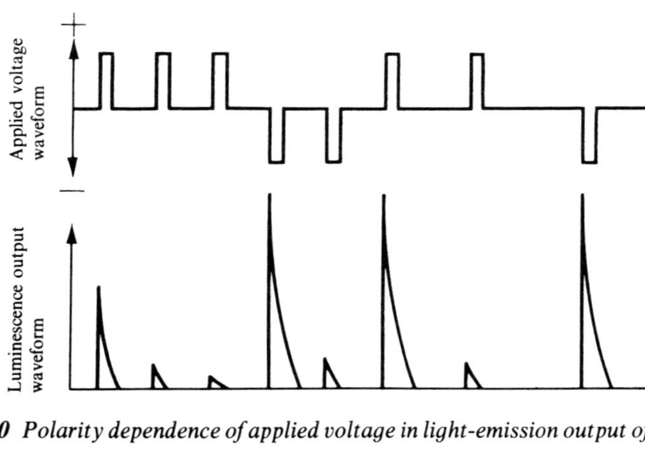

Figure 5.20 Polarity dependence of applied voltage in light-emission output of thin-film ELD

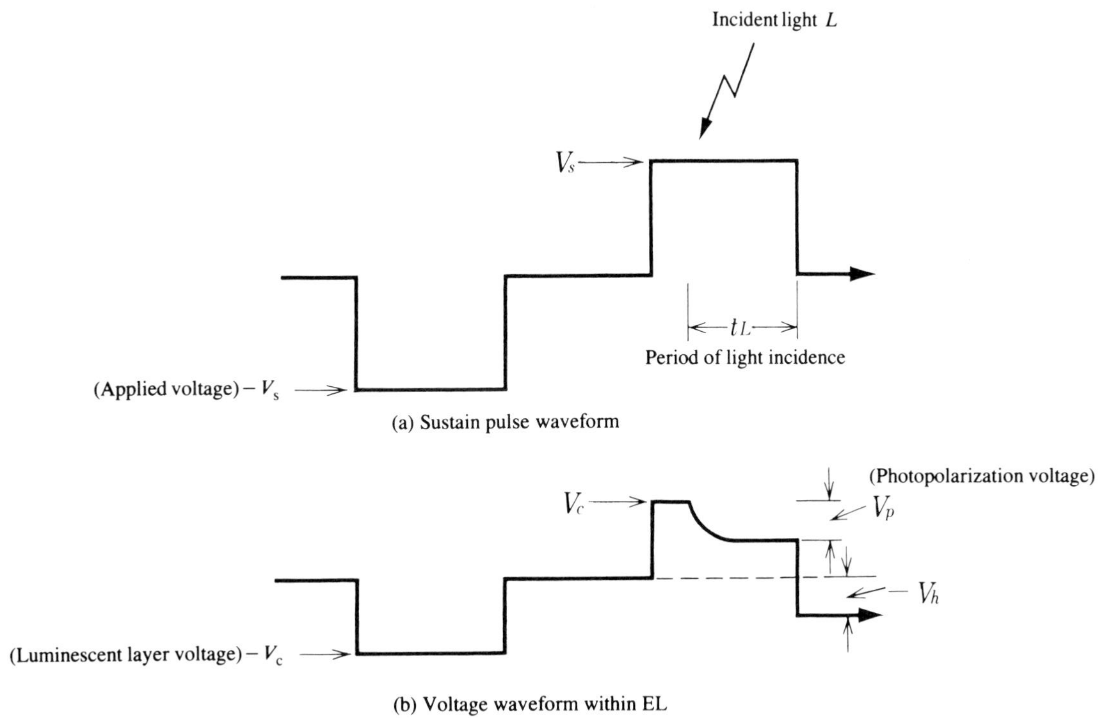

Figure 5.21 *Optical writing by photopolarization effect*

As mentioned above, a memory effect does not occur unless the Mn amount and thickness of the light-emitting layer are increased. However, as shown in Fig. 5.20, when a pulse of different polarity is applied, an intense light emission is seen, while when a pulse of the same polarity is applied, the light emission is weak. Furthermore, if the pulse of the same polarity continues to be applied, the light emission disappears. The reason for this is that if a pulse of different polarity is applied, the polarization voltage within the light-emitting layer is superimposed in the direction of increasing total voltage, whereas in the case of a pulse of the same polarity, the polarizing voltage within the light-emitting layer is of inverted polarity with respect to the applied voltage, so the total voltage that is applied to the light-emitting layer is decreased, and it no longer emits light.

In a memory effect thin-film ELD, the optical polarization effect and optical relaxation effect are present. Their respective actions are shown in Figs. 5.21 and 5.22. First of all, let us suppose that an alternating pulse $\pm V_s$ equal to the threshold voltage is applied. Next, if external light L is incident at the time of application of the pulse, the light energy causes an increase in the number of carriers in the light-emitting layer, making the phosphor emit light. Once it emits light, a polarizing voltage V_p is generated within the light-emitting layer, and this is superimposed on the next pulse, so that emission of light continues if pulses are successively applied. V_p is proportional to the intensity of the light, so an optical image can be written. On the other hand, once an ELD has emitted light, the external light irradiation L' during the space time $t_{L'}$ of the pulse voltage causes relaxation of the interior polarization charge, which consequently disappears.

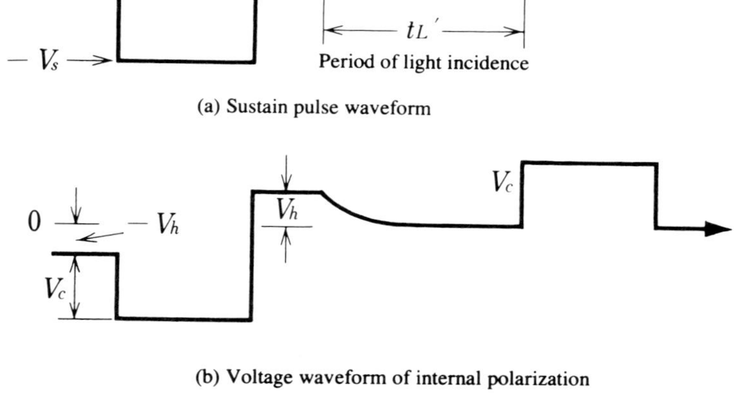

***Figure 5.22** Erasure using photorelaxation effect*

Blanking can therefore be achieved [17], by illumination with light, by using this optical relaxation effect.

5.4.4 *Thin-film DC-ELD*

In a thin-film DC-ELD, as shown in Fig. 5.23, a transparent electrode is evaporated onto a glass substrate, a light-emitting layer (ZnS:Cu) is evaporated onto this, and forming processing is then carried out. After this, a metal electrode is provided as a counterelectrode. As a result of this forming processing, a heterojunction is formed between the *n*-type ZnS region, wherein Cu has migrated in the light-emitting layer, and the *p*-type Cu_xS region, which has not been changed by the forming. As shown in Fig. 5.24, it seems that light emission occurs from this junction region [8].

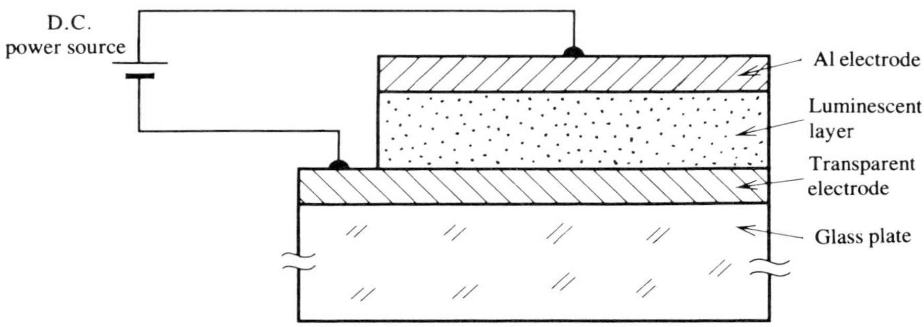

Figure 5.23 *Structure of thin-film DC-ELD*

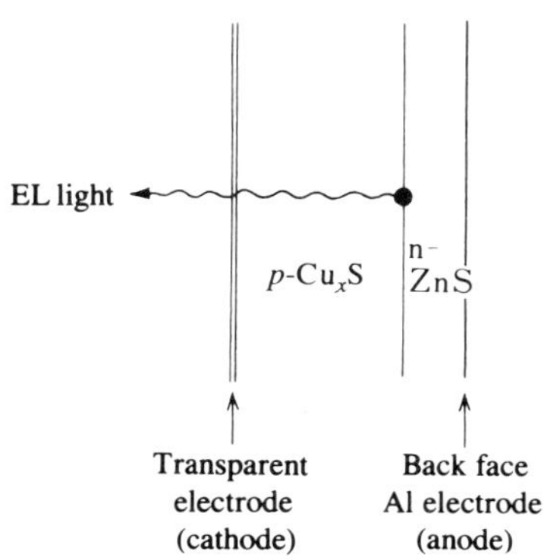

Figure 5.24 *Luminescence mechanism of thin-film DC-ELD*

5.5 VARIOUS DRIVE SYSTEMS FOR ELECTROLUMINESCENT DISPLAYS

5.5.1 Multiplex drive

The drive system of a typical three-layer thin-film a.c. electroluminescent matrix panel will be described.

Figures 5.25 and 5.26 show an external view including the drive circuitry of a 280 × 120 dot matrix electroluminescent panel and the drive circuit. In this electroluminescent panel, there are 280 transparent electrodes on the data side and 120 aluminium electrodes on the scanning electrode side, arranged on the other side of a triple electroluminescent layer. The arrangement is such that they cross over the data side electrodes at right angles. The drive circuit is constituted by connecting each group of electrodes to electrode terminals arranged alternatively above and below and to the left and to the right in a comb arrangement. Thus in Fig. 5.26, the upper data side consists of electrodes numbered $1, 3, \cdots, i-1, i+1, \cdots, 279$, whereas the lower electrode side consists of the electrodes numbered $2, 4, \cdots, i, i+2, \cdots, 280$. In addition, the left-side scanning electrodes are numbered $1, 3, \cdots, j-1, j+1, \cdots, 119$, and the right-side scanning electrodes are $2, 4, \cdots, j, j+2, \cdots, 118, 120$.

To drive this electroluminescent panel, a line sequential scanning system is used. A number of electrodes are selected at the same time on the data side, while a

Figure 5.25 External view of three-layer thin-film AC-EL matrix panel and drive circuitry

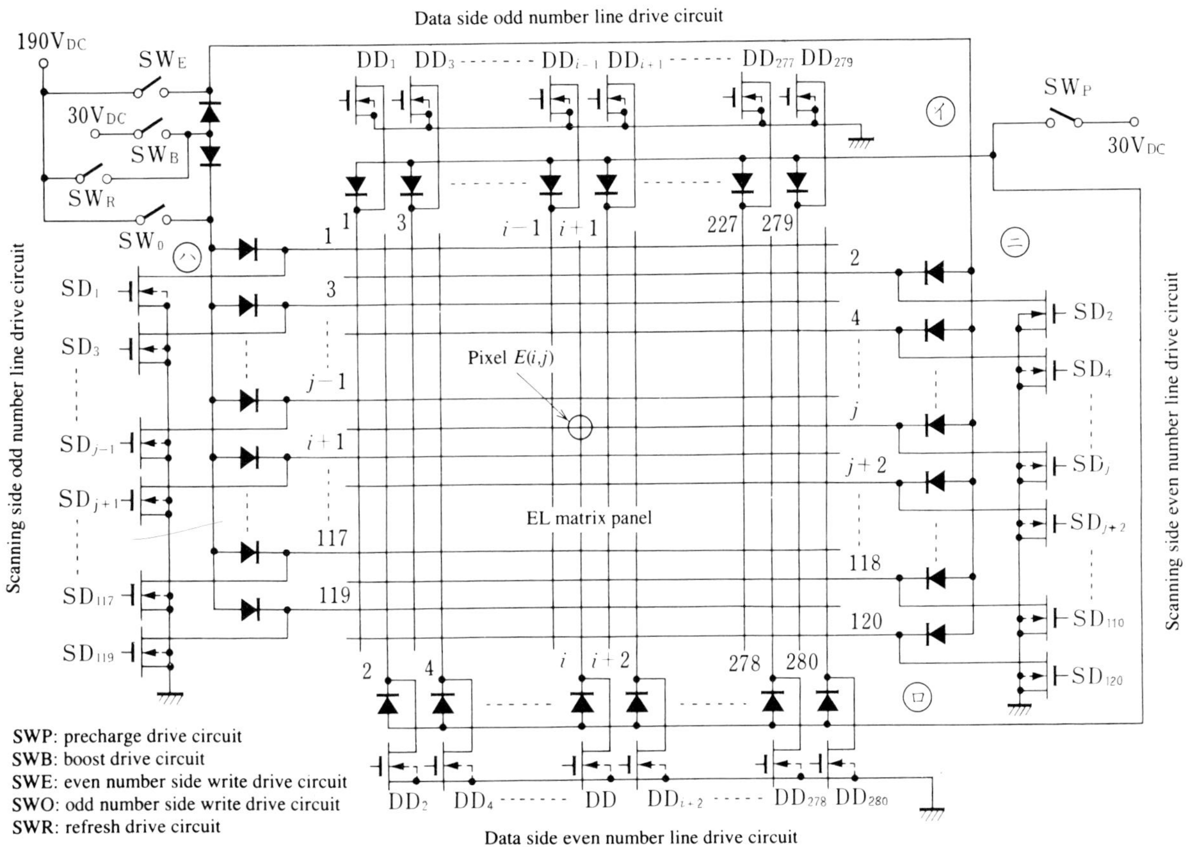

Figure 5.26 *Circuit diagram of three-layer thin-film AC-ELD drive circuit*

scanning electrode is sequentially selected on the scanning side. Accordingly, when the scanning of one picture takes place from scanning line 1 to scanning line 120, an inverse polarity pulse is applied simultaneously to all of the pixels of the panel. This completes one cycle, and scanning writing is then carried out afresh from scanning electrode 1. The repetition frequency of this alternating cycle is called the field frequency. A frequency of 60 to 120 Hz is chosen.

In order to understand the operation of the drive circuitry, the operation of writing a pixel E (i, j) will be described. The drive circuit consists of a combination of high threshold voltage MOS transistors and high threshold voltage diodes. As shown in Fig. 5.27, its action is carried out in three stages. In the first stage, precharging is performed. First of all, the high threshold voltage scanning side electrode MOS transistors SD_1, SD_2, $\cdots$, SD_{119}, SD_{120} are switched to the on state. At the same time precharging drive circuit switch SW_P is put into the on state. At this point, all the high threshold voltage MOS transistors DD_1, DD_2, $\cdots$, DD_{279}, DD_{280} on the data side are put into the off state; that is the scanning electrode is at earth potential (0 V), 30 V_{DC} is connected from SW_P through common wires (i) and (ii) to the data electrode, so that all the pixels are charged to 30 V. Charging is completed in a short time (about 10 μs).

In the second stage, in the period in which modulated discharge drive is performed, all the scanning side MOS transistors are turned off, the boost drive circuit switch SW_B is turned on, and 30 V is applied to the scanning electrode by common wires (iii) and (iv) on the data side. In this process, the intersections of the

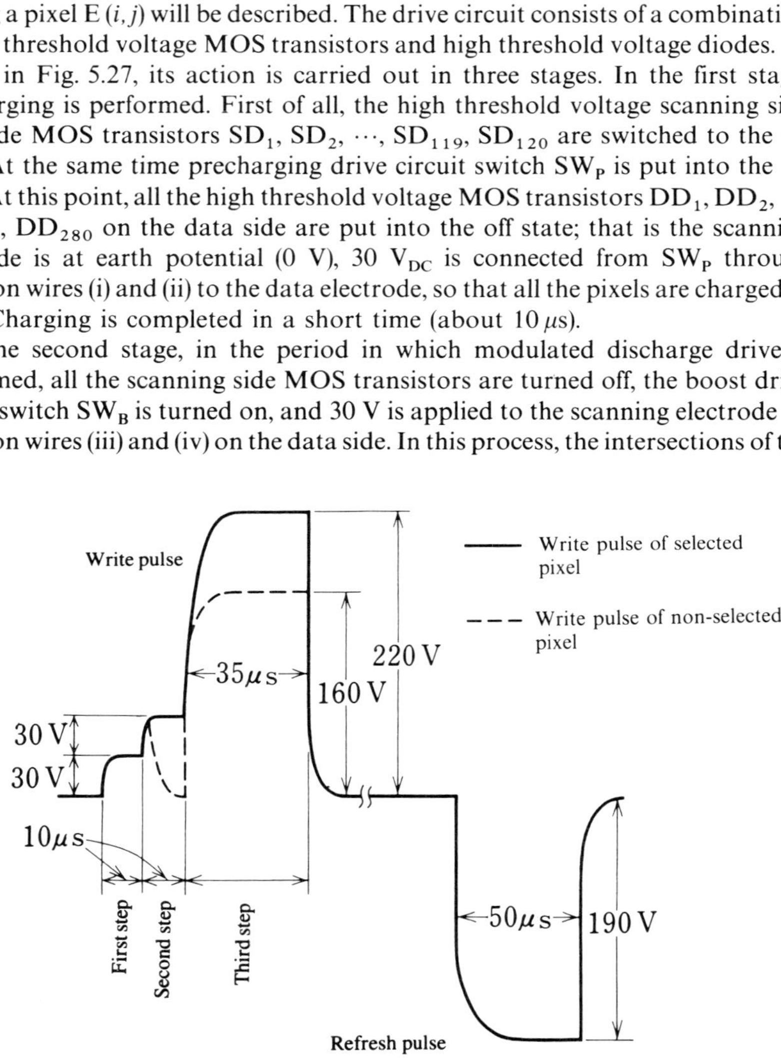

Figure 5.27 Drive panel waveform of thin-film AC-ELD produced by a three-step pulse

electrodes, i.e. the pixels, are capacitatively coupled, so all the data side electrodes are at 60 V. Thereupon, in order to write pixel E to make it emit light, only the MOS transistor DD_i of the ith electrode on the data side is turned off, all other data side electrode MOS transistors $DD_{k \neq i}$ being turned on, and therefore discharging their charge. The voltage waveform shown by the continuous line in Fig. 5.27 appears only at this electrode i, all the other electrodes following the voltage waveform shown by the dotted lines. This process is performed in a very short time (10 μs). None of the pixels yet appears to show any change, since they have not yet reached light-emitting voltage.

Next, writing is performed as the third stage; that is writing drive circuit switch SW_0 on the odd numbered side is turned on, and a voltage of 190 V is applied to common wire (iii). At this point, since the data side electrodes are capacitatively coupled, only electrode i becomes 220 V, the other electrodes becoming 160 V. Thereupon, the MOS transistor that is connected to scanning electrode j is turned on, and the other transistors are turned off. When this happens, a pulse voltage of 220 V is applied to pixel E(i,j), so that only pixel E emits light. The drive period for this operation is 35 μs. The firing voltage of this electroluminescent panel is about 190 V. Luminance of about 50 fL is obtained by repeated operation at 220 V, 100 Hz.

A drive operation as described above is performed for all the 120 scanning electrodes. On completion of one screen, all the data side MOS transistors are turned on, and at the same time the refresh drive circuit switch SW_R is turned on. This causes the diode that is connected to the scanning side to conduct, with the result that a refresh drive pulse of about 195 V of inverse polarity to the write voltage is supplied, causing only light-emitting pixel E(i,j) that is being written to emit light once more. This requires about 50 μs [7].

5.5.2 Active matrix drive

In an active matrix type electroluminescent panel, as shown in Fig. 5.28, an aluminium metal film electroluminescent display electrode and a thin-film transistor (TFT) are produced by the evaporation technique on a substrate. A common drain electrode and common gate electrode are arranged in positions intersecting on the same plane. Thus, to the source electrode of a first TFT (T_1), there are connected a capacitor and the gate of a second TFT (T_2). The source electrode is connected, together with the capacitor, to the earth line. The drain electrode of T_2 is connected to an ELD electrode. Such a combined unit is provided for each pixel, all these units being integrated on a single IC. In an electroluminescent panel, an electroluminescent light-emitting layer of, for example, the dispersed type is applied onto this active matrix circuit board, and a transparent electrode, consisting of PbO or the like, is evaporated onto it.

The method of driving the active matrix will now be described with reference to Fig. 5.29.

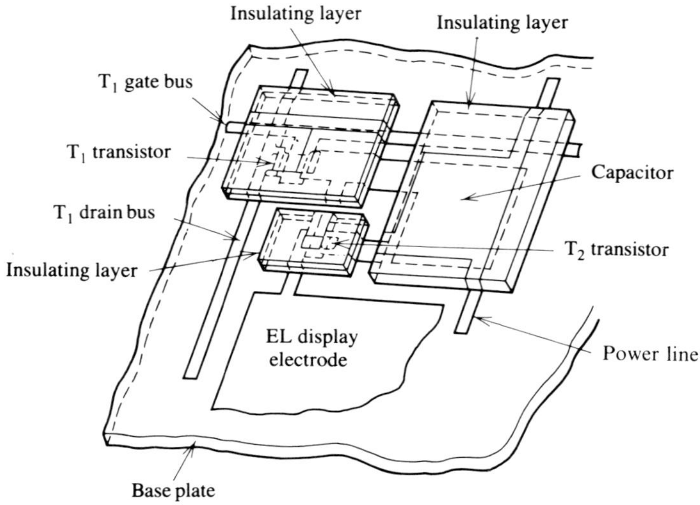

Figure 5.28 *Drive circuit of active matrix type EL panel (TFT layout diagram)*

When pulse voltage is applied to the selection intersection points of drain electrodes X_i, X_{i+1}, X_{i+2}, ... and gate electrodes Y_j, Y_{j+1}, Y_{j+2}, ..., T_1 is turned on and the capacitor is charged. The voltage to which it is charged is determined by the pulse width that is applied to the gate. If this charging voltage is sufficiently large, the gate voltage of T_2 becomes large, with the result that it is turned on, so the voltage that is applied to the upper transparent electrode is applied to the electroluminescent layer through the electroluminescent layer. As discharging of the capacitor proceeds, the gate voltage gradually decreases. The conduction

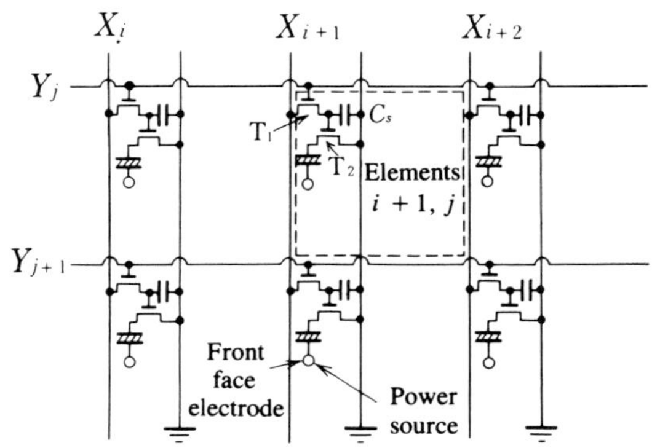

Figure 5.29 *Active matrix drive circuit*

time of T_2 is therefore determined by the charging voltage of the capacitor. The luminance of the ELD is also related to this conduction time. In this way, by successive scanning of the *X* electrodes and *Y* electrodes, the pixels at the points of intersection are selected and caused to emit light. Even in active matrix drive, usually the line sequential scanning method is adopted.

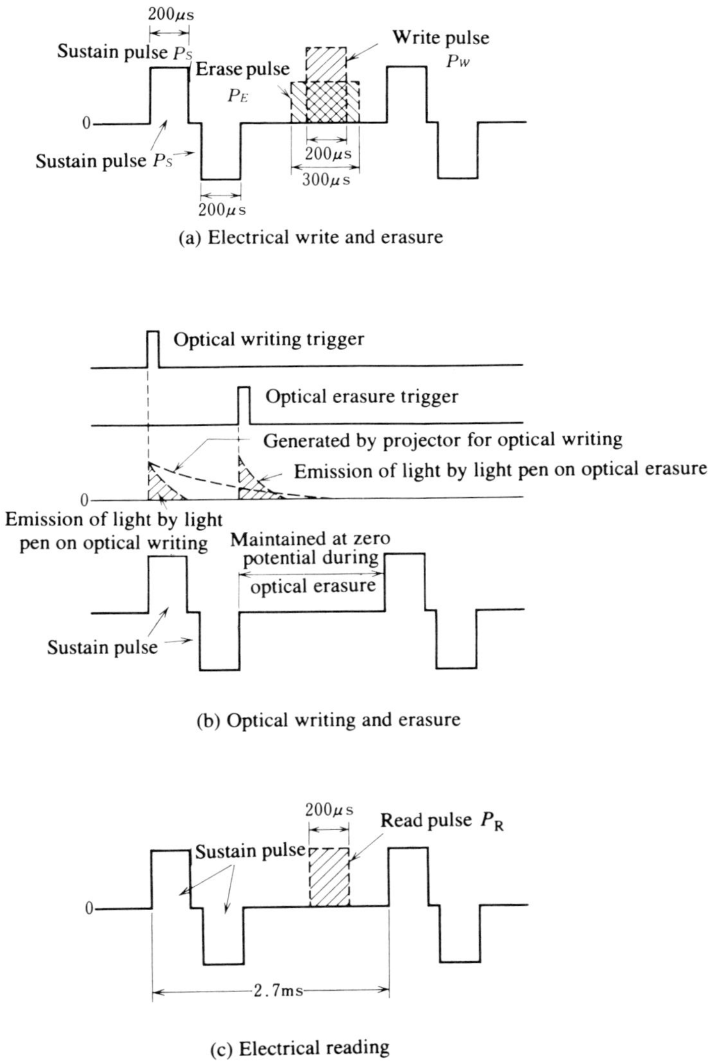

Figure 5.30 *Drive waveforms of memory type ELD*

5.5.3 Drive for optical writing, optical erasure, and electrical reading

A memory type electroluminescent matrix panel has the same construction as a non-memory ELD, but, as shown in Fig. 5.19, the relationship between its applied voltage and luminance follows a hysteresis curve. The method of electrical writing and erasure will therefore be described using the pulse waveform shown in Fig. 5.30(a).

If a sustain pulse P_S of voltage equal to the threshold voltage is simultaneously applied to all the pixels, and an electrical write pulse P_W is applied by point sequential scanning or line sequential scanning in the space period between one cycle and the following cycle, display with writing to an arbitrary pixel as shown in Fig. 5.31 can be achieved. Further, if an erasure pulse P_E of voltage equal to the V_E voltage shown in Fig. 5.19 is applied, the pixel that was up to then written and displayed can be erased.

Alternatively, as shown in Fig. 5.30(b), writing can be carried out optically using an active light pen to illuminate the panel with an optical write trigger pulse whilst the sustain pulse is being applied. An example of a display optically written using a light pen is shown in Fig. 5.32. If, on the other hand, an erasure trigger pulse is used to drive the active light pen light during the sustain pulse space period, optical relaxation takes place in the light-emitting layer of the ELD, causing the polarization charge to disappear and stopping light emission from

Figure 5.31 *Example display produced by electrical writing of memory type ELD*

Figure 5.32 Example display produced by optical writing of memory type ELD

Figure 5.33 Example of CRT display obtained by electrical reading memory type ELDs

the pixels that were emitting light hitherto. A negative display can therefore be achieved by driving the light pen in erasure mode to produce optical erasure.

In the case of direct electrical read-out, a read pulse P_R is applied by point sequential scanning in the space period P_S of the sustain pulse. This causes current to be generated in the region of the light-emitting pixels when the read pulse is applied; that is the light emission gives a light-emission current, which flows on the rising edge of the applied pulse. Those pixels that are not written to, however, do not provide any light-emission current, even though a read pulse is applied to them. The light-emission display data can be electrically read directly by distinguishing the difference between these two cases. Figure 5.33 shows an example of redisplay onto a CRT by transmitting a read electrical signal[19].

5.6 *APPLICATIONS OF ELECTROLUMINESCENT DISPLAYS*

5.6.1 *Digital displays*

Numerical and symbol displays such as measuring instruments etc. may be given as examples of applications of digital displays. The only way in which they are superior to phosphor display tubes is that it is possible to make them of smaller thickness and larger size. The uses of this are few.

As an example of a digital display ELD, the Rohiya Company have manufactured a prototype numerical display unit of the transmission segment type using only transparent electrodes. An example of its external appearance is shown in Fig. 5.34. The size of the characters is: height 2.5 cm, width 1.5 cm. The segment display has the form of a rectilinear figure-eight[20]. Apart from this NEC has commercialized a large-segment character display device using organic film.

5.6.2 *Analogue displays*

Dispersed type ELDs are often used to provide an analogue display in meters. A display of this type is currently under test for a car dashboard instrument panel. The ELD meter shown in Fig. 5.35 employs a thin-film green light-emitting electroluminescent panel of double insulating film construction. A transparent electrode is evaporated onto a glass substrate, and the high-permittivity material $BaTiO_3$ is evaporated onto this as an insulating film. $ZnS:TbF_3$ is used as the phosphor material of the light-emitting layer. The display is driven by applying an alternating voltage of 80 V, 1 kHz, to obtain a luminance of 140 to 150 fL. A display with a life of more then 5000 hours has been practically implemented[20].

Yellow–orange analogue display thin-film electroluminescent panels using $ZnS:Mn$ as the light-emitting layer are being produced on a trial basis for a meter as described above.

Figure 5.34 *Transparent type segment thin-film AC-ELD (Rohiya Company)*

5.6.3 Character displays

The Rohiya Company uses a thin-film a.c. electroluminescent panel in airport flight guide display boards. Figure 5.36 shows an example of such an electroluminescent guide display board. Each character is constructed of 8 × 11 dots, and has dimensions of 40 × 35 mm. The luminance is 115 cd/m^2. This gives a very good contrast ratio of 10:1, the ambient light being 5000 lx. An external view of a character ELD module is shown in Fig. 5.37.

The dimensions of the display board as a whole are 3 × 2.2 m, with a depth of 20 cm. It is composed of 16 rows with 45 characters in each row. The power required by the entire display is 1.2 kW. The colour of the letters is yellow–orange, giving a good contrast with a blackened layer used for the backing face.

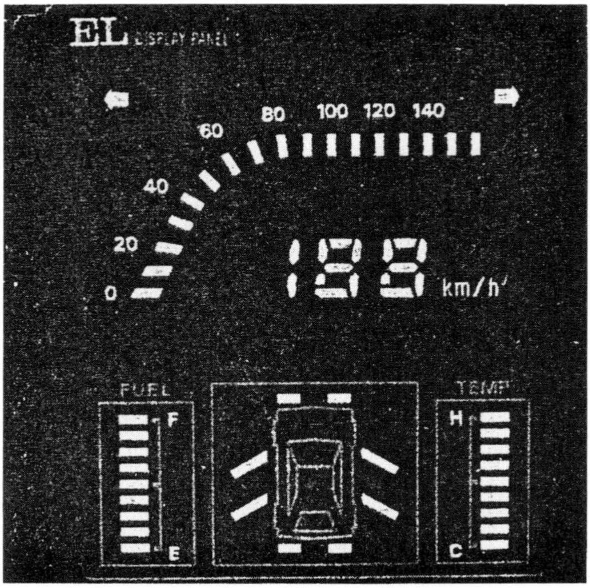

Figure 5.35 Double-insulated thin-film green EL unit (Nippon Seiki)

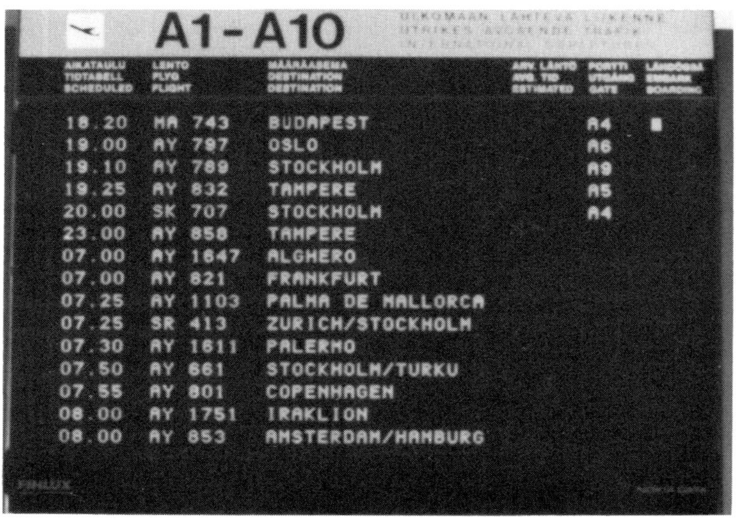

Figure 5.36 Airport guidance display board using thin-film AC-ELD (Rohiya Company)

Figure 5.37 *8 × 11 dot character ELD unit (Rohiya Company)*

5.6.4 Graphic displays

An example of graphic display using a thin-film type alternating current electroluminescent panel is shown in Fig. 5.38(a). The area of the display is 120 × 90 mm, with 512 × 256 dots, giving a resolution of 2.7 dots/mm. The colour of the display is yellow–orange.

The line sequential method is used with 60 frames displayed each second. The luminance is 25 fL. The circuit layout of this display unit is shown in Fig. 5.39.

A drive circuit embodied as an IC is connected to the scanning electrodes and data electrodes, where 32 lines can be driven per chip. A drive voltage of 215 V is used, for which a 30 V power source and a 185 V power source are provided. In addition, there are provided a modulating drive source of + 35 V and a power source for the logic circuitry, supplying + 7 V. The temperature range of operation is 0 to + 50°C and the range of storage temperature is − 20 to + 70°C. Apart from this, the 640 × 200 thin-film a.c. electroluminescent panel for message display shown in Fig. 5.38(b) has also been developed as a commercial product.

5.6.5 Picture displays

Attempts have been made to provide television picture displays using three-layer thin-film a.c. electroluminescent matrix panels. The electroluminescent panel in

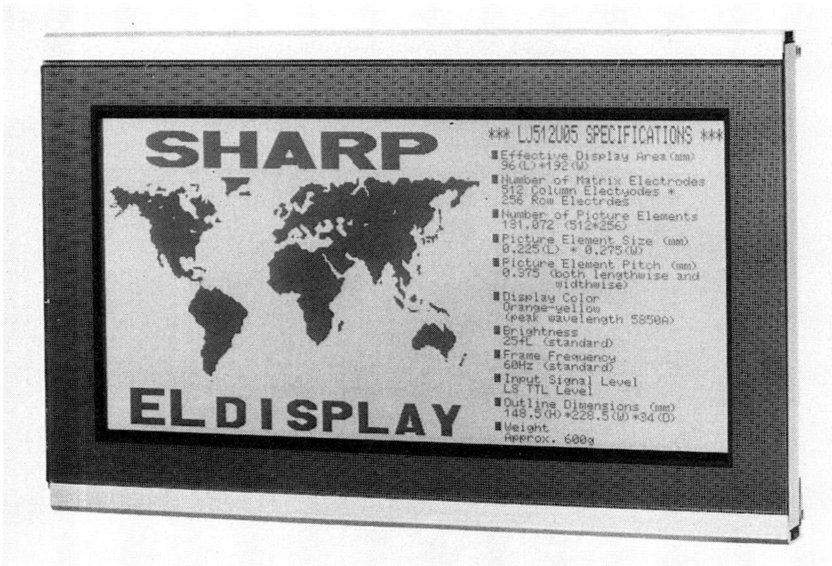

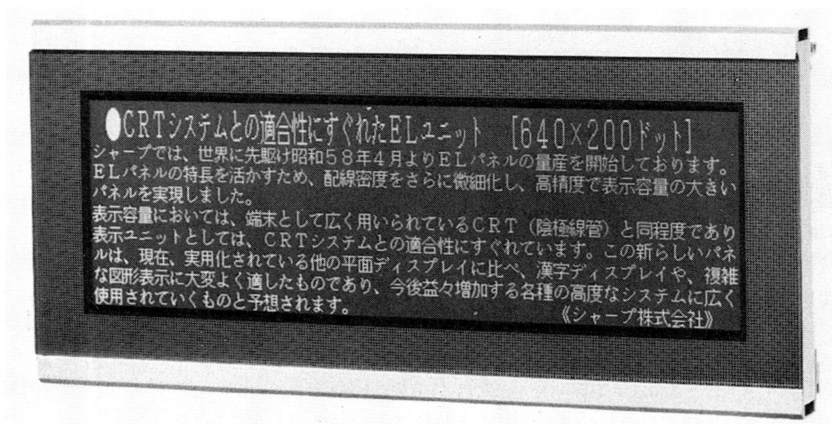

Figure 5.38 *Three-layer thin-film AC-ELD unit (Sharp Corporation)*
(a) 512 × 256 dots
(b) 640 × 200 dots

this case is the same as that used in the graphic display of Fig. 5.38(a). The amplitude of the modulating drive pulse is varied by the image signal, and the display is scanned by the line sequential method. The layout of this TV display device is shown in Fig. 5.40, and its external appearance is shown in Fig. 5.41. An advantage of this device is that the amplitude of the modulating drive pulses is

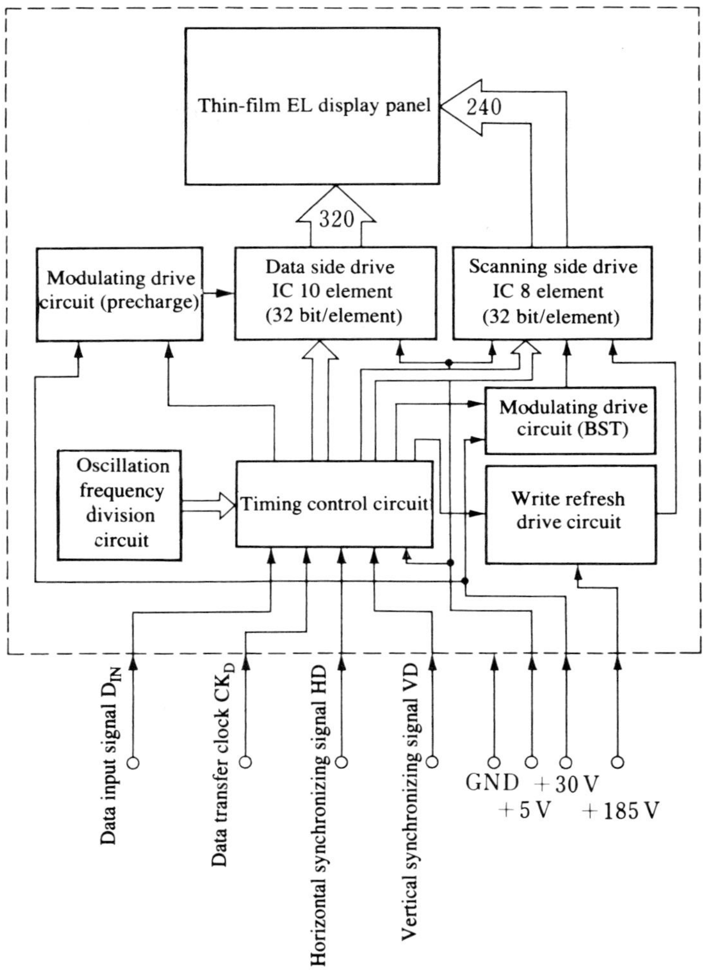

Figure 5.39 *Circuit layout of graphic display unit*

analogue modulated, so a gradation of 50 or more can be obtained. Analogue memories are provided for each line, utilizing the gate electrode capacitance of the drive MOS transistor. It can be used in two modes—hanging on a wall or as a desk top unit—thanks to a construction in which the display unit and tuner unit can be folded at an angle of 90°.

5.6.6 Colour displays

Trial manufacture is currently being carried out on yellow–orange (Mn doped) and green (TbF_3 doped) displays. Multicolour electroluminescent displays are

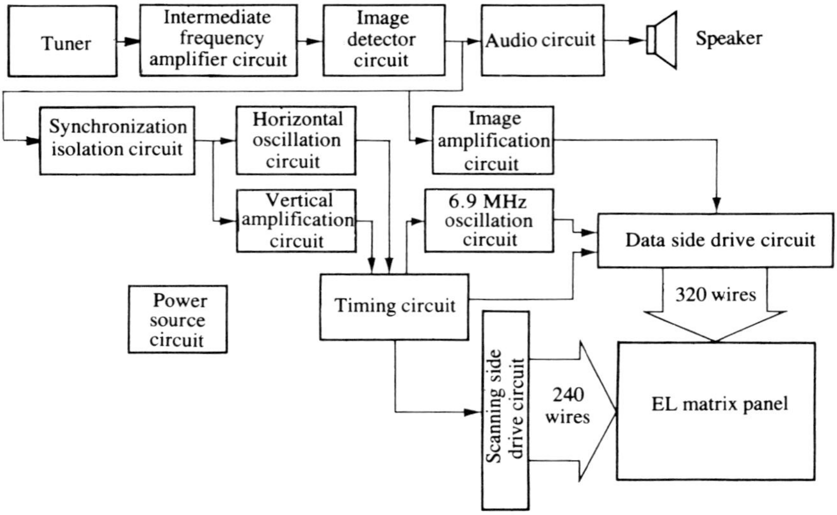

Figure 5.40 *Circuit layout of television receiver using thin-film AC-EL panel*

also being developed using yellow (DyF_3), red–orange (SmF_3), white (PrF_3), and Blue (TmF_3), etc., activators. In addition to the above, there are examples of three-colour displays (red, yellow–orange, green), obtained by varying the magnitude of the applied voltage, and examples of mixed colours, obtained by superimposing two electroluminescent layers, each provided with transparent electrodes[21, 22].

5.6.7 Other types of display

Since liquid crystal display elements are of the passive type, in order to make them easy to see at night, background illumination is required. For such applications, a dispersed type green light-emitting electroluminescent sheet has been commercialized, which employs a flexible organic film as the substrate. Its thickness is only about 1 mm. The drive voltage used is 30 to 60 V, 100 Hz to 2 Hz a.c. The luminance is 3 fL when driven at 100 V, 60 Hz. The life until the luminance is reduced by one half is about 10 000 hours, and the operating temperature is 10 to 50°C. Figure 5.43 shows an example in which this is used as the background of a liquid crystal clock.

5.7 OUTLOOK

Development of ELDs began earlier than development of other displays, but only some aspects of this development have been utilized in commercial products. The

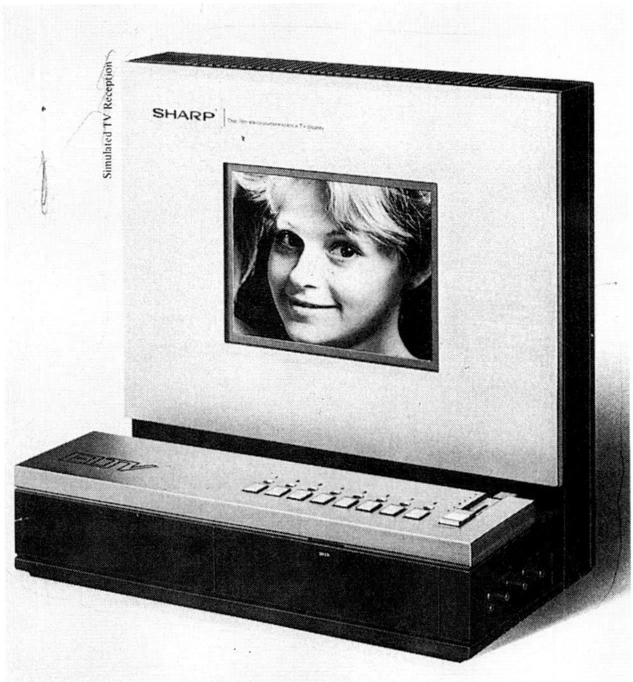

Figure 5.41 External appearance of prototype EL television receiver

reason for this is that there are still problems, mainly regarding luminance, life, obtaining different colours, cost, etc. Regarding luminance and life, it appears that these problems have now been solved, but the stage of cheap mass production has not yet been reached. Also the cost is two to three times that of a CRT. However, it is expected that the mass production effect will apply to the cost problem. Regarding the problem of obtaining different colours, research is at present proceeding into green and red colours.

When considering electroluminescent displays, it is often necessary to discuss the peripheral drive circuitry rather than the electroluminescent panel itself. At present, the drive circuitry is constructed of high threshold voltage MOS ICs. Efforts are required to make panels that can be driven by lower voltages. At the present stage of investigation, there are reports of panels that can be driven at less than 10 V, but, unless the electrooptical conversion efficiency is raised, even though the voltage is low, the current becomes large, making the drive IC difficult

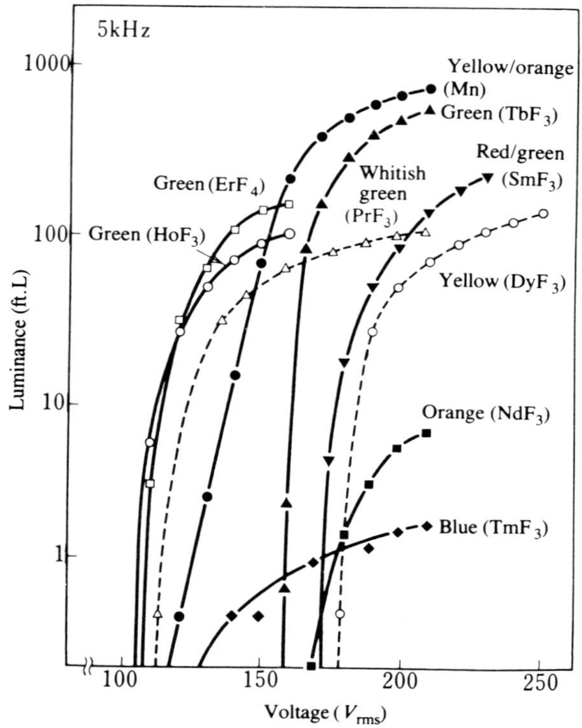

Figure 5.42 *Luminance versus voltage of ELs of various colours*

to manufacture. From the point of view of the drive circuitry, a DC-ELD, which can be made to emit light under d.c. drive, is desirable.

Regarding the problems of obtaining high resolution, increasing the number of pixels and increasing the size, at present, of displays with 5 pixels/mm and sizes 1000 × 1000 and 300 × 300 mm^2 have been practically implemented. As a target, by adding a memory function, it is hoped to double these numbers.

Regarding future prospects, reduction in the size and thickness of computer terminals is continuing, so there is an increasing requirement for making electroluminescent display devices in portable form. Also, their application to transportable display devices, using the fact that the mode of light emission of ELDs is completely solid state, is eagerly anticipated.

ELDs face their chief competition from PDPs and VFDs and display quality and cost will be the deciding factors in this comparison. Regarding cost, this should be comparable with LCDs provided with backlight lamp.

As a general trend, it may be noted that flat screen type displays such as ELDs are gradually invading the territory of CRTs.

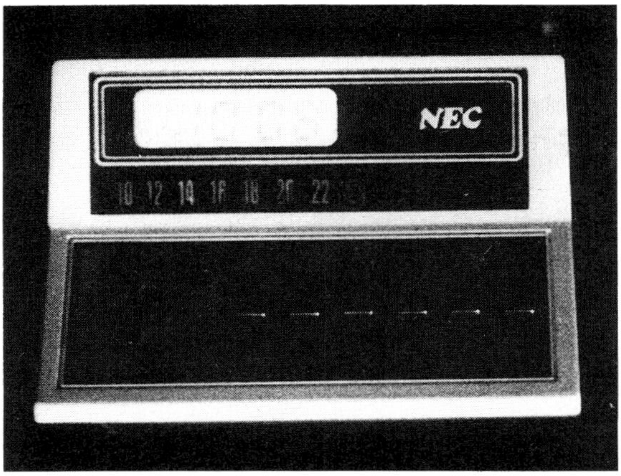

Figure 5.43 Use of EL back lighting in liquid crystal clock

BIBLIOGRAPHY

Inoguchi and C. Suzuki: Ko kido cho jumyo o tassei shita hakumaku EL paneru no kozo to tokusei (Structure and properties of thin-film EL panels displaying high luminance and long life), Nikkei Electronics, pp. 84–104 (1974-11-18).

Suzuki, C.: EL displays, *Shingaku Shi (Journal of Communications Association)*, **65** (7), pp. 741–7 (1982).

Suzuki, C., *et al.* Thin film EL displays, *Information Display*, **13**, Spring, 14–19 (1977).

Suzuki, C., *et al.*: Memori-gata san-so hakumaku EL paneru no denkiteki kakikomi shokyo (Electrical writing and erasure of memory type thin film EL panels), *Terebi Shi (Television Journal)*, **32** (11), 969 (1978).

REFERENCES

[1] S. Kawashima: *Electroluminescence*, Nikkan Kogyo Shinbun Sha (first publication 1965).

[2] J. Ibuki: *Kotai Hyoji Sochi (Solid-state Display Devices)*, Asakura Shoten (first publication 1972).

[3] M. Wada: *Insha Kogaku v Denshi Display (Printing and Projection Engineering V Electronic Displays)*, Sankyo Shuppan (1975).

[4] M. Yoshiyama *et al.*: DC-EL paneru niyoru terebi hyoji sochi (Television display devices using DC-EL panels), National Technical Report (1971-12).

[5] D. Kahng: Electroluminescence of rare-earth and transition metal molecules in II–VI compounds via impact excitation, *Appl. Phys. Lett.*, **13**, 210–12.

[6] T. Inoguchi *et al.*: Stable high-brightness thin film electroluminescent panels, *Digest of Technical Papers*, 1974 SID Int. Symp., pp. 84–5.

[7] Y. Kaneya *et al.*: Hakumaku electroluminescent moji display no jitsuyo-ka gijutsu (Thin film electroluminescent character displays), *Nikkei Electronics*, **209**, 118–42 (1979-04).

[8] M. Takeda *et al.*: Inherent memory effects in ZnS:Mn thin film EL devices, Proc. 6th Conference on *Solid State Devices*, Tokyo (1974).
[9] Y. Kaneya *et al.*: ZnS kei sanso hakumaku EL soshi no kobunkyoku koka, 1973 TV Zendai (Photopolarization of ZnS triple-layer thin film EL elements 1973 TV National Conference).
[10] T. Santola *et al.*: Atomic layer epitaxy for producing EL thin film, *SID 80 Digest*, **11**, 108–9 (1980-04).
[11] M. Abdalla *et al.*: Low voltage DC electroluminescence in ZnS (Mn, Cu) thin film phosphrs, *SID 78 Digest*, **10**, 130–1 (1978).
[12] T. Mishima *et al.*: Molecular beam epitaxial ZnSe:Mn DC electroluminescent cell with low threshold voltage, *J. Appl. Phys.*, **52** (9), 5797–9 (1981).
[13] T. Suzuki: Electroluminescence Zairyo (Electroluminescent Materials), *Kogyo Zairyo (Industrial Materials)*, **29** (2), 42–7 (1981).
[14] M. Yoshida *et al.*: AC thin film EL device that emits white light, *Digest of SID*, pp.106–7 (1980).
[15] Vecht: DCEL dot matrix displays in a range of color, *Digest of SID*, pp.110–11 (1980).
[16] A. Murase: EL display no saikin no shinpo (Recent progress in EL displays), *TV Kai Shi (Journal of the TV Association)*, **36** (11), 970–976 (1982).
[17] C. Suzuki *et al.*: Optical writing and erasing on a thin film EL panel with inherent memory, *Digest 1976 SID*, pp.52–3 (1976).
[18] M. Abdalla et al.: Performance of DCEL coevaporated ZnS:Mn, Cu low voltage devices, *IEEE Trans*, **ED-28** (6), 694–7 (1981).
[19] C. Suzuki *et al.*: Direct electrical read-out from thin film EL panels, *Digest of Technical papers*, 1978 SID Int. Symp., pp.134–5.
[20] H. Hayashi: Honkakuteki na jitsuyoki ni hairu moji/graphic yo EL display (Character/graphics EL displays approaching genuine practicality), *Nikkei Electronics*, **297**, 89–104.
[21] T. Suyama *et al.*: Multi-coloring of thin film electroluminescent device, *Jpn. J. Appl. Phys. Suppl.*, **21** (1), 383 (1981).
[22] Y. Oishi *et al.*: Tunable color EL device, *Proc. 3rd Int. Display Res. Conf.*, pp. 570–3 (1983) (Japan Display '83).

Chapter 6

Vacuum fluorescent displays (VFDs)

6.1 INTRODUCTION

6.1.1 Development and progress of vacuum fluorescent displays

Vacuum fluorescent displays (VFDs) are a type of display device that was developed in Japan and is even now produced only in Japan. They are supplied world-wide, new types are continually being developed, and their range of applications continues to expand. There have been three great changes in VFDs since their development in 1967. First of all, a single-digit tube was developed that displayed only one digit, using a round bulb. This was used for the display of numerals and some symbols in desk-top calculators. This suffered from many drawbacks both from the customers' and the manufacturers' point of view. As shown in Fig. 6.1, this display was fixed in position on a PC board, and a large number of leads had to be soldered into place. It was also necessary to use displays of different luminance separately.

In 1972, a round multidigit tube was developed, capable of displaying six to thirteen digits using round vacuum enclosures. This had many advantages from the point of view of both the customer and the manufacturer. Thanks to this device, it was possible to cope with the mushrooming demand for VFDs resulting from the boom in desk-top calculators. Figure 6.2 shows a round multidigit tube mounted on a PC board. Separate alignment of each digit is unnecessary. There are also very few leads that need soldering. Separation of elements of different luminance is also unnecessary. However, since this device, like the single-digit tube, employed an assembly of electrodes on a ceramic plate, material costs were high, and it was not possible to cope with the strong demand for cost reduction.

In response to this, the flat type of display device that was required by the market was developed in 1974. This is the flat multidigit tube. The main electrodes are manufactured by thick-film printing onto a glass plate, or of a mixture of thin-film technology and thick-film printing. The various electrodes are assembled onto this, and the assembly is enclosed in a vacuum enclosure,

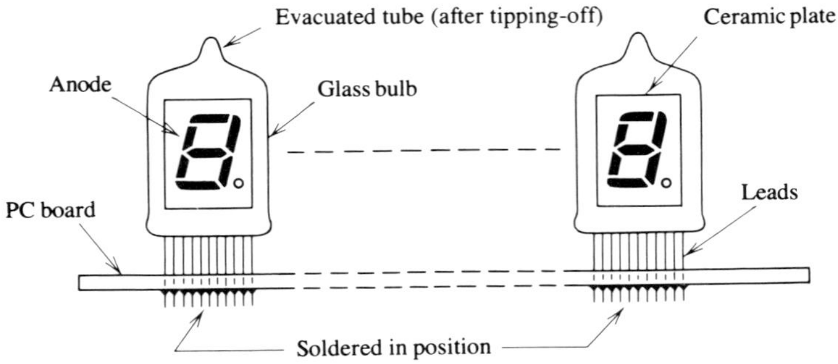

Figure 6.1 Single-digit tube

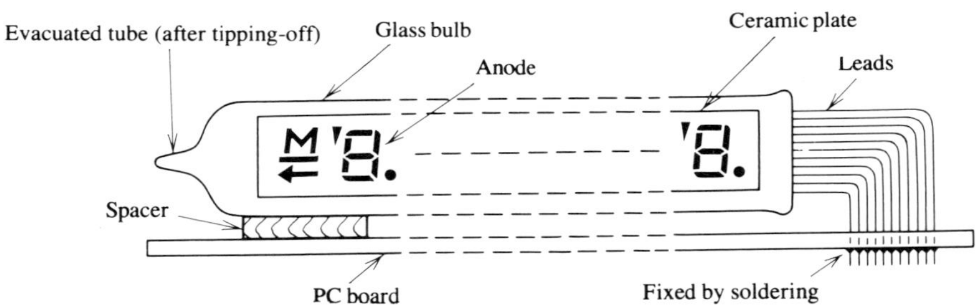

Figure 6.2 Round multidigit tube

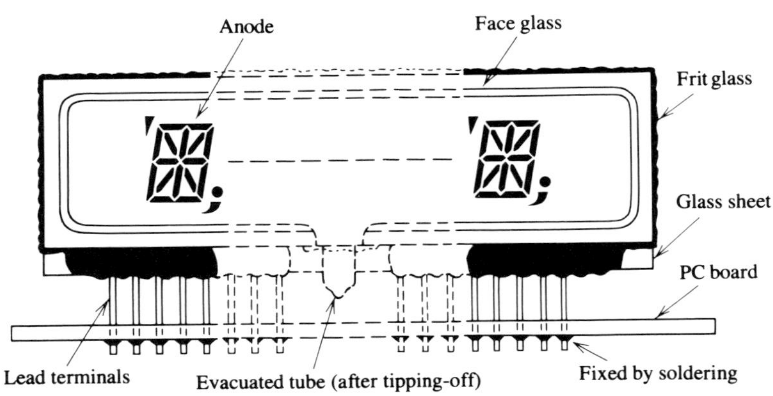

Figure 6.3 Flat multidigit tube

which is sealed using a face glass. This device has a degree of freedom regarding its shape, and it has been possible to produce a large number of display patterns. Figure 6.3 shows how it is mounted on a PC board.

6.1.2 *Characteristics of vacuum fluorescent displays*

When VFDs first appeared, they consisted of low-voltage drive, high luminance green display tubes. They were prized as monochrome display elements emitting bluish light. Once flat multidigit tubes were developed, the degrees of freedom of the shape and display pattern increased, and the types of light-emitting colours available also increased, making it possible to expand their field of application as multicolour display elements. As will be described, a VFD is a triode, wherein the display is effected by control of thermions produced by a narrow filament (cathode). Consequently, most of the power consumption is used in heating this cathode. For this reason, as an application to handy gadgets it is inferior to other display elements of low power consumption, and the space factor of the enclosure also imposes limitations on the fields of use of the display. Efforts are being made to overcome these problems by use of thinner filaments and improved face glasses.

6.2 *CONSTRUCTION OF VACUUM FLUORESCENT DISPLAYS AND THEIR MAIN COMPONENTS AND MATERIALS*

6.2.1 *Construction of VFDs*

The construction of a VFD will be explained with reference to a flat multidigit tube.

Figure 6.4 shows an example of the cross-section of a VFD. A VFD is a type of

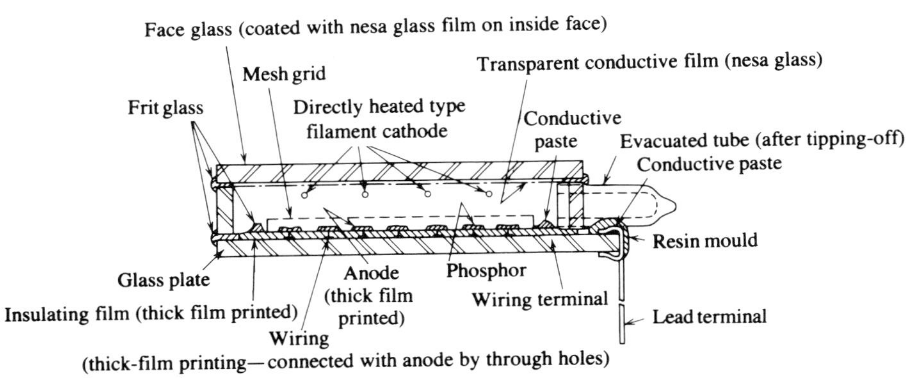

Figure 6.4 *Cross-sectional view of vacuum fluorescent display tube*

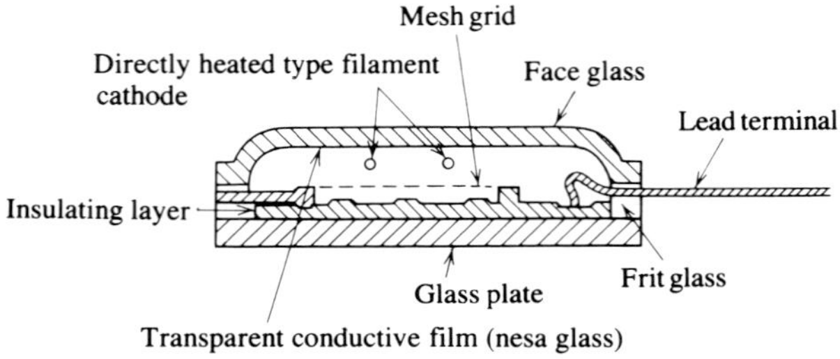

Figure 6.5 *Example of lead-out of vacuum fluorescent display tube*

triode[1] wherein an anode, to which phosphor is applied, and a cathode and grid are sealed into a vacuum enclosure that is transparent on at least one side. In this example, the lead terminals are connected outside the tube. However, as shown in Fig. 6.5, there are also types of display in which the lead terminals are directly led out from within the tube[2].

6.2.2 Main components and materials of VFDs

In flat multidigit tubes, the vacuum enclosure is formed by a glass substrate, a face glass, and an exhaust tube. The display is operated by applying drive voltage from the lead terminals to electrodes formed by thick-film printing onto the glass substrate, or by a combination of thin-film technology and thick-film printing. The main components and materials used in the construction will be described below.

(1) Glass substraate

The glass substrate is one of the most important components of a VFD, since the VFD electrodes and other main components are formed onto it. Figure 6.6 shows thick-film printing onto a glass substrate. In this case, since the display pattern is comparatively simple, the wiring is not complicated. The wiring, insulating layer, and anode are built up on the glass plate by repeated thick-film printing and baking. For these printing steps, a paste is used of which the principal

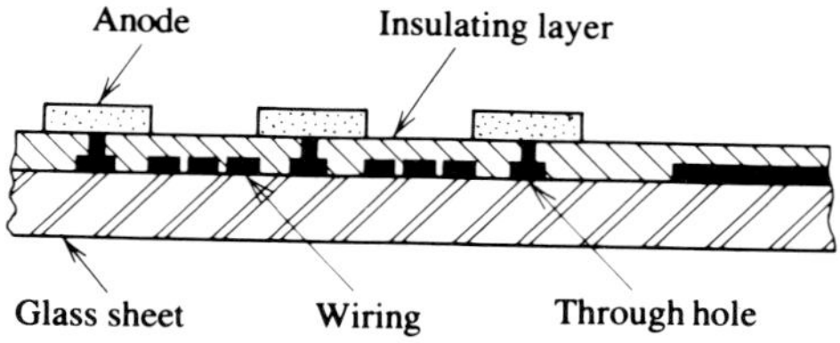

Figure 6.6 *Cross-sectional view of thick-film glass plate*

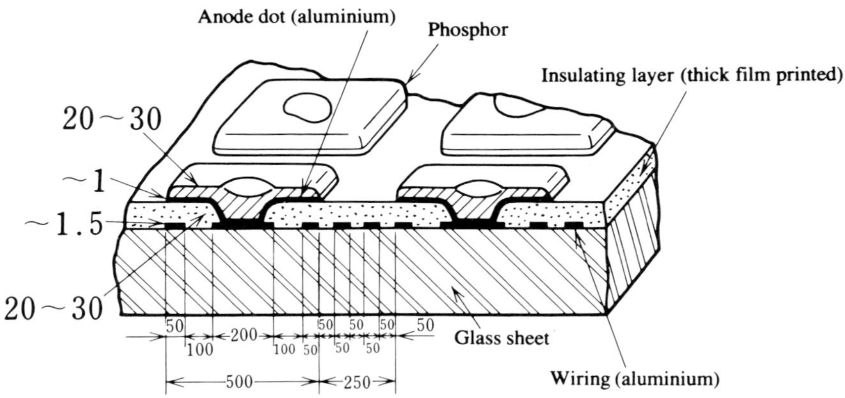

Figure 6.7 *Partial cross-sectional view of glass plate produced by combination of thin-film and thick-film printing techniques*

Table 6.1 *Comparison of thick-film wiring and thin-film wiring (units: μm)*

	THICK-FILM WIRING	THIN-FILM WIRING
Film thickness	10	1
Wire width	100–200	30–50
Wire separation	150–300	30–50
Wiring material	Silver	Aluminium

constituents are a silver-based material, a material consisting of a mixture of powdered glass with a suitable pigment, and a carbon-based material.

Complicated display patterns involve high wiring densities so multilayer wiring must be produced by thick-film printing on a glass plate of standard dimensions. This causes problems both from the point of view of manufacturing cost and drive conditions. This is combatted[3] by using a combination of thin-film techniques and thick-film printing on a glass substrate, as shown in Fig. 6.7. The wiring and anode are formed of aluminium thin film, while the insulating layer is formed by thick-film printing. A comparison of thin-film wiring and thick-film wiring is shown in Table 6.1.

(2) Grid

This consists of a metal mesh made of an ultrathin sheet of stainless steel formed by photoetching in a tortoise shell or lattice shape. Its transmissivity affects the

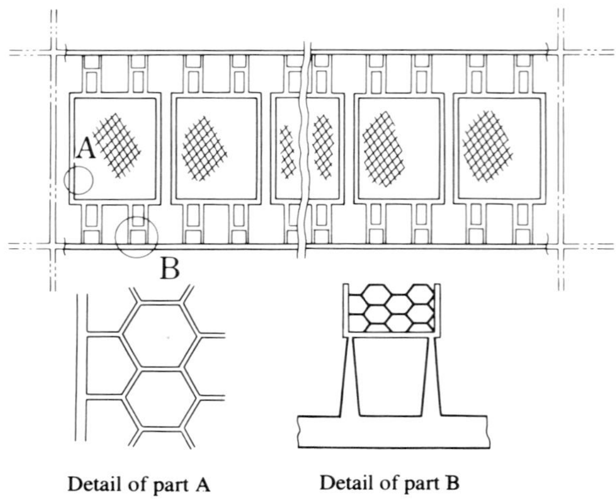

Figure 6.8 *Example of grid*

current distribution, luminance, and display quality, etc. Figure 6.8 shows an example of a tortoise-shell-shaped grid.

(3) Cathode

This is produced by applying carbonates of three alkaline earth metals onto a cathode made of ultrafine pure tungsten wire, heating this cathode by passing current through it immediately before completion of the evacuation process in the tube manufacturing step, and then activating it as follows:

$$(Ba + Sr + Ca)CO_3 \rightarrow (Ba + Sr + Ca)O + \uparrow 3CO_2$$

That is, by driving off and evacuating the carbon dioxide by heating, the oxides of the three alkaline earth metals are left behind on the pure tungsten wire, providing an electron source. The suitability or otherwise of the activation conditions has a very considerable effect on the performance of the VFD.

(4) Phosphor

It is no exaggeration to say that it is the fact that ZnO:Zn phosphor emits light of blue–green colour with slow electrons that has made possible the development of VFDs with a green display tube. As will be explained below, there has been an increasing demand in various fields for provision of VFDs of multiple colours. Phosphors providing a large number of light-emission colours are used in various applications, but they do not provide luminance of a practically useful level when used with slow electrons. In fact it might be more accurate simply to say that under these conditions they do not emit light. This is because the resistivity of

Table 6.2 *Data of various types of phosphor*

Phosphor composition	Emitting colour	Peak wavelength (nm)	Colour coordinates		Emitting efficiency (lm/w)	Luminance ratio (%)
			X	Y		
ZnO:Zn	Blue green	505	0.25	0.44	5–7	100
(ZnCd)S:Ag + In_2O_3	Red	665	0.67	0.33	0.5–1	10
ZnS:AuAl + $^{In}{}_2O_3$	Lemon	555	0.39	0.56	1–1.5	40
ZnS:Mn + In_2O_3	Yellow	585	0.52	0.47	0.3–0.6	20
ZnS:Ag + In_2O_3	Blue	450	0.16	0.12	0.1–0.3	13
ZnS:CuAl + In_2O_3	Green	520	0.25	0.59	1–1.5	—

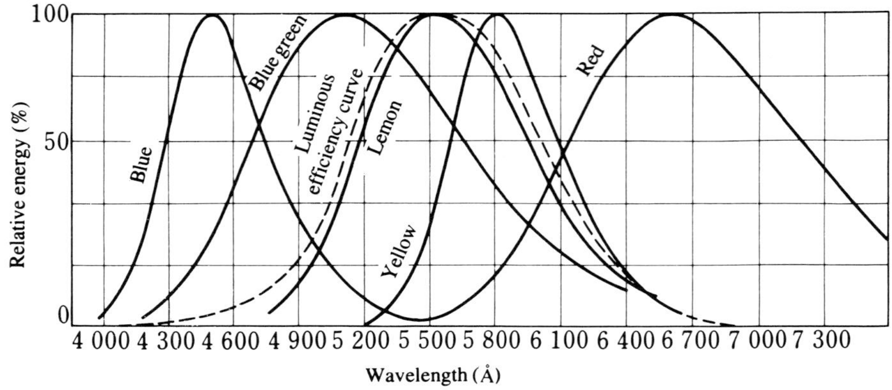

Figure 6.9 Spectral distribution of various types of phosphor

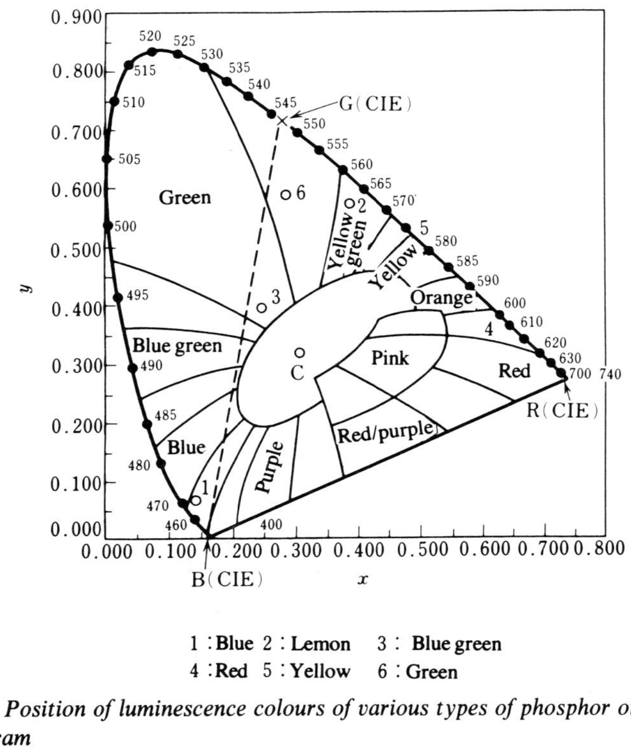

1 : Blue 2 : Lemon 3 : Blue green
4 : Red 5 : Yellow 6 : Green

Figure 6.10 *Position of luminescence colours of various types of phosphor on chromaticity diagram*

these phosphors is extremely high (10^2 to $10^{12}\,\Omega\cdot$cm). (For comparison, the resistivity of ZnO:Zn phosphor is 10^3 to $10^4\,\Omega\cdot$cm.)

Accordingly, various schemes for lowering their resistivity have been tried[4]. At present, the resistivity has been lowered to 10^4 to $10^7\,\Omega\cdot$cm by adding a specific weight percentage of indium oxide, followed by treatment by the passage of current. When the luminance of such phosphors is measured by driving them under the same conditions as ZnO:Zn phosphor, the luminance obtained is only 15 to 45 per cent that of the ZnO:Zn phosphor. However, due to the effect of visual sensitivity, this does not cause much of a practical problem.

Table 6.2 shows data for various types of phosphor. Figure 6.9 shows their spectral distributions. Figure 6.10 shows the positions of the colour of light emitted by various phosphors on a chromaticity diagram. ZnO:Zn phosphor has high luminance, but it can be seen that its light emission is of low colour purity.

(5) Face glass

This may be produced either by forming a glass plate by hot-working or by assembling a combination of short glass plates. These processes are shown in Figs. 6.11 and 6.12. In the types of product that are presently being developed, the latter method is most often used. This is because it contributes to reduction of the

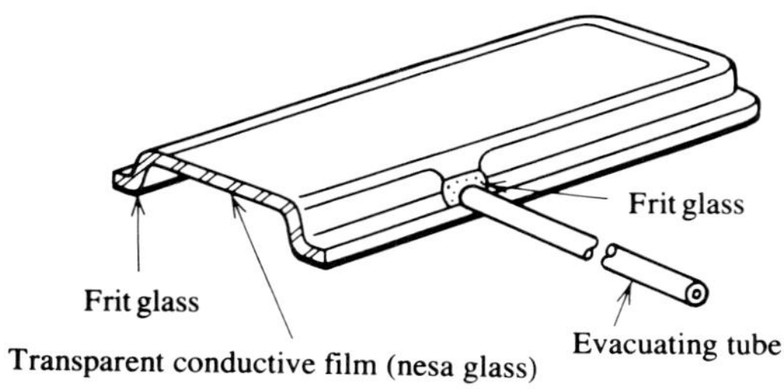

Figure 6.11 *Moulded face glass*

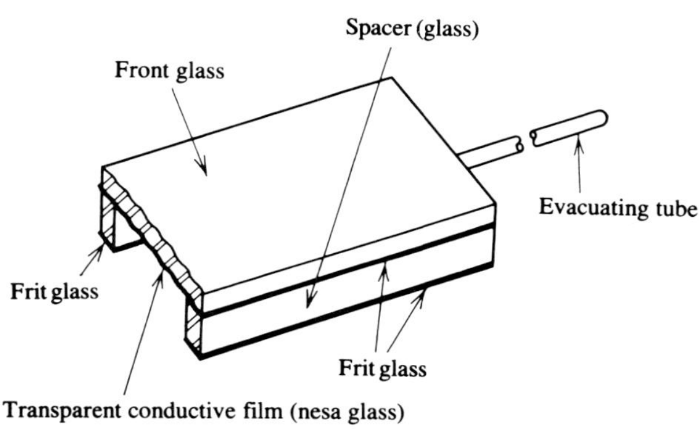

Figure 6.12 *Spacer face glass*

external dimensions relative to the display dimensions. In both cases, a vacuum enclosure is produced by sealing a glass substrate and exhaust tube by means of frit glass, a transparent conductive film being applied to the inside face of the front glass and maintained at cathode potential to act as a shield to external electrical fields.

(6) Other components

The getter is a component that is used in most electron tubes. It is used to stabilize operation by maintaining the degree of vacuum. There are various forms of getter, but in a VFD the getter shown in Fig. 6.13 is used. This consists of $BaAl_4$ and Ni powder introduced into a pure iron channel-shaped ring and fixed, being heated by radio frequency heating after exhaustion, so as to evaporate barium metal onto the inside surface of the face glass. In a broad sense of the term, it may be regarded as a type of vacuum pump.

Apart from the above, there are the anchor and two types of filament support. These are metal components for supporting the cathode in the prescribed position and for keeping it under a suitable tension. The former are connected to the filament wiring by conductive paste and fixed to the glass substrate by frit glass. The frit glass is used in the form of a paste made using a flux or the like, and consists of powdered glass of low melting point and having an expansion coefficient that is close to that of the face glass and glass substrate. Apart from being used in forming the vacuum enclosure, it is also used for fixing the grid and filament supports.

Apart from these, conductive paste is used to make connections to the glass substrate wirings of the grid and filament supports respectively, lead terminals are used for connection between the external power source and the various electrodes in the tube, solder or conductive paste is used to connect these to the glass

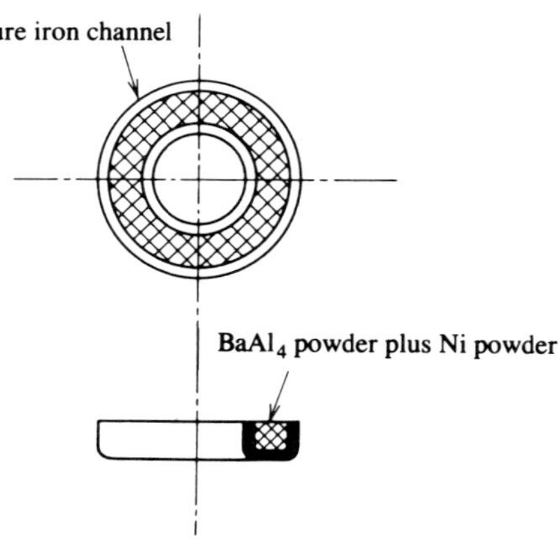

Figure 6.13 *Ring getter*

substrate wiring, and resin is used for fixing and to protect against the ingress of moisture between the wiring terminals.

6.3 OPERATING PRINCIPLES AND CHARACTERISTICS OF VACUUM FLUORESCENT DISPLAYS

A VFD is a light-emitting display including a grid, to which a positive voltage is applied, and an anode, to which a phosphor is applied. This phosphor is excited by thermoelectrons emitted from the cathode. The principle of operation and characteristic of a VFD, and also its drive circuitry, will be described in detail below.

6.3.1 Principle of operation of VFDs

The principle of operation of a VFD will now be described, with reference to Fig. 6.14.

When the prescribed voltages are applied to each electrode, the cathode is heated to about 650°C, causing thermoelectrons to be emitted. Grids provided between the cathode and anode have the effect of either (a) uniformly diffusing and accelerating thermoelectrons from the cathode towards the anode or (b) preventing thermoelectrons from reaching the anode. These actions are respectively obtained by holding the grid potential positive or negative with respect to the cathode. The grids consist of metal mesh, so some of the thermions flow into them, producing grid current, while the rest pass through the grids and

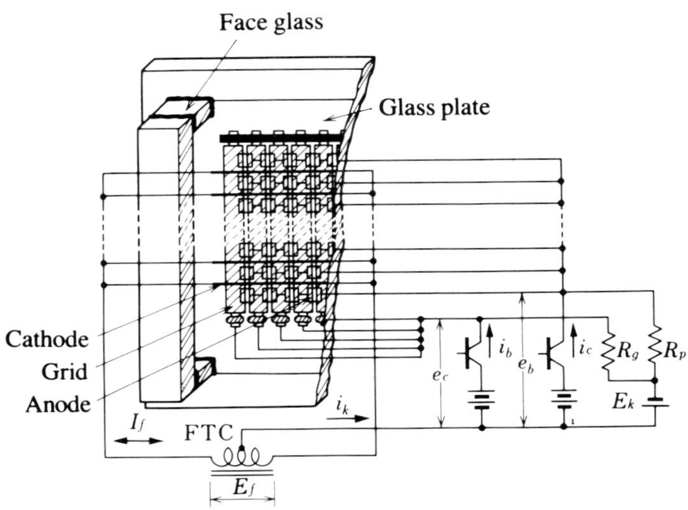

Figure 6.14 Operating principle of vacuum fluorescent display

arrive at the anode, exciting the phosphor and causing it to emit light, producing an anode current. In this case, a positive voltage must be applied to the anode too. Thus fluorescence only takes place when a positive voltage is applied to both grid and anode at the same time[5]. Negative voltage is applied to anode dots that are not to emit light or to the grids of digits wherein light emission is not to take place. To this end, bias voltage E_k is applied through grid resistance R_g and anode resistance R_p, as shown in Fig. 6.14.

Since a VFD is a triode, the relationship between applied voltage and current at each electrode can be expressed by the following equation in the practical voltage range[6]:

$$i_k = K\,(e_b, e_c)^n \tag{6.1}$$

where i_k is the cathode current, K is the perveance* (a constant determined by the construction of the electrodes), e_b and e_c are the anode voltage and grid voltage (usually the same), and n is a constant (3/2 in the case of an ideal parallel flat plate electrode and 1.5 to 2.0 in the case of a VFD).

The cathode current i_k is the sum of the anode current i_b and the grid current i_c. The following relationship exists:

$$i_k = i_b + i_c = i_b(1 + d) = i_c\left(\frac{1 + d}{d}\right) \tag{6.2}$$

where $d = i_c/i_b$ is called the distribution factor and is a value determined by the construction of the electrodes; $d = 0.8$ to 3.

The following relationship exists between e_b, e_c, and the luminance L:

$$L = \eta.Du.e_b{,}e_c \cdot i_b = \eta.Du\overleftarrow{\ }k'(e_b{,}e_c)^{n+1} \tag{6.3}$$

where η is the light-emitting efficiency of the phosphor, Du is the duty factor, $k' = K/(1 + d)$. Usually, L is proportional to the 2.5 to 3.0-th power of e_b, e_c.

$$(i_h = K(e_b, e_c)^n$$

$$i_k = i_b + i_c = i_b(1 + d) = i_c\left(\frac{1 + d}{d}\right), \qquad d = \frac{i_c}{i_b}$$

$$i_k = i_b + di_b = i_b(1 + d)$$

$$i_k = \frac{i_c}{d} + i_c = \frac{i_c + di_c}{d} = \frac{i_c(1 + d)}{d} = i_c\left(\frac{1 + d}{d}\right)$$

$$L = \eta \cdot Du \cdot e_b{,}e_c\,\dot{}\,i_b$$

$$= 8\eta \cdot Du \cdot k(e_b{,}e_c)^{n+1}, \quad K = \frac{K}{1 + d} \qquad n = 3/2$$

*In general the relationship between the anode voltage E_a and anode current I_a in a diode tube is expressed by: $I_a = KE^{3/2}$, where K is called the perveance and is a constant determined by the dimensions and structure of the electrodes. The relationship expressed by this equation is called the 3/2 power law of L. Langmuir.

$$L = \eta \cdot Du \cdot e_b, e_c \cdot i_b = \eta \cdot Du \cdot e_b, e_c \cdot \frac{i_k}{1+d}$$

$$= \eta \cdot Du \cdot e_b, e_c \cdot \frac{K(e_b, e_c)^n}{1+d} = \eta \cdot Du \frac{K}{1+d}(e_b, e_c)^{n+1}$$

$$= \eta \cdot Duk(e_b, e_c)^{n+1}$$

In the case of a vacuum fluorescent display, usually the grid and anode are used at the same voltage, so the above equation can be applied to the diode constituted by the cathode and grid.

6.3.2 Characteristics of VFDs

A VFD has various characteristics similar to a triode. The characteristics of typical types will be described.

(1) Cathode rise characteristic

As explained above, when it is heated to the prescribed temperature, the cathode emits electrons. There is a correct heating temperature for the emission of electrons at a necessary and sufficient rate.

As shown in Fig. 6.15, the filament current I_f, L, i_b, and i_c increase with an increase in the filament voltage E_f. L, i_b, and i_c show a saturation characteristic with respect to E_f. However, I_f continues to rise. Although not shown in the figure, the cathode temperature T_f also rises with E_f. If E_f is set to a value lower than the saturation region, the change of L in response to power source fluctuation becomes large, resulting in a lowering of display quality. On the other hand, if E_f is too high, there will be an unnecessary increase in power

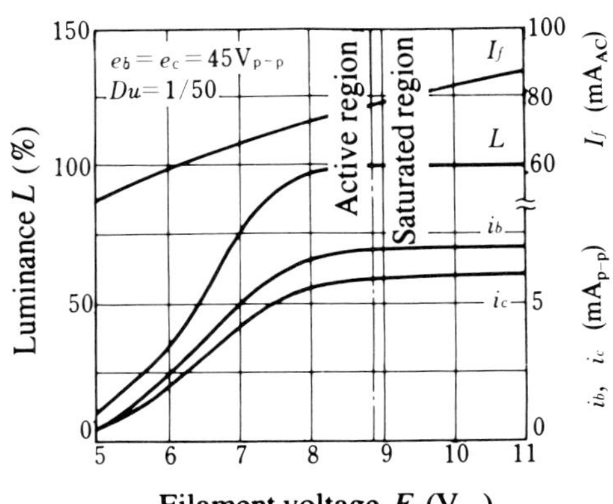

Figure 6.15 *Cathode rise characteristic*

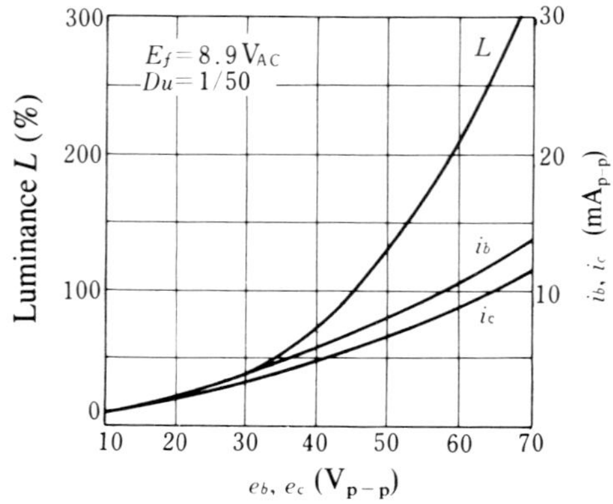

Figure 6.16 Grid and anode voltage characteristic

consumption without an increase in L. Also, T_f will become high, affecting the life. Consequently, the best plan is to set E_f to the lowest possible value in the saturation region.

(2) Voltage characteristic

The greatest effect on the luminance of a VFD is exerted by e_b and e_c. A VFD can be efficiently driven by optimum selection of these values.

Figure 6.16 shows their voltage characteristics. Apart from the special case, to be discussed later, of a dot matrix VFD, e_b and e_c are used with the same voltage. In the practically used voltage range, for a rise in e_b and e_c there is a continuous rise in L, i_b, and i_c, with no saturation characteristic. Consequently, the cathode current I_k ($= i_b + i_c$) also increases. So if e_b and e_c become too high, the limit of electrode emission of the cathode is exceeded, causing damage to the cathode and shortening the life of the device. Therefore e_b and e_c are set at a value such that sufficient luminance is obtained, but not so high as to cause cathode damage.

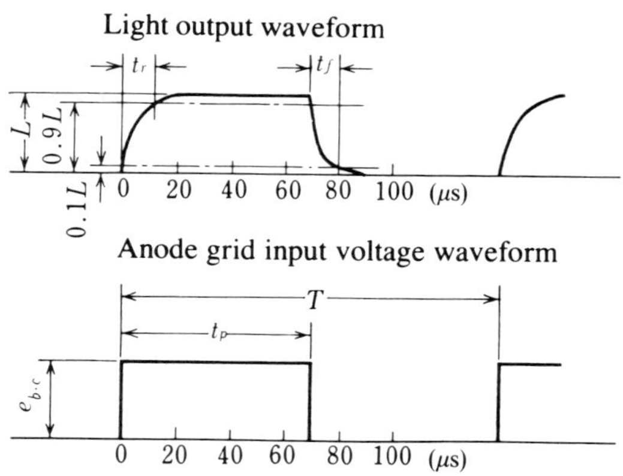

Figure 6.17 Pulse response characteristic of ZnO:Zn phosphor

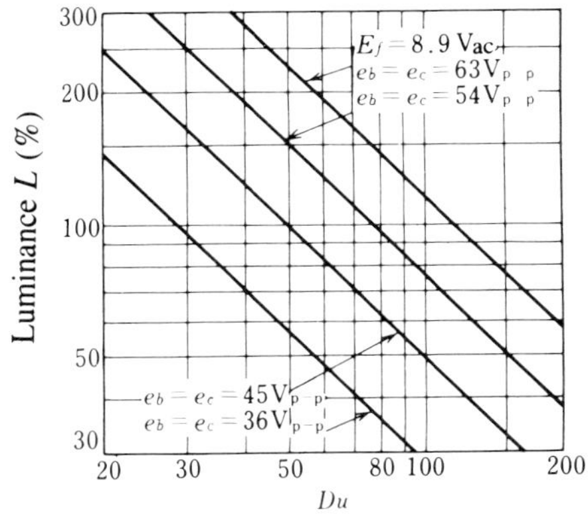

Figure 6.18 *Du–luminance characteristic*

(3) Duty factor and luminance

In the dynamic drive VFD, to be described later, the duty factor *Du* is a further important factor determining luminance and display quality. The duty factor is determined by the blanking time and excludes the effects of the number of digits of the VFD and the rise and fall characteristics of the phosphor.

The pulse response characteristic of a ZnO:Zn phosphor is shown in Fig. 6.17; t_r and t_f are the rise and fall characteristics of the phosphor, and are both about 10 μs. Consequently, a pulse width t_p of at least 20 μs must be used. The luminance of the VFD is inversely proportional to *Du* (Fig. 6.18), irrespective of the pulse width, provided that the pulse width is at least 20 μs. *Du* is given by the following equation:

$$Du = \frac{t_p}{T} \tag{6.4}$$

where T is the frame time. In Figs. 6.15, 6.16, and 6.18, measurements were made with $t_p = 100\ \mu s$.

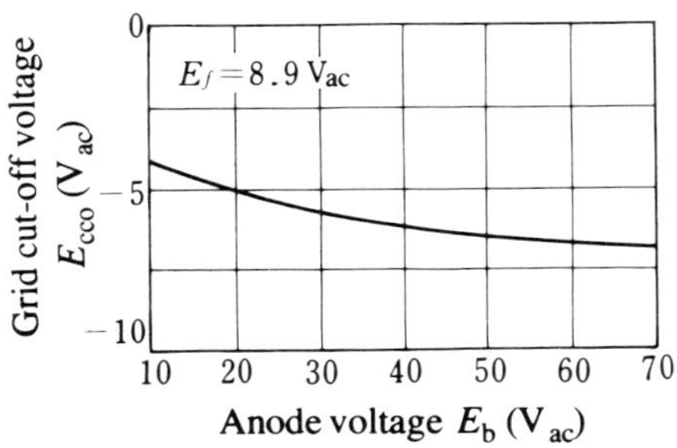

Figure 6.19 *Grid cut-off characteristic*

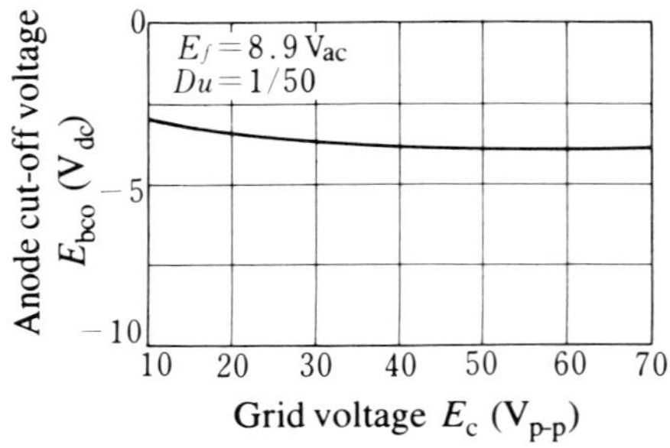

Figure 6.20 *Anode cut-off characteristic*

(4) Cut-off characteristic

In the use of a VFD, some digits and anode segments are made to emit light, while other digits and anode segments are not. As described above, bias voltages are applied to the digits and anode segments that are not to emit light. Their characteristics are shown in Figs. 6.19 and 6.20.

All these characteristics were obtained for a d.c. display VFD (DC 405A2) to be described. However, apart from special devices, all VFDs show practically the same characteristics. The conditions of measurement were: $E_f = 8.9$ V, $e_b = e_c = 45\ V_{pp}$, $Du = 1/50$, $t_p = 100\ \mu s$.

6.3.3 Drive of VFDs

The method of cathode heating and grid scanning frequency will be described.

(1) Method of cathode heating

The cathode is driven by applying E_f such that it reaches the prescribed temperature. This may be done by direct current heating, or alternating current or square wave heating. In the case of the latter, the effective value must be the prescribed voltage. These methods are explained below.

(a) Direct current heating method The principle of heating the cathode by applying direct current voltage is shown in Fig. 6.21. The potential relationship of the various electrodes is as shown in Fig. 6.22.

A potential difference in E_b and E_c (corresponding to E_f) appears at the digits at the two ends of the VFD, producing a gradient in luminance. Since, as explained above, the luminance is proportional to the 2.5 to 3.0-th power of E_b or E_c, a large luminance difference between the digits at the two ends is unavoidable. To compensate for this, the countermeasure of applying a gradient, corresponding to this potential gradient, to the values of the gap widths G_{gk} and G_{gk} between the cathode and grid is taken. Specifically, where the potential is low, the gap is made

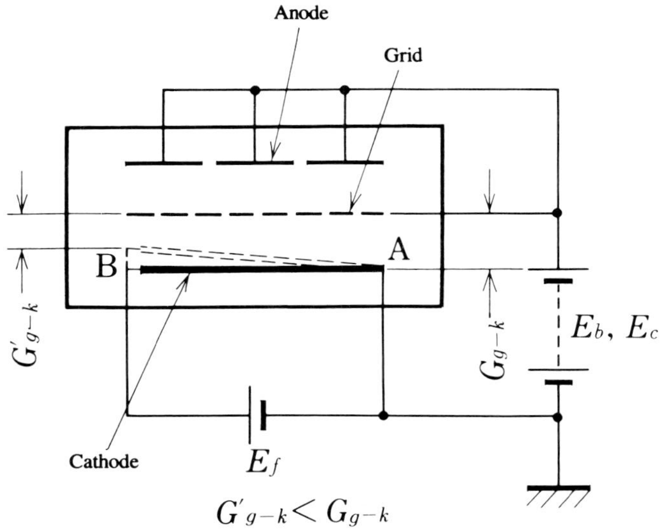

Figure 6.21 *Principle of d.c. heating*

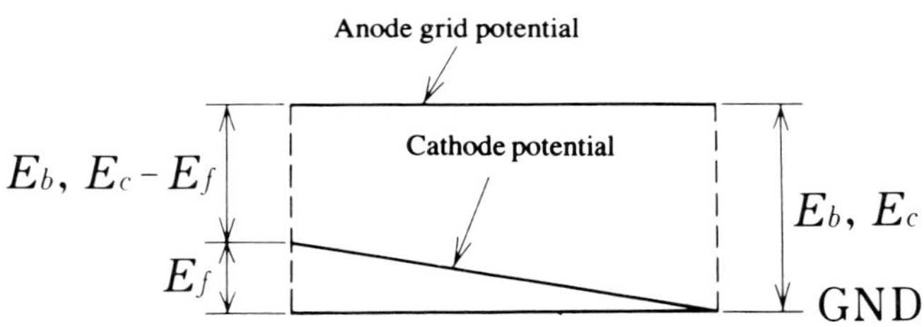

Figure 6.22 *Potential relationship of d.c. heating*

narrow, to increase the anode current I_b and restore the balance of the luminance. This can be explained as follows.

Assuming that the cathode and the grid form a diode valve, the relationship between the emission current i_e from the cathode and the electrode gap is expressed by the following equation [7]:

$$i_e = \left(\frac{K}{d^2}\right) E_c^n \tag{6.5}$$

where i_e is the i_k shown in equation (6.2), K is a constant determined by the electrode structure, and n is a constant equal to 3/2; that is i_k is inversely proportional to the second power of the gap between the cathode and grid. In the same way, i_b is contained in equation (6.2) and varies in accordance with i_k.

This state of affairs is shown by the dotted line in Fig. 6.21. This makes it essential to designate which is the plus side and which the minus side of the filament terminals. This method of heating becomes difficult to put into practice

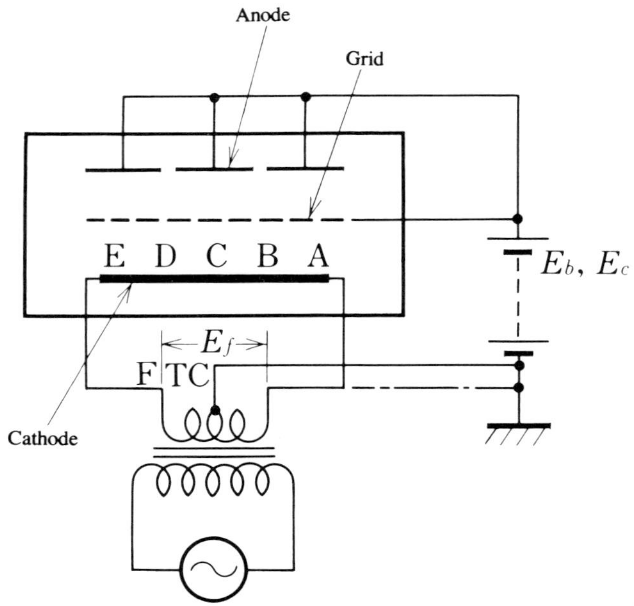

Figure 6.23 *Principle of a.c. heating (FTC: filament transformer centre tap)*

as the size of the device increases, so it is only adopted in restricted applications, such as for clocks etc. of about four digits, for use in vehicles.

(b) Alternating current heating method In the case of a VFD, the cathode is generally heated by alternating current, but, when, as in Fig. 6.23, a filament transformer centre tap (FCT) is connected to earth, the time variation of the cathode potential at positions A to D on the cathode is as shown in Fig. 6.24. Consequently, E_b and E_c at the respective positions vary, so strictly there is a variation of luminance. To visual observation, the difference in luminance is not noticeable. However, as shown by the chain dotted line in Fig. 6.23, if one side is connected to earth, the time variation of the cathode potential is as shown in Fig. 6.25, giving a marked difference in luminance, as in the case of direct current heating, making this method impracticable.

When the cathode is used with alternating current heating, beats occur between the power source frequency of cathode heating and the grid scanning frequency, causing vibration of the cathode. This may result in flickering of the display, so attention has to be paid to this possibility.

(c) High-frequency heating When a d.c./d.c. convertor is used, the power source frequency is about 20 kHz. This frequency is remote from the grid scanning frequency of 100 to 1000 Hz, so beats are not produced, and flickering of the display caused by vibration of the cathode can be prevented. Also, in the case

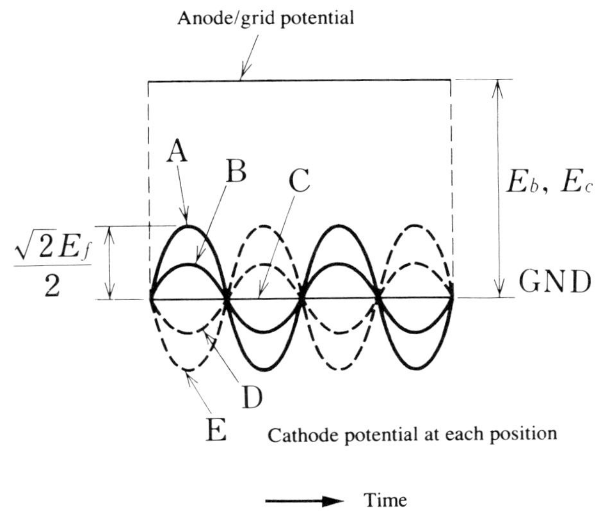

Figure 6.24 Potential relationship of a.c. heating (where FTC is earthed)

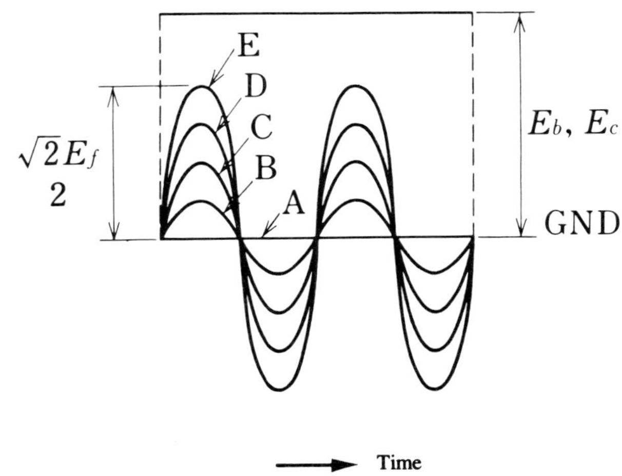

Figure 6.25 Potential relationship of a.c. heating (one side earthed)

when the cathode is pulse heated, the FCT of the power source is earthed in use (Fig. 6.26). A typical pulse waveform in this case is shown in Fig. 6.27.

(2) Grid scanning frequency

Display quality may be impaired by flickering produced in the display by various factors, such as variations in the gap between the cathode and the grid due to cathode vibration, producing variation of the luminance with time. The cathode

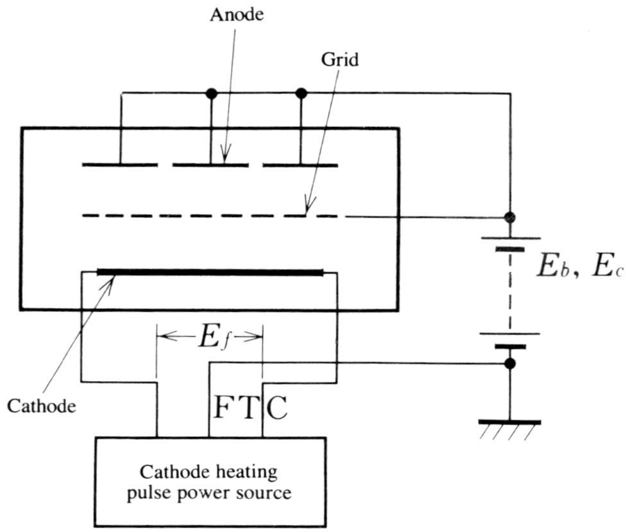

Figure 6.26 *Principle of pulse heating*

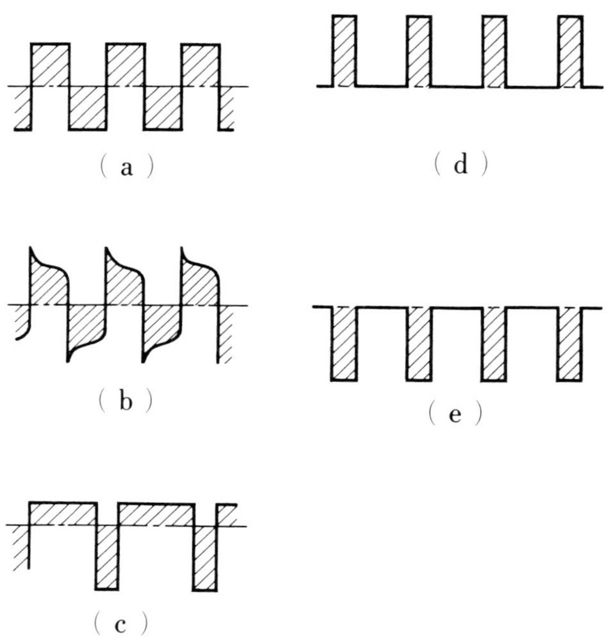

Figure 6.27 *Example of waveform of cathode pulse heating*

is made of very fine wire of pure tungsten, and is arranged with a prescribed gap from the grid, being stretched by a tension determined by its tensile strength and thickness, which is applied to it by an anchor made of springy material. It therefore forms a sort of chord, which vibrates in response to external force. Its characteristic frequency v_0 is given by the following equation:

$$v_0 = \frac{1}{2l}\sqrt{\frac{980T}{\sigma}} \tag{6.6}$$

where l is the cathode length (cm), T is the cathode tension (g), and σ is the mass per unit length of the cathode (g/cm).

The half-amplitude a is given by the following equation:

$$a = \frac{A}{4\pi^2}\frac{1}{v_0^2} \tag{6.7}$$

where A is a constant depending on the force that is applied to the VFD and on the mass of the cathode.

Increasing the v_0 of the cathode can make a small, but, in order to make v_0 large, T must be made large by reducing l and σ. σ and T are in a proportional relationship determined by the design, so the only possibility left is to make l small. To this end, since in the case of a long tube l will also be long, an intermediate support must be provided in the middle of the cathode. By this means, l can be reduced to essentially one half, enabling v_0 to be doubled, and reducing a.

Cathode vibration is caused by various factors apart from external forces. One of the most significant of these factors is the occurrence of beats between the cathode heating power source frequency f_F and the grid scanning frequency f_G. The relationship of f_F and f_G such that such beats do not occur is as follows:

$$f_F - 30 < f_G < f_F + 30$$

or (6.8)

$$\frac{f_F - 30}{n} < f_G < \frac{f_F + 30}{n}$$

(Units: Hz; $n = 2, 3, 4, \ldots$)

If mains power is used for the cathode heating power source, the minimum value of f_G is 90 (or 80) Hz, but, from the point of view of safety, it is desirable to use at least 110 Hz. If a circuit capable of adjusting f_G by about $\pm$ 50 Hz is provided and if flickering is produced by f_G being an integral fraction of f_F, it can be removed by adjusting this circuit. However, if as mentioned above, a d.c./d.c. converter is used for the cathode power source, since its frequency is at least 20 kHz, which is much higher than $f_G = 100$ to 1000 Hz, there is no possibility of the display quality being impaired by flickering.

6.4 VARIOUS DRIVE SYSTEMS FOR VFDs

The basic drive system of a VFD may be either a static drive or a dynamic drive. Depending on which of these is used, the construction of the VFD itself will also differ. Recently active drives have also been developed. The details of these are explained below.

6.4.1 Static drive

This is a drive system for displays with a small number of digits. As shown in Fig. 6.28, in the case of the grids, prescribed direct current voltage is applied in common to each digit, whereas each anode segment has an independent terminal attached to it. Any desired time can therefore be selected for the display digit and display pattern simply by applying a signal to the corresponding anode segment. The reason why this drive system is suited to a display with a comparatively small number of digits is that, if the number of digits is increased, the number of anode segment terminals increases steeply.

An example of a static drive circuit is shown in Fig. 6.29. The peripheral circuitry is simple, the drive voltage being 10 to 15 V_{DC}, but an anode segment selection circuit is required for each digit. This type of drive is often used in automobile clocks, etc.

6.4.2 Dynamic drive

Dynamic drive is used in displays where the number of digits is large. As shown in Fig. 6.30, the anode segments in the same position of each digit are connected within the tube and are connected to common terminals. Consequently, the number of segment terminals is not related to the number of digits, but is determined by the number of anode segments per digit. The grid is divided for each digit, terminal leads being provided for each grid subdivision.

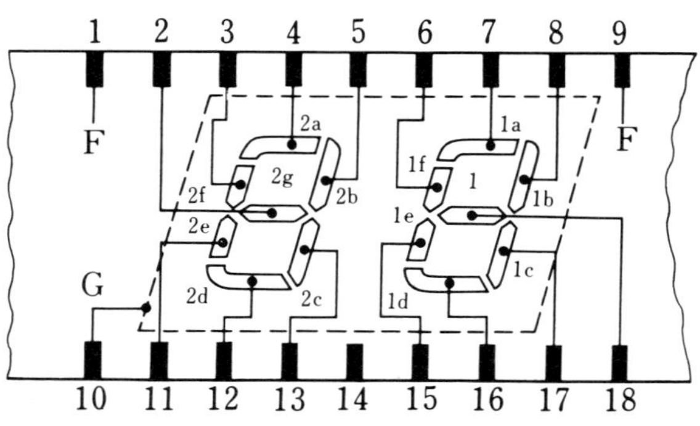

Figure 6.28 *Electrode connections of static drive VFD*

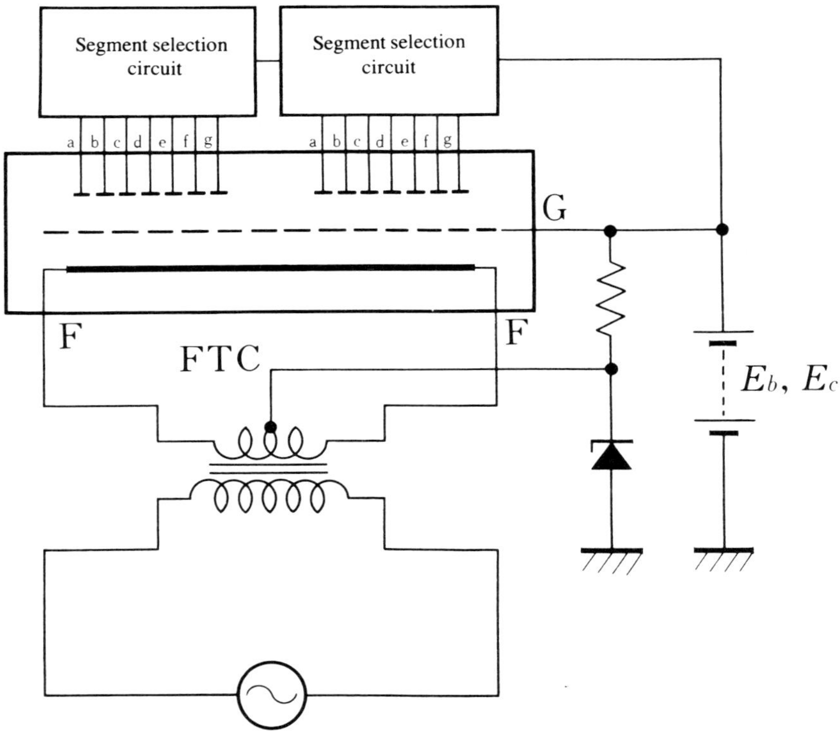

Figure 6.29 Example of static drive circuit

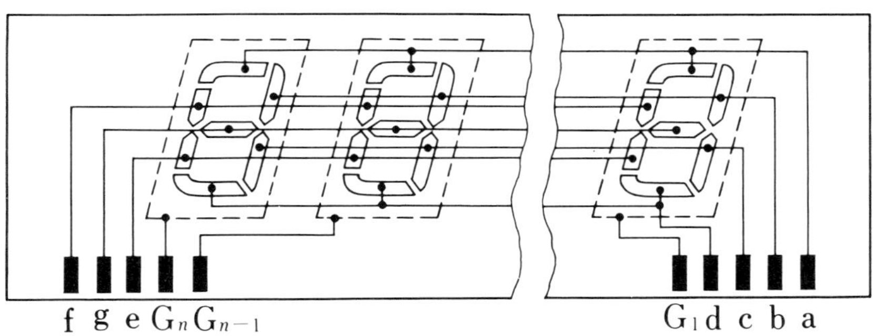

Figure 6.30 Electrode connections of dynamic drive VFD

An example of a dynamic drive circuit is shown in Fig. 6.31. The segment selection circuit may be a single circuit, irrespective of the number of digits, but a grid scanning circuit is required. In a display using a dynamic drive, the anode segments are selected in accordance with the timing of the digit signal (grid signal). Figure 6.32 is a timing chart of the anode segment signal.

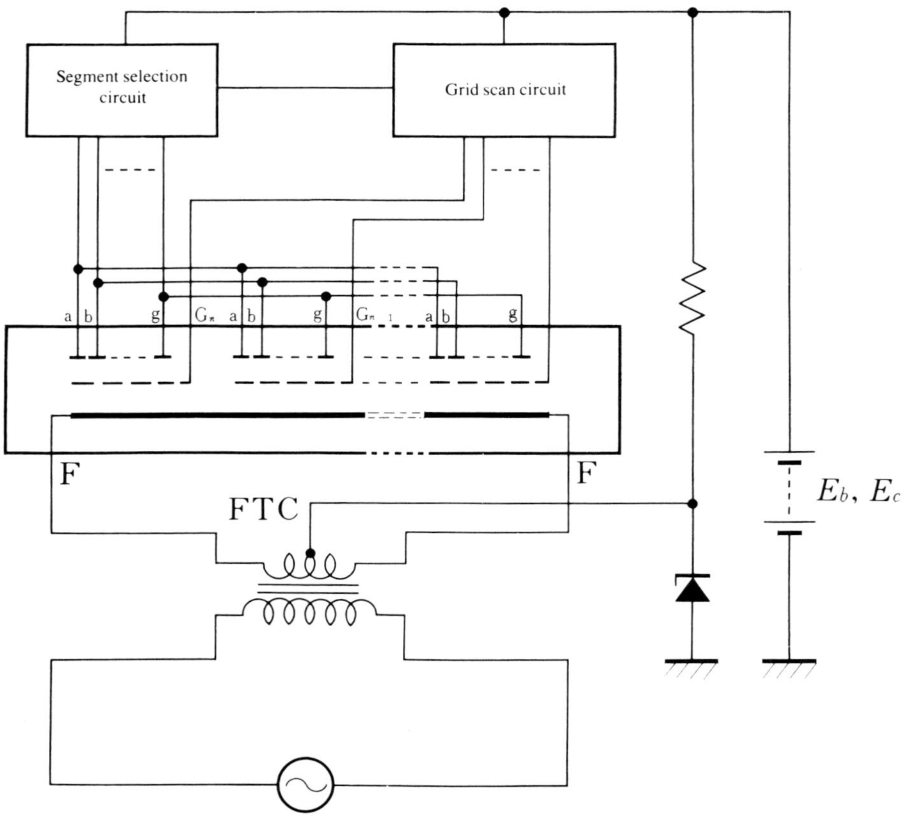

Figure 6.31 *Example of dynamic drive circuit*

6.4.3 Active matrix drive

In a VFD with a construction as described up to this point, it was not easy to solve the problems of cost caused by the requirement for high drive voltage and for peripheral circuitry, associated with an increase in pixel density and number of pixels. In order to deal with these, from 1976 onwards, active matrix drive systems have been researched[8].

The essential principles of their operation will be described with reference to Fig. 6.33.

Each pixel consists of a phosphor layer applied onto the drain electrode of a MOSFET. The construction includes a screen cathode, which is common to all the pixels, and grids, which uniformly accelerate and diffuse the thermions emitted from thermion-emitting cathodes[9].

To achieve a fine, high-density display, research has been carried out into the surface potential of the phosphor layer and into ways of making the internal resistance due to the screen cathode uniform. Reliability of the operation of the

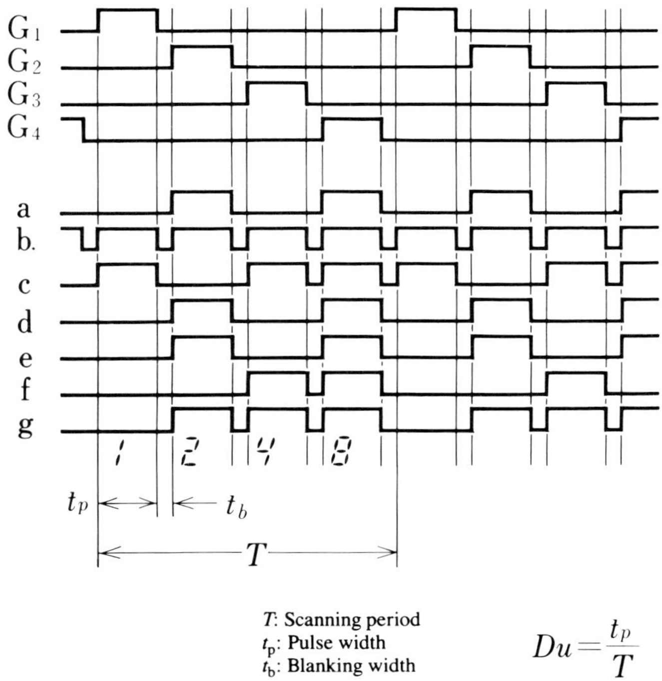

Figure 6.32 *Timing chart of grid and anode segment signals (displaying the four digits '1248')*

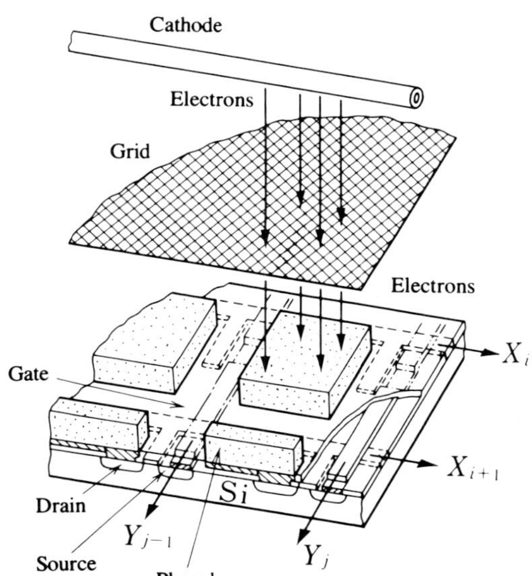

Figure 6.33 *Basic operating principle of active matrix drive*

silicon chip under irradiation by slow electrons, matching of the VFD production process and the silicon process, etc., have also been investigated[9, 10]. As a result, active matrix drive VFDs having the following features are beginning to be practically implemented:

1. Since they use light emission of high efficiency (7 to 8 lm/W), excited by slow electrons at low voltage (less than 15 V), a silicon LSI chip can be employed.
2. Since a silicon LSI can be utilized, it is possible to provide each pixel with a memory circuit, eliminating problems such as lowering of luminance and increase of drive voltage through the duty cycle.
3. The area of the silicon chip can be made small, and lowering of costs can be achieved by using a number of small-area chips.
4. As the silicon LSI chip is maintained in high vacuum, it shows high reliability irrespective of the environment.
5. Since the drive circuitry is incorporated, a simplification of the peripheral circuitry can be achieved, making it possible to reduce the overall size of the system.

6.5 TYPES OF VACUUM FLUORESCENT DISPLAY AND THEIR APPLICATIONS

Starting with displays of numbers and symbols using seven or eight segments, VFDs have progressed to displays of various types of symbols and capital letters using fourteen or sixteen segments. In future, displays of many symbols such as small letters, Japanese katakana symbols and most symbols using 5×7 or 5×12 matrices will be possible, and page displays are being approached by increasing the number of digits. In addition, by using a dot matrix to increase the range of character display to Japanese hiragana symbols and Japanese or Chinese characters, etc., graphics displays and image displays will also become possible. On the other hand, analogue displays using bars or blocks are employed in industrial meters or level meters for audio equipment, and combination displays of screen patterns are used in hobbies such as electronic games etc.

Table 6.3 shows an outline of the types of VFD and displays, and the major applications. A tree diagram of VFD applications based on the various technologies is shown in Fig. 6.34. The details of the respective classifications are introduced below.

6.5.1 Segment display

In a seven-segment (7SG) display, numerals and simple symbols are displayed by the pattern shown in Fig. 6.35 By adding a large number of letters and symbols as in Fig. 6.36, they can be used in digital displays of various types of industrial measuring device. Furthermore, their range of application has been expanded by producing combined displays, as will be described later.

Table 6.3 *Classification of vacuum fluorescent displays, diagram of display, applications*

CLASSIFICATION	DIAGRAM OF DISPLAY	MAIN APPLICATIONS
Seven-segment display		Calculators; industrial meters, electronic weighing machines; ECRs; timers; vehicle clocks
Fourteen-segment display		POS terminals; conversation aids for deaf and dumb persons; educational equipment; OK monitors; display devices of various kinds
Dot displays		POS terminals; data terminal devices; bank window systems; monitoring and control devices; seat reservation systems; OA systems
Dot matrix displays; scrolling displays	20 × 20, 40 × 41, 64 × 64, 80 × 80, 128 × 128, 256 × 256, 320 × 240 dots; 256 × 26, 256 × 64, 256 × 128 dots	Graph displays; image displays; electronic game machines; data terminal devices; OA systems
Bar displays		Analogue display industrial meters; level meters; speedometers; revolution counters; thermometers; medical equipment
Composite displays	Combinations of two or more types	Vehicle apparatus; educational equipment, POS terminals; various types of display devices
Image displays	Combinations of images	Electronic game machines; educational equipment

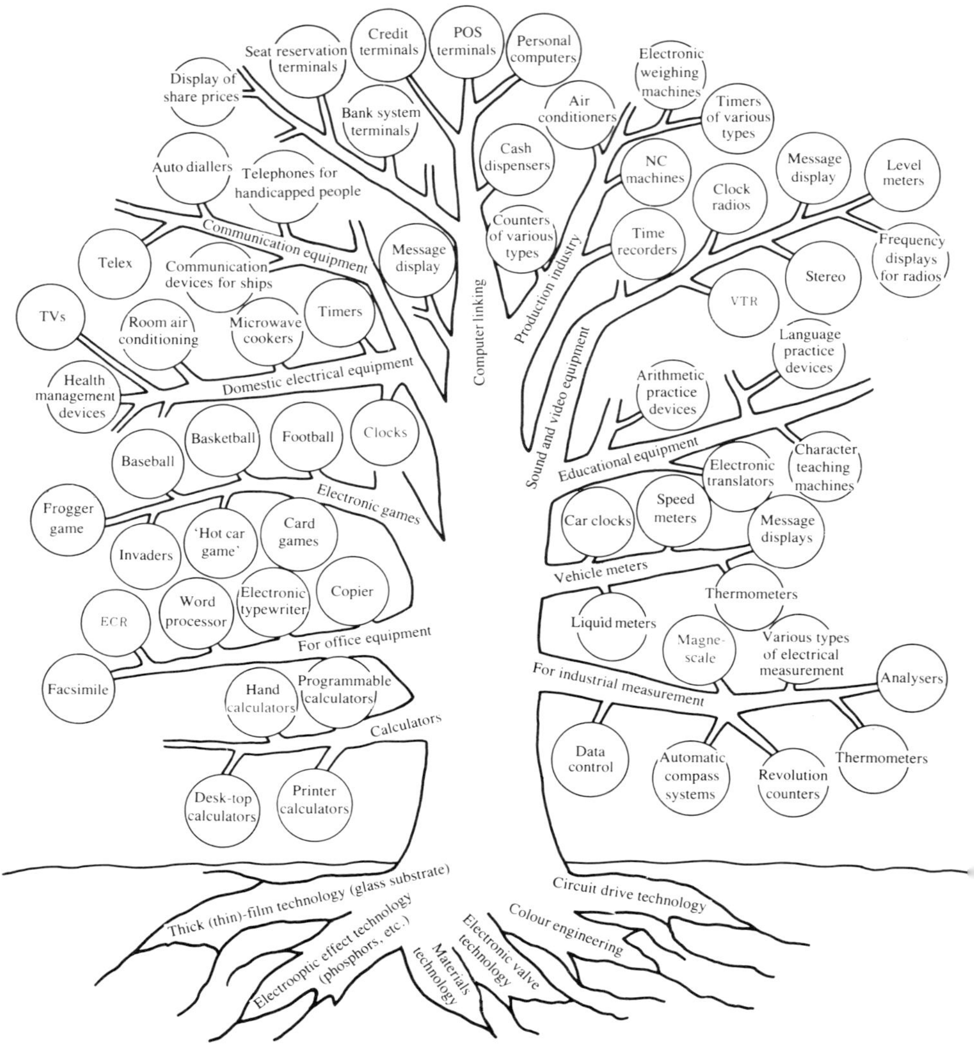

Figure 6.34 *Tree showing applications of VFDs*

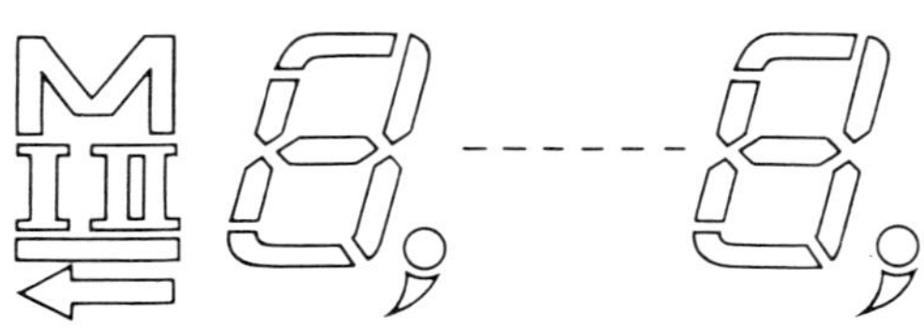

Figure 6.35 *Example of pattern of seven-segment display*

Figure 6.36 Examples of eight-segment VFD with additional display of symbols and letters

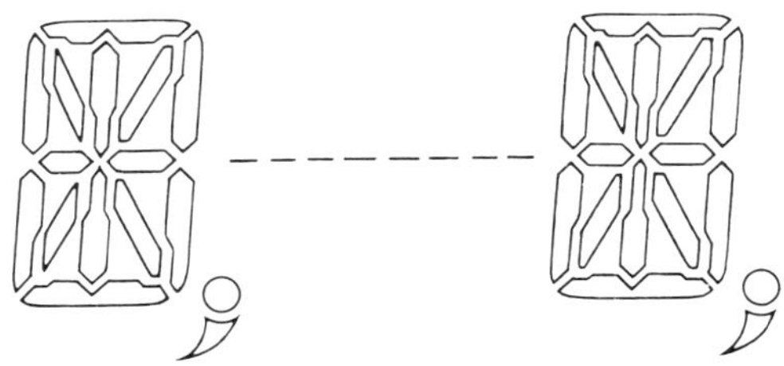

Figure 6.37 Example of pattern of fourteen-segment display

Figure 6.37 shows the pattern of a 14SG display. 16SG display patterns can also be obtained by dividing the segments into upper and lower portions. As shown in Fig. 6.38, capital letters and various kinds of symbols can be displayed by this pattern, but numerals or letters are somewhat distorted in shape. An example of a 14SG VFD is shown in Fig. 6.39.

7SG displays are often used in various devices such as desk-top calculators, ECRs, electronic weighing machines, and watches, etc. There is a strong demand for lower voltage and lower power consumption while maintaining present levels of luminance.

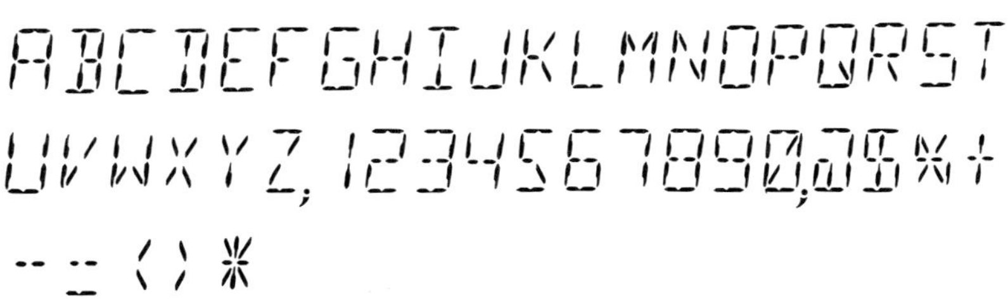

Figure 6.38 Display pattern produced by fourteen segments

Figure 6.39 Example of fourteen-segment VFD

Next, we introduce a method of calculating the power consumption P_t per digit[1]:

$$P_t = P_F + P_B + P_C \tag{6.9}$$

where P_F, P_B, and P_C are respectively the cathode, anode, and grid power consumption, and are calculated as follows:

$$P_F = \frac{E_F \cdot TI_f}{n}$$

$$P_B = e_b i_b \cdot Du$$

$$P_C = e_c i_c \cdot Du$$

where E_f, I_f, e_b, e_c, i_b, i_c, and Du are as defined above (Section 6.3), and n indicates the number of digits.

14SG displays and 16SG displays are used in equipment of many types in all countries. For example, large numbers of such displays of the desk-top calculator type are used in educational equipment such as alphabet training machines, and they are also used in devices for enabling deaf people to carry out telephone conversations.

6.5.2 Dot character displays

The development of dot character displays (DC displays) has made it possible to display large amounts of information at the same time. This is an important role

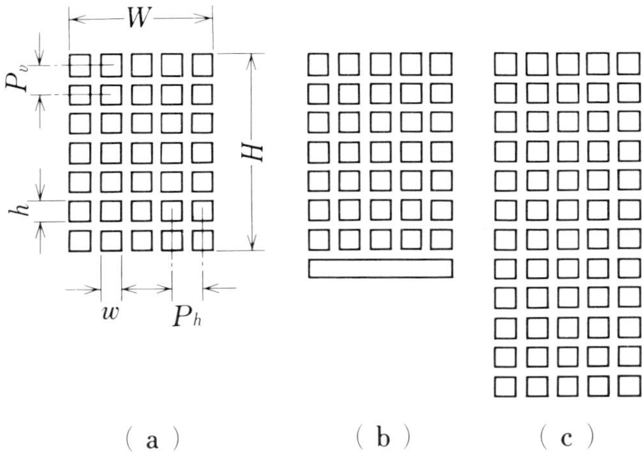

Figure 6.40 Types of dot pattern

of display devices, which has greatly increased the range of applications of VFDs.

As shown in Fig. 6.40, there are many types of dot character display: 5 × 7 dots, 5 × 7 dots with cursor, and 5 × 12 dots. As shown in Fig. 6.41, Japanese katakana can be displayed, as well as capital letters and small letters.

An example of a VFD dot character display is shown in Fig. 6.42. At present, this is the most common display quantity, and displays can be produced with a letter height of 5 mm, with 5 × 7 dots with cursor, 40 digits and 12 rows, i.e. 480-digit displays. In the VFDs of such dot character displays, it is necessary to provide a large number of displays in restricted dimensions, so the wiring is formed of aluminium thin film. Figure 6.43 shows a view of this wiring to a larger scale. Since the development of dot character displays, VFDs have tended to become larger, but, in the type wherein 40 digits and 4 rows, i.e. a 160 digit display, are provided, the number of outgoing leads is large, so the dot character display VFD is fixed to the PC board. These leads are mounted by fixing them in position with solder, so that they can be connected to, for example, another PC board by a commercially available PC joiner.

There are various methods of driving the VFDs of a dot character display[11, 12]. An example of a drive circuit for a 40-digit, 6-row, i.e. 240-digit, display variety is shown in Fig. 6.44. An example of the display is shown in Fig. 6.45. This is an example of a display module to be described below.

The range of displays that can be achieved by VFDs when used in dot character displays is wide, and a large number of digits can be displayed, so this is one of the most important fields of application of VFDs. Regarding digit number, displays have been developed with a large number of rows and up to 480 digits. The maximum number of digits per row is 80 digits, and there are two types of letter height: either 9 or 6 mm. In both cases, the display is a 5 × 12 dot display. The former is the type that gives the largest overall length. The chief applications are

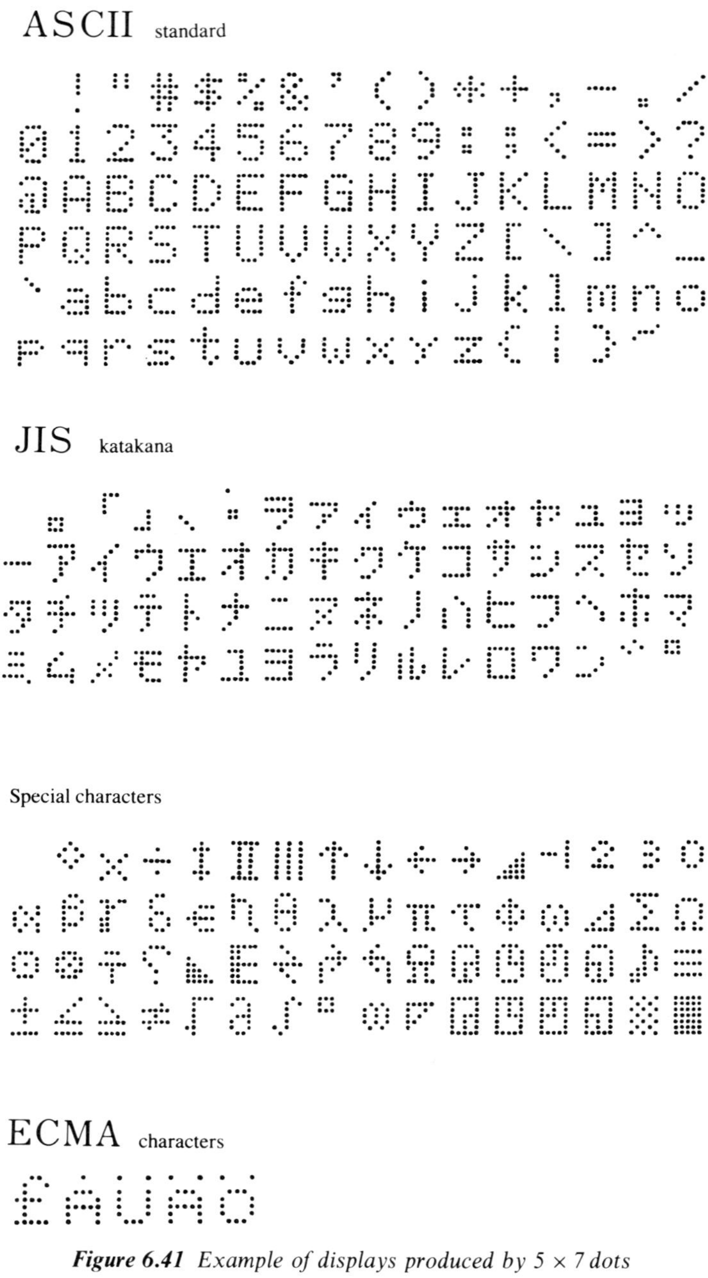

Figure 6.41 *Example of displays produced by 5 × 7 dots*

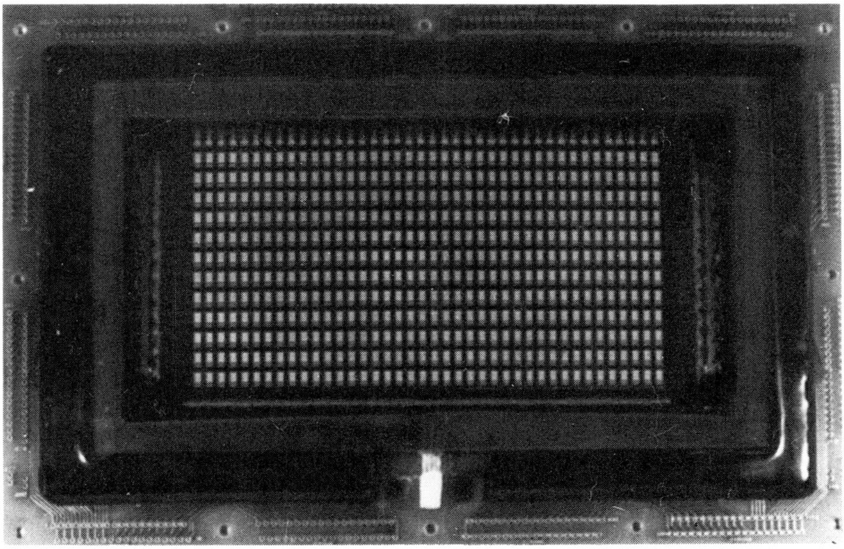

Figure 6.42 480-digit display VFD

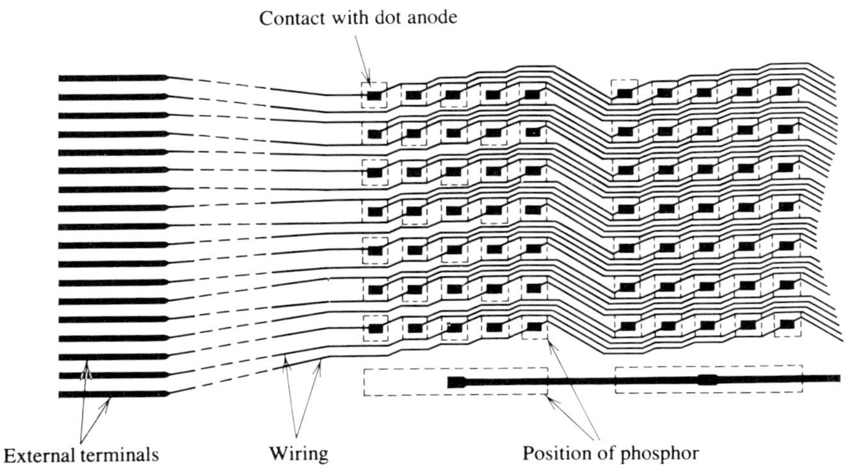

Figure 6.43 Detail of wiring of aluminium thin-film glass substrate

shown in Table 6.3. However, apart from this, 5 × 7 dot displays provided with an annuniciator are mounted on high-grade ECRs, etc. Multirow types are used in the control units of various types of environmental test equipment, and they are considered to be useful for condition setting and monitors.

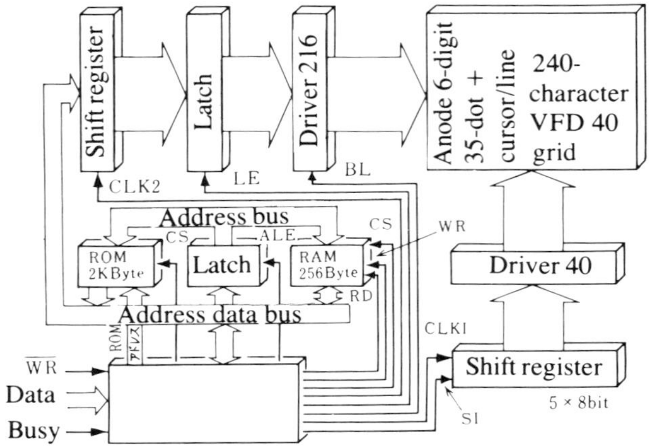

Figure 6.44 Example of drive circuit of 240-digit display VFD

6.5.3 Dot matrix displays

In segment displays and dot character displays, the columns and rows are partitioned from each other and are discontinuous, so previously the type and the amount of the display were limited[13]. This is overcome by the dot matrix display (DM display). This can display not only numbers, various types of symbol, Japanese katakana symbols, and capital and small letters of the alphabet, but also Japanese hiragana symbols or Chinese characters and other complicated letters, or even graphics displays or TV images[14].

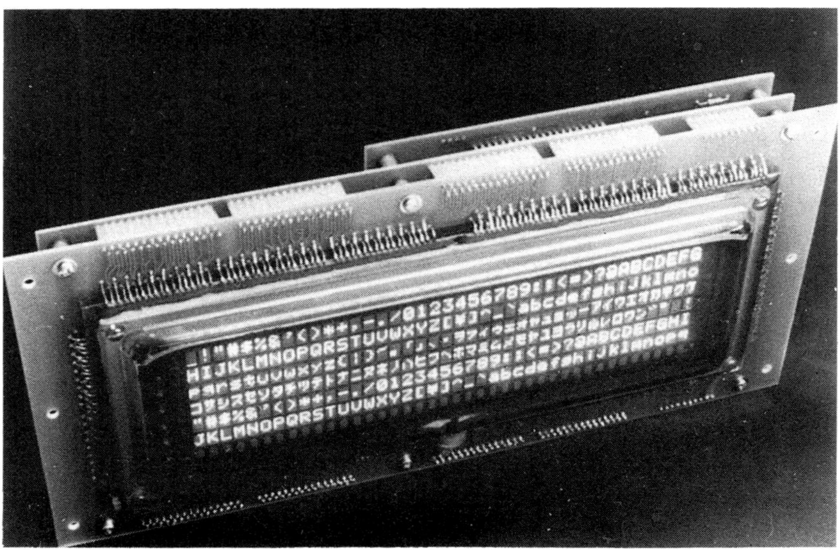

Figure 6.45 Example of 240-digit VFD display

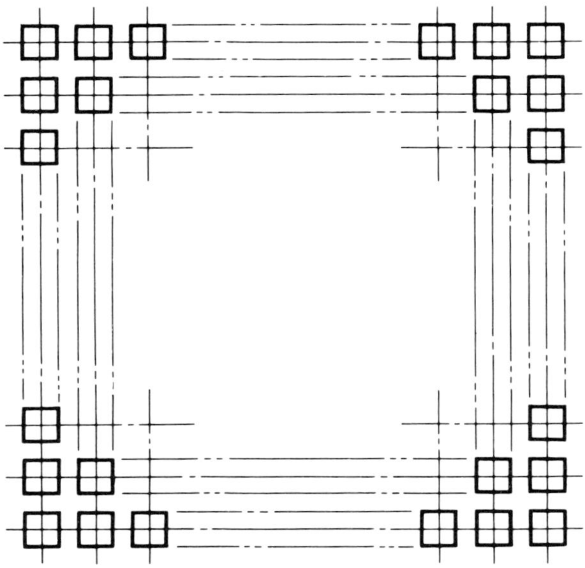

Figure 6.46 Arrangement of dot anodes of dot matrix display VFD

Figure 6.47 Arrangement of dot anodes of scrolling display VFD

In a dot matrix display, the dots are arranged with equal vertical and horizontal pitch. Figures 6.46 and 6.47 show two different types of dot arrangement. In the former, approximately the same number of dots are arranged in both the vertical and horizontal directions. In the latter, the vertical number is much less than the horizontal number. Hereafter, the latter will be called the scrolling display (SL display) and will be dealt with separately.

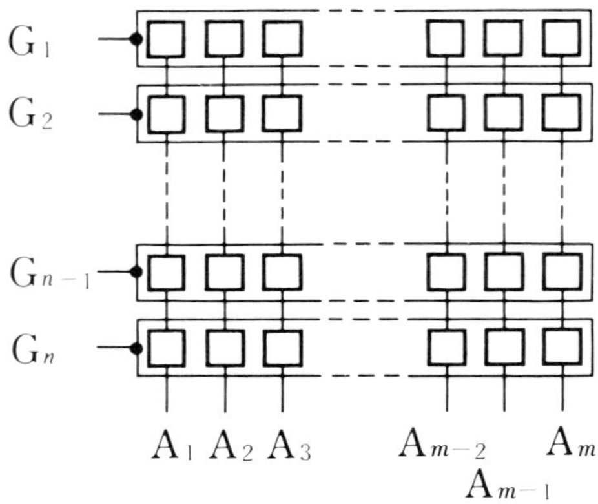

Figure 6.48 *Basic arrangement of grids and connections of dot anodes*

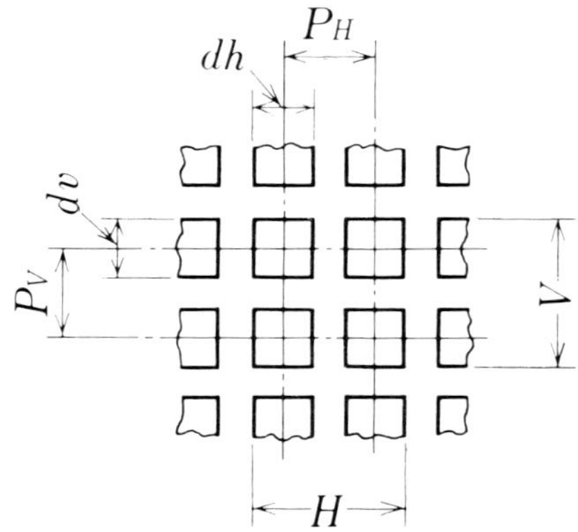

Figure 6.49 *Dot dimensions*

(1) Dot matrix display

Both the dot matrix display and scrolling display are based on the triode construction shown in Fig. 6.4. However, the lead-out method is different. Basically, as shown in Fig. 6.48, a matrix is formed by arrangement of the dot anode connections and of the grids, so that when a positive voltage is applied to any particular pair of these two, the dot anode at their intersection is caused to emit light. In a dot matrix display or scrolling display, the degree of resolution is vital.

For calculation of the degree of resolution, the dot relationship dimensions shown in Fig. 6.49 are used. From this figure, H and V can be expressed by the following equations:

$$H = P_H + dh$$
$$V = P_V + dv \tag{6.10}$$

where P_H and P_V are the horizontal and vertical pitch of the dot anodes and dh and dv are the horizontal and vertical dimensions of the dot anodes. In general, $P_H = P_V$ and $dh = dv$, so $H = V$.

As is clear from the figure, since H or V is the dimension of two dot anodes, the resolution of a dot matrix display VFD is defined as follows:

$$\frac{2}{H} \quad \text{or} \quad \frac{2}{V} \tag{6.11}$$

The number of dot anodes per millimetre calculated by dividing the H or V dimension by two dot anodes is called the relative resolution[15].

To obtain a dot matrix display VFD with good resolution, the dot size and the dot pitch must be made small. In the case of the pure matrix shown in Fig. 6.48, reduction in the pitch entails reducing the width of the grid. However, if this is done, the electrons from the cathode cannot be satisfactorily controlled and are affected by the bias applied to the adjacent grid, with the result that uniform light emission from all of the dot anodes is not obtained, making it impossible to achieve the necessary luminance.

To deal with this problem, the dot anode connection and grid arrangement shown in Fig. 6.50 (b), with adjustment of the drive voltage and bias voltage, may be adopted. The single matrix of Fig. 6.50 (a) shows a simple matrix as illustrated in Fig. 6.48, in which, due to the effect of the bias voltage of the adjacent grid, no light emission was obtained from the light-emitting dot anodes at the two ends. However, in the double matrix shown in Fig. 6.50 (b), the dot anodes are divided into two portions which are separately connected and provided with separate outgoing leads and consequent adjustments made to the drive voltage and the bias voltages of the grid and dot anodes. The result of these modifications is that the entire light-emitting dot anode surface emits light, and the necessary luminance is obtained. In a double matrix display, drive voltage is applied to the

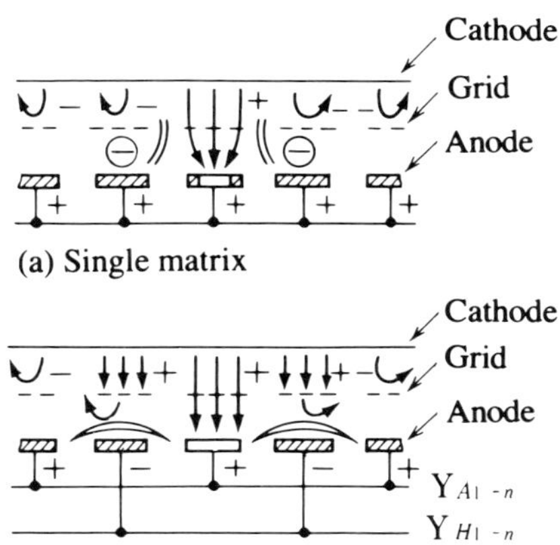

(a) Single matrix

(b) Double matrix

Figure 6.50 *Electrode connections and voltage application in dot matrix display*

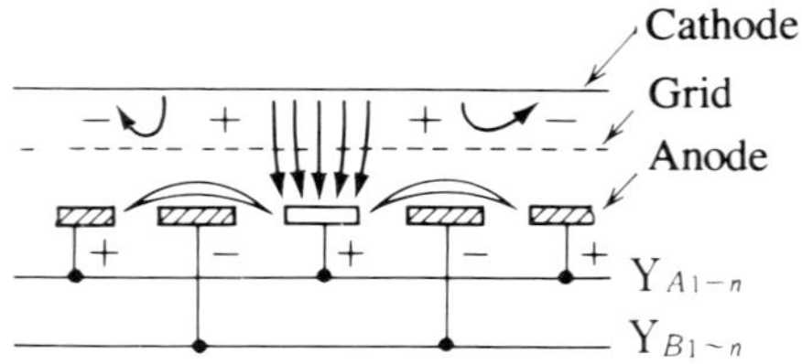

Figure 6.51 *Arrangement with grids offset by one half-pitch with respect to the dot anodes*

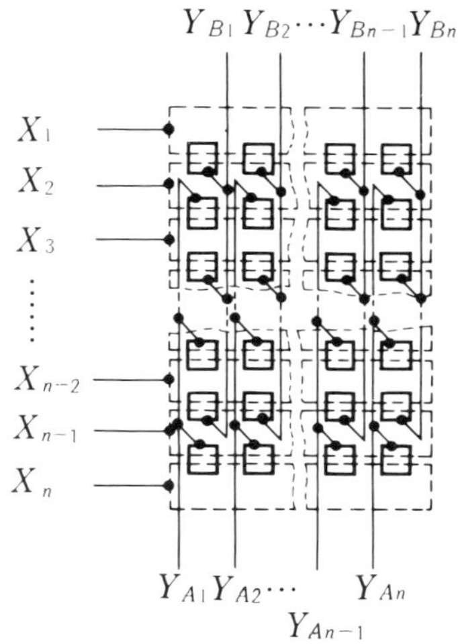

Figure 6.52 *Improved dot anode connection and grid arrangement*

dot anode adjacent to the middle of the display and to the three adjacent grids, bias voltage being applied to the other grid anodes.

As a means of obtaining the same effect as reduction of the grid current, the grids may be arranged as shown in Fig. 6.51, offset by one half-pitch with respect to the dot anodes, drive voltage being applied to the middle dot anode and to the adjacent two grids. In this way, a good display is obtained[17]. The connection of the dot anodes and arrangement of the grids is as shown in Fig. 6.52.

Since the development of a dot matrix display VFD of pitch 4.0 mm, dot anode dimension 1.5 mm^2, and relative resolution 0.31 dots/mm with 20 × 20 dot anodes, a 320 × 240 dot anode display (DM 320 × 240 B) has been developed. This display is of pitch 0.4 mm, with 0.2 mm × 0.25 mm dot anodes, relative resolution 3.3 dots/mm and 3.1 dots/mm[18]. With such a dot anode construction, a practically useful range has been reached. A prototype three-colour display has been manufactured[19] by applying, as in Fig. 6.53, two types of phosphors (bluish-green and red), using these two colours and a mixture of them. The main products that have been developed are introduced below.

Figures 6.54 and 6.55 show a graphics display and image display using a dot matrix VFD (DM 128 × 128C), wherein 128 dot anodes of dimensions 0.35

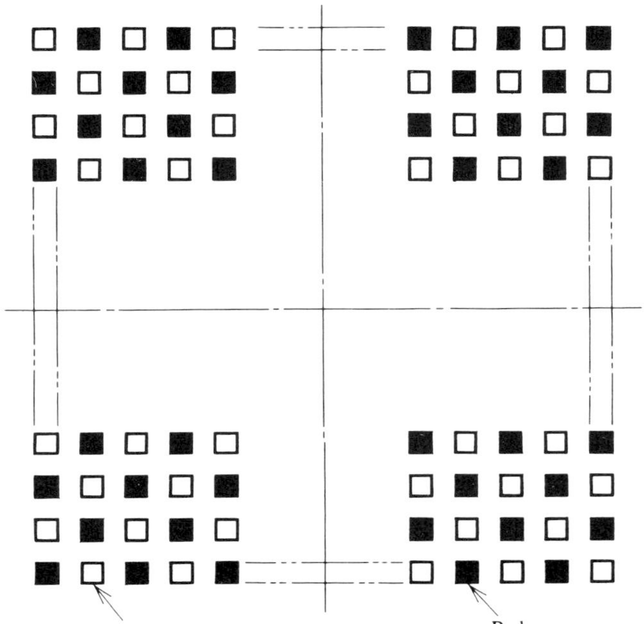

Figure 6.53 *Two-colour dot anode arrangement*

× 0.4 mm are arranged in the vertical and horizontal directions with a pitch of 0.6 mm in each case. The relative resolution is 2.1 dots/mm. The latter of these two is driven by the circuit shown in Fig. 6.56.

An example is shown in Fig. 6.57 of a graphics display using 256 dot anodes (in both the horizontal and vertical directions), of dimensions $0.2\,mm^2$ (DM256 × 256C), arranged with a pitch of 0.4 mm. The relative resolution in this case is 3.3 dots/mm.

Figure 6.58 shows an example of a graphics display pattern obtained using DM 320 × 240 C. Table 6.4 shows the dot dimensions, electrical ratings, and characteristics of this DM 320 × 240 C display.

(2) Scrolling display

The method of displaying a pattern while shifting it to left or right is called scrolling display (SL display). The pattern is shifted by one pitch at a time, and if this is repeated at constant speed, the pattern appears to move smoothly. The quality of the display can also be improved in this case by double-matrix connection. This method of connection is shown in Fig. 6.59.

Figure 6.60 shows an example of display produced by a scrolling display VFD (DM 256 × 64 B) with 256 (horizontally) × 64 (vertically), 0.4 × 0.4 mm dot anodes pitched at 0.65 mm. The example is part of an animated cartoon picture,

Figure 6.54 *Graphic display using DM128 × 128C*

Figure 6.55 *Image display using DM128 × 128C*

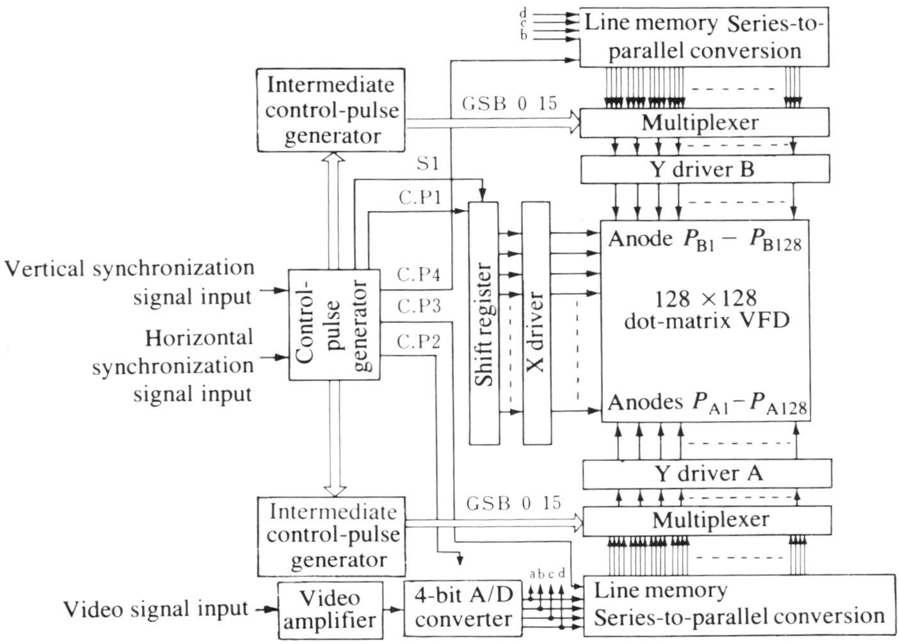

Figure 6.56 *Block diagram of image display using DM128 × 128C*

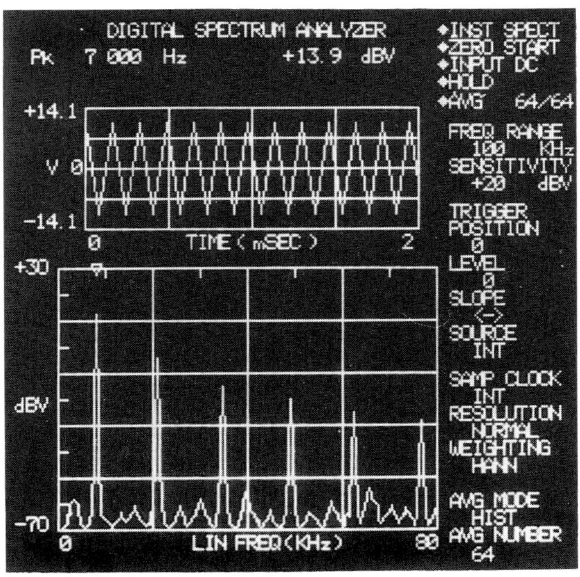

Figure 6.57 *Graphic display by DM256 × 256C*

Figure 6.58 Graphic display by DM320 × 240C

Table 6.4 Details of DM 320 × 240 C

Dot dimensions	Dot count	320 × 240 (= 76 800)		
	Dot size	0.2 mm (horizontal) × 0.25 mm (vertical)		
	Dot pitch	0.4 mm (both horizontally and vertically)		
	Display size	127.8 mm (horizontal) × 95.85 mm (vertical)		
Electrical ratings	Filament voltage	E_f	Typ.	$5.7\,V_{rms}$
	Anode voltage	e_b	Typ.	$180\,V_{p-p}$
	Grid voltage	e_c	Typ.	$70\,V_{p-p}$
	Cut-off voltage	E_{bc0}, E_{cc0}	Typ.	$-10\,V_{DC}$
Characteristics	Filament current	I_f	Typ.	1152 mA
	Anode current	i_b	Typ.	$0.05\,mA_{p-p}$/dot
	Grid current	i_c	Typ.	$10\,mA_{p-p}$
	Luminance	L	Typ.	100 fL

and the display is driven according to the block diagram shown in Fig. 6.61. Figure 6.62 shows a display of Japanese hiragana symbols and Chinese characters by another type of scrolling display VFD. The dot dimensions for this display are the same as those for DM 256 × 64 B, except that there are only 26 dots in the vertical direction.

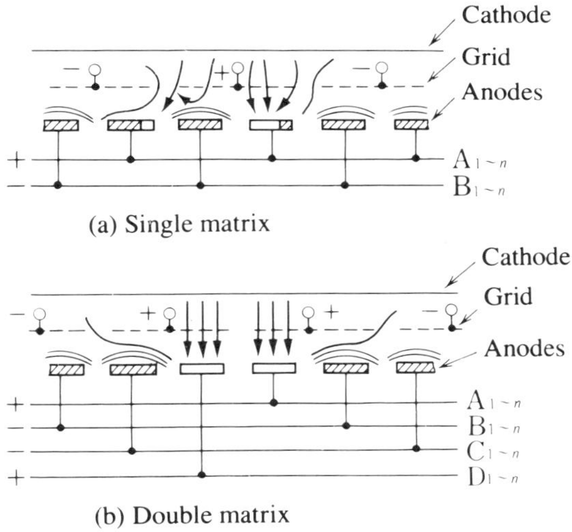

Figure 6.59 *Electrode connections and voltage application in scrolling display fluorescent display tube*

Figure 6.60 *Example of display by DM256 × 64B (part of animated cartoon)*

6.5.4 Analogue displays

Analogue displays for audio equipment and industrial measuring devices consist of an arrangement of bars or blocks. The number of bars or blocks that fluoresce varies according to the signal applied, allowing direct appreciation of changing physical quantities.

Figure 6.63 illustrates a bar display VFD for an industrial measuring device. In addition to the simple types consisting of bars, digital displays, etc., some types have widened their display range and include stationary alphanumerics or

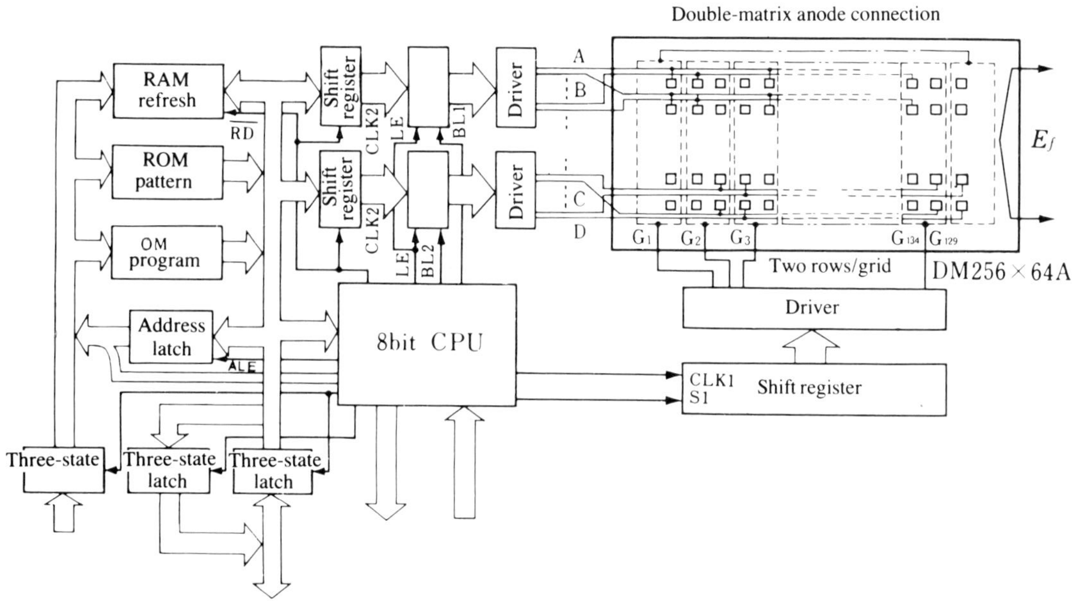

Figure 6.61 *Block diagram of DM256 × 64B drive*

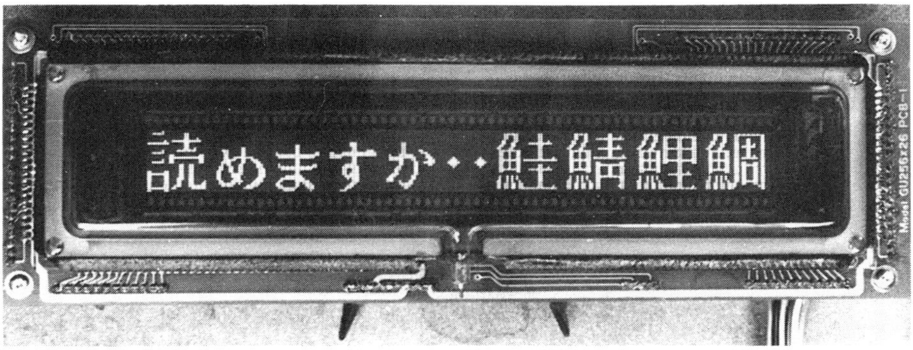

Figure 6.62 *Display of Japanese hiragana symbols and chinese characters by DM256 × 26B*

Figure 6.63 *Bar display VFD for industrial measurements*

graphic images. Multicoloured VFDs for use in level meters, in which the colour of the display changes when a certain level is reached, are becoming more common. The different sections of these are coated with phosphors that emit light of different colours.

6.5.5 Composite displays

Many types of compact, multifunction VFDs have been developed for use in household electrical appliances, office equipment, and automobile equipment. In

Figure 6.64 Example of VFD for copying machine

addition to the dark-green segment displays used for the clock, timer, and temperature displays of microwave ovens, displays employing other colours such as red can be used in other types of household electrical appliance to show alphanumerics (in English, Japanese, etc.) or graphic symbols illustrating settings, ranges of use, etc. In VTRs, graphic displays are used mainly for clocks and timers, but also for showing the day of the week, channel, tape count, and other operating information. Displays are thus being improved from both the functional and fashion aspects. An example of the use of VFDs in office equipment is the multicoloured displays used in the operating panels of copiers. Figure 6.64 shows an example of such a display. In car equipment, VFDs are used in digital clocks and various other types of dashboard indicator. Digital watches usually employ bluish-green monochrome displays, while other types of display may be multicoloured to answer the needs of legal regulations and fashion. Although not as bright, red displays are used in preference to dark-green ones in warning indicators.

6.5.6 Fixed-image displays

Large numbers of fixed-image displays are used in electronic games. Figure 6.65 shows an example of such a display, consisting of an array of multicoloured fixed images in various combinations. By changing the position at which they are made to fluoresce at intervals, the animals and strange monsters can be made to seem as if they are moving. For functional and fashion purposes, there is a strong demand

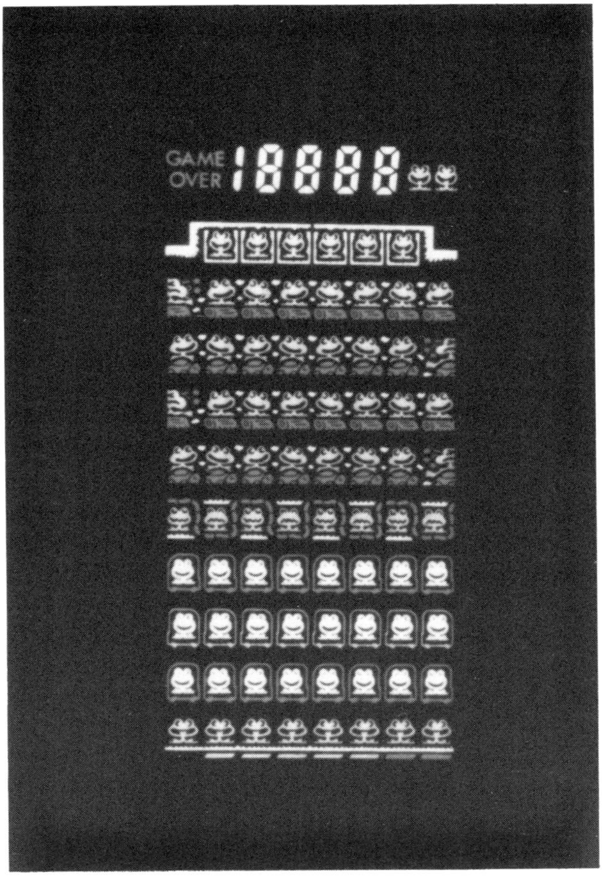

Figure 6.65 *VFD for use with 'Frogger' electronic game*

for fine-pattern displays that can give such images accurate, detailed expressions. The range of colours displayed can be increased by the use of filters, and enlarged displays can be obtained by using Fresnel lenses.

6.5.7 Active matrix displays

A prototype VFD for use as a TV display was announced in 1979[21], and an ultracompact image display VFD for use in video camera viewfinders was announced in 1982[22]. These VFDs had 172 × 108 pixels on an 8 × 6 mm screen. Since each of the pixels had its own dynamic memory, a luminance of approximately 10 fL could be obtained with a display voltage of 15 V. In viewfinders, hardly any external light enters when the display is viewed. For this reason, more importance was placed on uniformity and half-tone display

Figure 6.66 *Example of active matrix display*

Figure 6.67 *Detail of display image*

Figure 6.68 *External appearance of active matrix image display element*

Table 6.5 Particles of active matrix image display elements

Chip size	8.95 mm (horizontal) × 6.31 mm (vertical)
Display range	7.912 mm (horizontal) × 5.832 mm (vertical)
Pixel number	172 (horizontal) × 108 (vertical)
Pixel pitch	46 μm (horizontal) 54 μm (vertical)
Dimensions of area of phosphor application	Approx. 30 μm diameter
External dimensions	29.0 mm (length) × 18.0 mm (width) × 7.0 mm (thickness) 16 pin, dual in line fitted with exhaust tube
Drive voltage	− 9 V (chip, grid) − 15 V (cathode) 1.0 V (filament)
Power consumption	About 200 mW About 800 mW (including peripheral circuitry)
Fluorescent screen luminance	About 10 fL
Grey scale	11 grades
Contrast	30:1

characteristics than on luminance. Also, since the display image is seen enlarged by a lens, the display is of gridless construction. In addition, to make focusing easier, the central part of the display is enlarged by peripheral circuitry, allowing the effective number of pixels to be quadrupled in certain parts. Figures 6.66 and 6.67 show examples of such displays.

DIP type packages are employed wherein the number of terminals is reduced to sixteen pins and incorporating pixel arrays and shift registers in the silicon chip. Their external appearance is shown in Fig. 6.68. Their specification is shown in Table 6.5. The power consumption is about 200 mW, which is low. It is only about 400 mW even when a viewfinder system is constructed including a drive IC.

6.5.8 Reverse view displays

In a VFD, a phosphor is stimulated by electrons to emit light. Thus they have the advantage that a display of high luminance can be obtained with low voltage by viewing from the direction in which this stimulation takes place. As is clear from Fig. 6.4, since fluorescence takes place at a position further back from the face plate, there was the inconvenience that parallax was produced between letters and symbols recognized at a filter mounted on the front face of the VFD. In order to solve this problem, research is proceeding into a system whereby what is at present the display face is viewed from the back, i.e. a reverse view system.

The principle of a reverse view VFD is shown in Fig. 6.69[23]. The light emitted by the phosphor in response to the electron passes through the phosphor layer and the anode, through the glass substrate, and is observed from the rear face. Consequently, the anode need not be of good transmissivity. In general, a nesa glass film is used, but, since its resistance varies in response to temperature conditions, it is not practical for the anode. For this reason, the anode is constructed of a hexagonal or lattice mesh made of thin aluminium film[24] (Fig. 6.70). Rear face light emission by such an anode can attain 60 to 70 per cent of the luminance of front face light emission. However, it has not yet reached 80 per cent in the case of a nesa glass film. On the other hand, the thickness of the phosphor has a considerable effect on the rear face light emission; that is if the phosphor is thick, transmitted light is absorbed, so that the light transmitted to the rear face is reduced. However, if the phosphor is too thin, gaps appear in the phosphor, reducing the surface area of light emission and lowering the amount of such light emission. As shown in Fig. 6.71, which shows the relationship between luminance and the thickness of the phosphor layer, there is an optimum value for the thickness of a reverse view phosphor layer. If this value is exceeded, the luminance falls. However, the width of this peak is extremely narrow, because it is

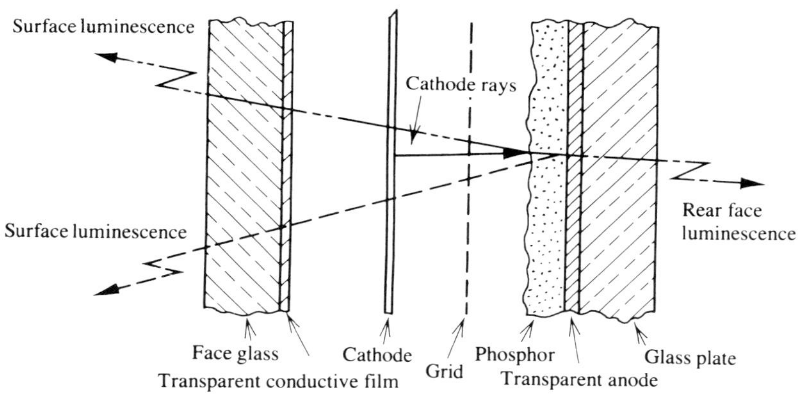

Figure 6.69 *Principle of reverse view VFD*

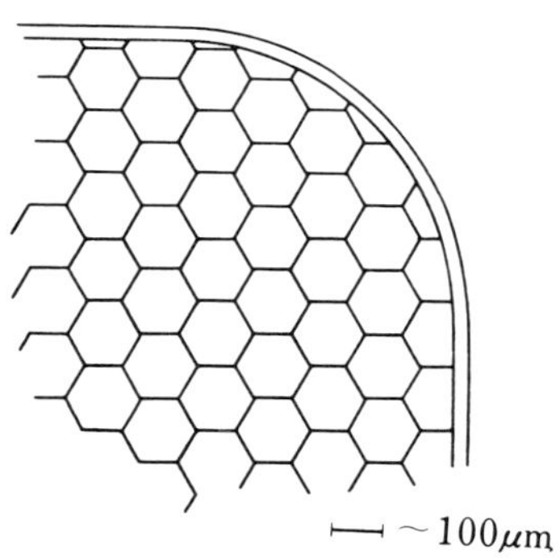

Figure 6.70 *Part of thin-film anode*

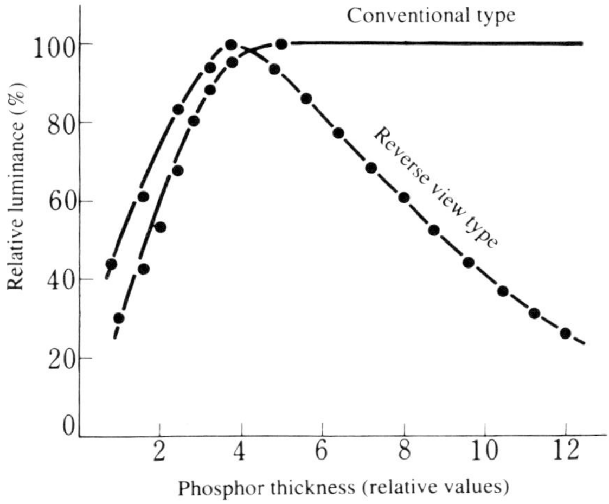

Figure 6.71 *Relationship between phosphor thickness and relative luminance*

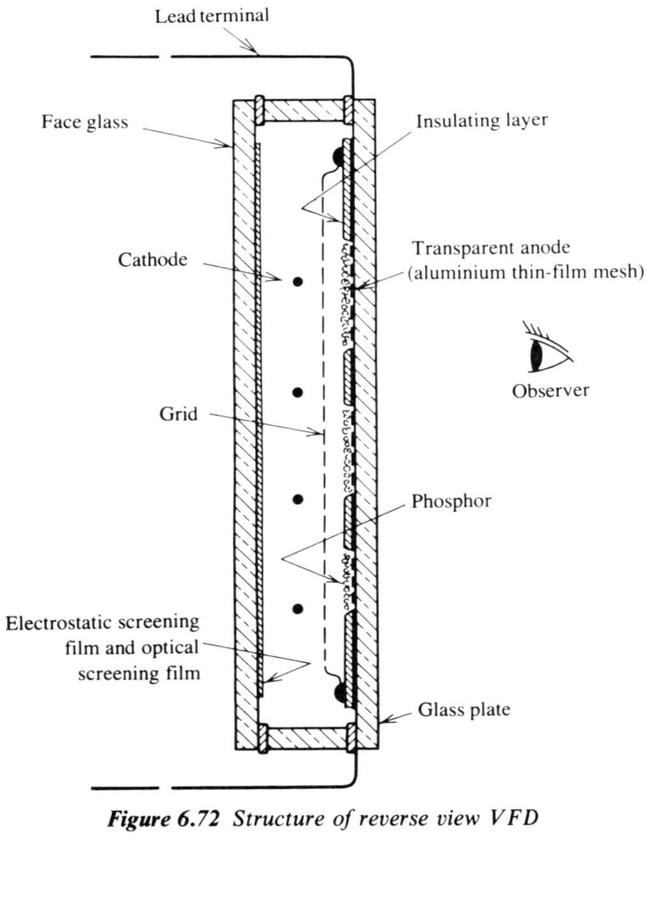

Figure 6.72 *Structure of reverse view VFD*

due to the transmissivity falling more than the amount of light emission. When a reverse view VFD is operated under the same conditions, a luminance of only 60 to 70 per cent that of the conventional type is obtained. The construction of a reverse view VFD is shown in Fig. 6.72.

6.6 DISPLAY MODULE USING A VACUUM FLUORESCENT DISPLAY

Dot character display VFDs or dot matrix display VFDs have found many applications and have made a considerable contribution in the field of information processing technology. However, driving these devices is not easy. Specifically, providing a large number of drivers that can withstand high voltages, and dealing with character generation, interfaces, and power sources, etc., waste time in troublesome design problems and require fairly specialist knowledge. Various types of display module are therefore being developed that can be provided for practical use and that incorporate VFD drivers, grid scanners, anode data latches, character generators, refresh RAMs, interface circuits, and d.c./d.c. converters. They can be used in many fields simply by connecting them to a + 5 V power source and host system.

6.7 OUTLOOK

It appears that great improvements in segment display VFDs are unlikely. However, it is confidently expected that their range of applications will be expanded by adding various display functions, increasing the amount of information by the use of multiple rows and partial use of multiple colours.

There is a tendency for the size of dot character display VFDs to increase, due to an increase in the amount of information. Further increase in the display capacity is required to approach that of a CRT. This will require scaling-up components, materials, and VFD fabrication processes.

Dot matrix display VFDs give most freedom of application. 320 × 240 dot displays have already been developed and progress is being made towards development of full-colour displays; at present two colours can be employed. To achieve full-colour display, development of drivers able to withstand high voltage will be required.

In VFDs for bar displays, composite displays, and fixed-image displays, function and display are being improved by miniaturizing the display pattern and using more colours. However, more uniform luminance balance between the light-emission colours is required.

Increased use of modularization of VFDs and accelerated implementation of active matrix display VFDs or reverse view VFDs are needed. In particular, modularization of dot matrix display VFDs will be important in expanding the range of application of VFDs.

BIBLIOGRAPHY

Masuda: Nishoku hyoji dot-matrix keiko hyoji kan (Two-colour dot-matrix VFD), *Eizo Joho (Image Information)*, **15**(24), 45–50.
Miyaji: Taito suru paneru gata display (The leading types of panel display), *Eizo Joho (Image Information)*, **12**(4), 27–32.
Sakamoto: Denshikan yo keikotai to sono oyo (Phosphors for electronic tubes and their application), *Electronics Ceramics '80*, Winter Issue, pp. 65–9.
Sakamoto: Denshikan yo keikotai to sono oyo (2) (Phosphors for electronic tubes and their application), *Electronics Ceramics '80*, Spring Issue, pp. 65–70.
Uchiyama, Nakamura, Nakashima, Masuda, Shimojo and Kiyozumi: Nishoku graphic keiko hyoji kan (Two-colour Graphic VFD), Terebijon Gakkai Hokoku (Reports of the TV Association) FD 741, IPD 82-2.

REFERENCES

[1] Hyojikan gijutsu iinkai hen (compiled by Display Technology Comittee): Keiko hyojikan no tsukaikata (The use of VFD), (Sha) Nihon Denshi Kikai Kogyo Kai (Electromechanical Industries Association of Japan) (1977-3).
[2] Morimoto: Keiko hyojikan no doko (trends in VFD), Denshi Kogyo Gijutsu Daikai (General Conference of the Association of Electronic Industrial Technology), p. 1c-2b-2(1981).
[3] Morikawa, Hattori, and Shimojo: Al hakumaku niyoru keiko hyojikan yo kiban no sakusei hoho (Method of manufacturing substrate for VFDs using Al thin film), Ise Denshi Giho (Ise Electronics Reports), No. 3, pp. 10–18.
[4] Hayashi: aka/midori/ao nado no teisoku denshi sen yo keikotai no kenkyu susumu (Advances in slow electron phosphors for R/G/B etc.), Nikkei Electronics, pp. 60–6 (1978-12).
[5] Nakamura and Kiyozumi: Keiko hyojikan to sono oyo (VFD and their applications), Nikkan Kogyo Shinbun Sha, Sho. 52-11 (1977-11).
[6] Morimoto: Keiko hyojikan to sono kaihatsu doko (Trends in the development of VFDs), Denshi Zairyo (Electronic Materials), pp. 66–73 (1977-11).
[7] Ito: Kiso denshi kogaku (Basic electronic engineering) (I), p. 172, Shokoto.
[8] S. Uemura, U.S. Patent 408/716.
[9] S. Uemura and K. Kiyozumi: *IEEE Trans. Electron Device*, **ED-28**, 749–55 (1981).
[10] S. Uemura: *The Electrochemical Society Extended Abstracts*, Vol. 83-2, Washington, D.C. 9–14 Oct., pp. 719–20 (1983).
[11] Kasano, Masuda, Simojo, and Kiyozumi: A 240-character vacuum fluorescent display and its drive circuitry, *Proc. SID*, **2**(1/2), 107–11 (1980).
[12] Iwade and Kasano: 240 moji keiko hyoji kan no kudo hoho (Method of driving a 240 character VFD), Ise denshi giho (Ise Electronics Reports), No. 3, pp. 10–18.
[13] Ise denshi ga kanji hyoji mo kano na atarashii keiko hyokan no kaihatsu (Development of a novel VFD capable of displaying Chinese/Japanese characters by Ise Electronics), Denshi Zairyo (Electronic Materials), p. 19 (1979-10).
[14] Iwade, Kasano, Masuda, and Nakamura: Vacuum fluorescent display for TV video image, SID International Symposium, *Digest of Technical papers*, **XIII**, 136–7 (1981).
[15] Tamura: Keiko hyojikan no kaihatsu jokyo to oyo tenkai (Development and Application of VFD), Denshi Zairyo (Electronic Materials) pp. 60–70 (1980-02).
[16] NE Report: 128 × 128 gaso no graphic hyoji yo keiko hyokan no kaihatsu (Development of 128 × 128 pixel graphic VFD), Nikkei Electronics, pp. 130–138 (1982-3).

[17] Uchiyama and Masuda: Komitsudo X-Y matrix keiko hyojikan (High density X-Y matrix VFD), Terebijon Gakkai Gijutsu Hokoku (Technical Reports of Television Association), Sho. 56-11 (1981-11).

[18] Uchiyama, Nakamura, Nishioka, Uemura, and Masuda: DM 320 × 240 B, No Syoukai (introduction of DM320 × 240B) Ise Denshi Giho (Ise Electronics Technical Reports), No. 8.

[19] M. Uchiyama, T. Nakamura, K. Nakajima, S. Uemura, M. Masuda, T. Shimojo, and K. Kiyozumi: P.1.3 Two color vacuum fluorescent image panel, Proc. 3rd International Display Research Conference, Kobe, 3–5 Oct., pp. 116–19 (1983).

[20] Nakanishi, Iwade: DM 320 × 240 B drive circuit, Ise Denshi Giho (Ise Electronics Technical Reports), No. 8.

[21] S. Uemura and K. Kiyozumi: Institute of TV Engineers of Japan Report IPD-46-15, pp. 91–6 (1979) (in Japanese).

[22] S. Uemura, K. Kiyozumi, M. Yoshimura, K. Fujii, Y. Torimaru, R. Miyake, and T. Nakamura: 7.2 MOS addressed image display on a chip, Proc. 3rd International Display Research Conference, Kobe, 3–5 Oct., pp. 238–41 (1983).

[23] K. Morimoto: On front luminous vacuum fluorescent display. *JEE*, pp. 46–8 (1983-10).

[24] Okubo, Uemura, and Masuda: Backview gata keiko hyojikan (Backview type VFD), Ise Denshi Giho (ISE Electronics Technical Reports), No. 8.

[25] Kasano: Keiko hyojikan display module no tsukaikata (Use of VFD display module), *Denshi Gijutsu (Electronic Techniques)*, **24**(3), 97–103.

Chapter 7

Light-emitting diode displays (LEDs)

7.1 INTRODUCTION

7.1.1 What are light-emitting diodes?

Light-emitting diodes (LED) are solid state functional devices that convert electricity to light. They are photon-emitting semiconductor elements that utilize the radiative recombination phenomenon* which occurs when minority carriers are injected into a *pn* junction made up of *p*-type and *n*-type semiconductor crystals arranged adjacent to each other. By changing the semiconductor material used, the wavelength of the emitted light can be altered. 'Visible LEDs', which emit light in the visible region, are mainly used for display purposes whilst 'IR LEDs', which emit in the infrared region, are used for optical communication and other forms of information transmission or processing.

The advantages of LEDs are (a) they provide small light sources of high reliability, (b) they require a low drive voltage (about 2 V) and so are readily integrated in semiconductor drive circuits, and (c) they are capable of high-speed modulation.

The advantages of visible LEDs when used as display elements lie in the fact that they consist of small solid state active light source elements that can be endowed with special features not shown by other types of light-emitting display devices such as CRTs, PDPs, VFDs, electroluminescent panels, and the like. One drawback of LEDs is the fact that power consumption per element is rather large, ranging from several milliwatts to several tens of milliwatt, but there is a considerable improvement in power consumption compared with conventional tube or bulb type display units, and since LEDs are also greatly superior from the point of view of reliability, they have come to occupy a well-established position as electronic components.

In both visible LEDs and IR LEDs, light emission in response to current

*Also known as injection electroluminescence

injection occurs at the *pn* junction face. The *pn* junction face normally has an area of about 0.3 mm square, and in the case of LEDs used for optical communication this corresponds to about 1/50 of the overall surface area. These LEDs are in all cases small single-crystal semiconductor elements utilizing substantially uniform light emission from a surface area of given size. Various types of LED devices are produced which make use of this surface light emission.

Figure 7.1 illustrates some of the more common types of LED used in displays

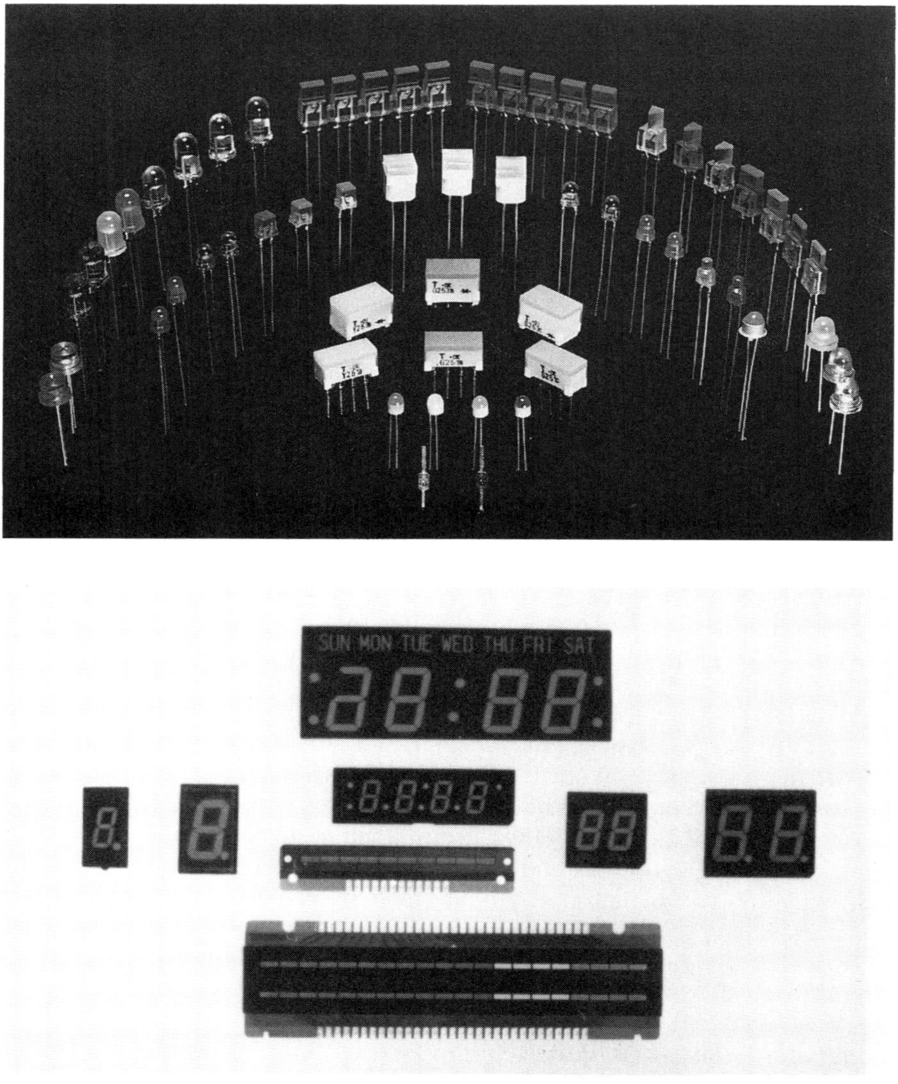

Figure 7.1 *Types of LED most widely used for displays*

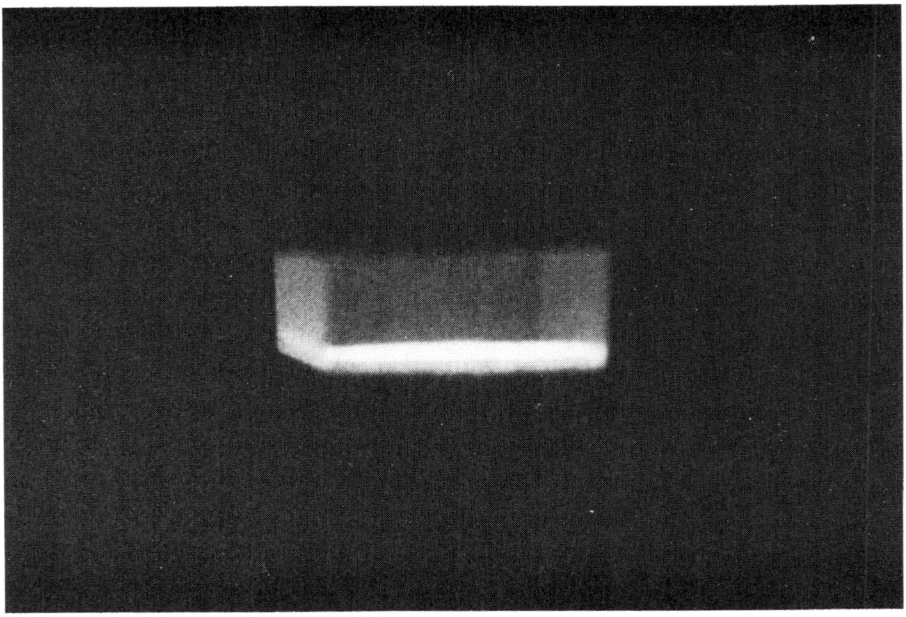

Figure 7.2 Light-emitting semiconductor crystal LED

and Fig. 7.2 shows the appearance of light emission from a semiconductor crystal. In Fig. 7.2, the LED single-crystal die is seen from the side, clearly showing that the light emission takes place from the *pn* junction. The pellet has approximate dimensions $0.3 \times 0.3 \times 0.2\,mm^3$.

7.1.2 History of the development of light-emitting diodes

The development of LEDs started in 1923 when Lossew observed the relationship between the rectifying action and light emission at the *pn* junction of silicon carbide (SiC)[1]. However, an accurate interpretation of injection electroluminescence was first given in a report in 1955 by Haynes[2] relating to the *pn* junction of germanium. Basic research was carried out by Welker[3] who indicated the possibility of using compounds of elements of groups III–V of the periodic table, such as gallium phosphide (GaP), gallium arsenide (GaAs), and so on, as semiconductors. Electroluminescence was first observed in GaP[4] and it was also observed that GaAs emits light with high efficiency[5]. Techniques for growing crystals of group III–V compound semiconductors and for utilizing these to produce electronic components were developed in the 1960s.

Against the background of general use of LEDs as semiconductor elements, the development of techniques for the mass production of visible LEDs for display units made rapid progress at the start of the 1970s. As a result of technical developments aimed at obtaining methods for synthesizing compound semiconductor materials, for producing single crystals, obtaining *pn* junctions, and so on, visible LEDs utilizing semiconductors consisting of compounds of elements of groups III–V of the periodic table are widely used. However, although the wavelengths of the light emitted by visible LEDs extend from the red to the green region, no blue LED has so far been perfected. For practical utilization of a blue LED, new materials, different from the group III–V compound semiconductors, will also have to be used. Development of the basic techniques is already in progress.

In the present decade, advances in the communications field have been accompanied by rapid advances in data-processing techniques involving the use of light. Broadening of the wavelength region of LEDs and improving their performance for the purpose of providing vital components for interfaces between human operators and machines or between machines and other machines will continue.

The principles of operation of LEDs, materials and techniques used in their fabrication, and their applications will be discussed below.

7.2 OPERATING PRINCIPLES AND CHARACTERISTICS OF LIGHT-EMITTING DIODES

Figures 7.1 and 7.2 show photographs illustrating the nature and appearance of LEDs. Figure 7.1 shows LED displays used for audio equipment, electrical appliances, and automatic ticket-vending machines for stations and the like. It can be seen that they are of different forms and appearance, but they all function by utilizing the light emission from small semiconductor crystals of about 0.3 mm square, as shown in Fig. 7.2. When used in pilot lights, LED dies having an area of 0.3×0.3 mm are moulded with transparent resin as illustrated in Fig. 7.3 (a), whilst in the case of a digital display element, one LED die (or several, depending on the particular case) of 0.3×0.3 mm area is mounted surrounded by a reflector plate as shown in axial section in Fig. 7.3 (b). This small (0.3×0.3 mm) crystal is light-emitting diode consisting of a *p*-type region showing p-type conductivity and an n-type region of the opposite conductivity type.

7.2.1 Light-emission mechanism and wavelength of emitted light

The luminescence of semiconductor crystals is produced by radiative recombination between two levels related to the energy states of electrons, as shown in Fig. 7.4. Accordingly, the rate of this recombination is determined by the product of the density of electrons in the high energy level E_2, the density of empty states in the low energy level E_1, and the probability of a transition between these two levels. Since thermal equilibrium is expressed approximately by the Boltzmann distribution, the relationship between the electron densities n_1 and n_2 at the two

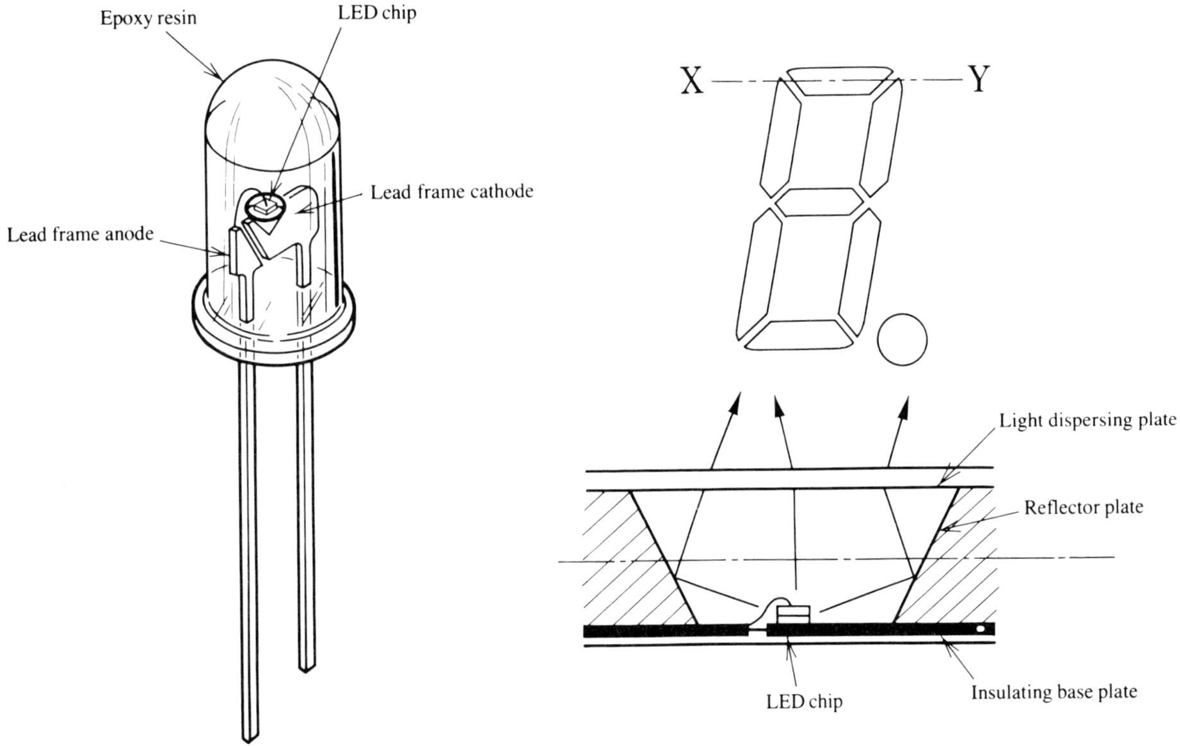

(a) Pilot lamp

(b) Digital display element

***Figure* 7.3** *Construction of an LED display*

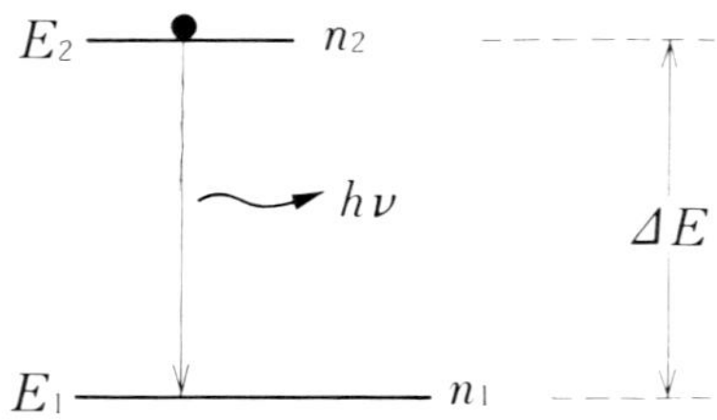

Figure 7.4 *Radiative recombination between two levels*

levels is expressed by the following equation:

$$\frac{n_2}{n_1} = \exp\left(\frac{-\Delta E}{kT}\right) \tag{7.1}$$

where k is the Boltzmann constant ($= 1.38 \times 10^{-23}$ J/K), T is the absolute temperature, and ΔE is the energy difference between the two levels.

If the electron density in the E_2 level becomes excessive for some reason, seen from the point of view of thermal equilibrium, the system will try to emit light and return to thermal equilibrium. This is known as radiative recombination. When this occurs, the Bohr condition, expressed by the following relationship between the frequency ν of the emitted light, its wavelength λ, and ΔE is established:

$$\nu = \frac{c}{\lambda} = \frac{\Delta E}{h} \tag{7.2}$$

where c is the velocity of light and h is Planck's constant ($= 6.625 \times 10^{-34}$ J·s).

Thus it can be seen that from the point of view of thermal equilibrium, some process that excites the substance to a higher energy state is essential for producing luminescence. A well-known excitation process of this type is light absorption. In the case of a semiconductor crystal, the luminescence produced by photoexcitation is referred to as photoluminescence. Cathode luminescence, which is produced by excitation by cathode rays, is a similar phenomenon. In the case of semiconductor crystals, this type of excitation state is characteristically obtained by injection of electrons or positive holes at a *pn* junction. The light emission of an LED is obtained by this process.

In actual semiconductor crystals, the gaps between adjacent atoms are small (approximately 5 Å) so the wave functions of the electrons of the different atoms interact, forming energy bands having a certain width. As a result of this, the luminescence given off by the semiconductor crystal does not produce a line spectrum but instead has a width, generally of several hundred ångströms. Apart from transitions between the conduction band and the valence electron band, transitions that proceed via impurity levels formed between these two bands* also place an important role in the light-emission process in a semiconductor

*The region between these bands is known as the forbidden band.

crystal. As discussed above, the wavelength of the luminescence light from a semiconductor crystal is determined by the width of the forbidden band of the semiconductor crystal (band gap: E_g) and the depth of the impurity level (luminescence centre) in the forbidden band, as can be seen from equation (7.2).

Rewriting equation (7.2) we obtain

$$\Delta E(\text{eV}) = \frac{1.23981}{\lambda\,(\mu\text{m})} \tag{7.3}$$

From equation (7.3) it will be evident that in order to obtain a visible LED ($\lambda < 0.7\ \mu$m) the semiconductor crystal requires a forbidden bandwidth of at least 1.77 eV and to produce a blue LED ($\lambda < 0.49\ \mu$m) we require a semiconductor crystal with a forbidden bandwidth in excess of 2.53 eV.

7.2.2 Current injection and light emission

When a voltage is applied to a *pn* junction in the forward direction, i.e. a positive voltage in the case of a *p*-type crystal and a negative voltage in the case of an *n*-type crystal, as seen in Fig. 7.5 (b) and (c), the potential of the electrons in the *n* region rises, and the height of the potential barrier that hitherto prevented electrons from flowing out into the *p* region is reduced. Therefore, electrons diffuse (are injected) into the *p* region and positive holes into the *n* region, so that, (as shown in Fig. 7.5 (d), excess of electrons and positive holes are present in these regions relative to the numbers that would be present under thermal equilibrium. These excess electrons and positive holes are the minority carriers produced by injection.

The number of these minority carriers are reduced by recombination with majority carriers in the respective regions, and a region (indicated in Fig. 7.5(d)) determined by the diffusion lengths L_e and L_h determined by the recombination rate and the diffusion constant becomes the principal light-emitting region. This diffusion length is of the order of 1 to 2 μm. Assuming that the lifetime of the minority carriers is 10^{-9}s, passage of a current of 10 A/cm^2 through an LED will result in an excess carrier density of about 10^{15} cm^{-3}. This is much larger than the density of minority carriers under conditions of thermal equilibrium and produces radiative recombination.

The forward current flowing as a result of applying a voltage in the forward direction is the ordinary diode current, which is expressed by the following equation:

$$I = I_0\left[\exp\left(\frac{eV_j}{\beta kT}\right) - 1\right] \tag{7.4}$$

where V_j is the voltage applied to the *pn* junction and has a value obtained by subtracting the voltage drop produced by the series resistance of the ohmic electrode portion from the forward voltage applied to the diode. The value of β is

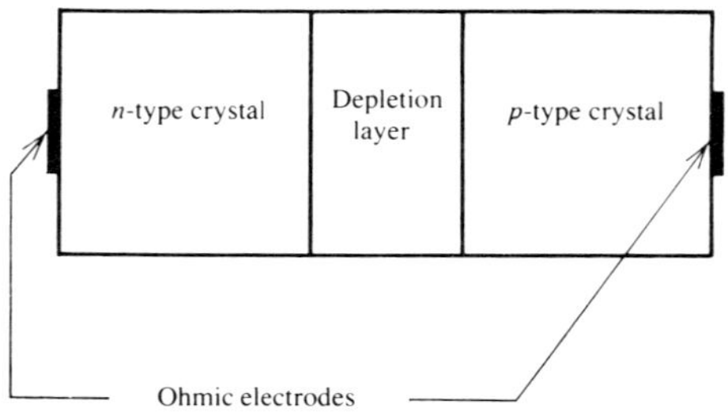

(a) Cross-section of diode structure

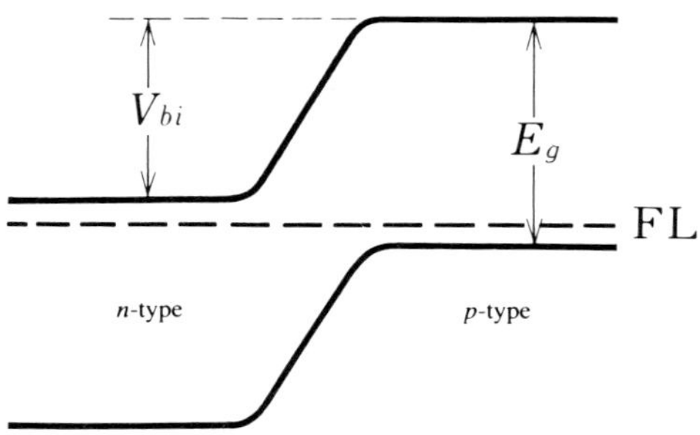

(b) Band structure when no voltage has been applied

E_g: Forbidden bandwidth

V_{bi}: Internal potential difference at *pn* junction

n_e: Electron density

n_h: Positive hole density

L_h: Positive hole diffusion length

***Figure** 7.5 Schematic diagram of pn junction*

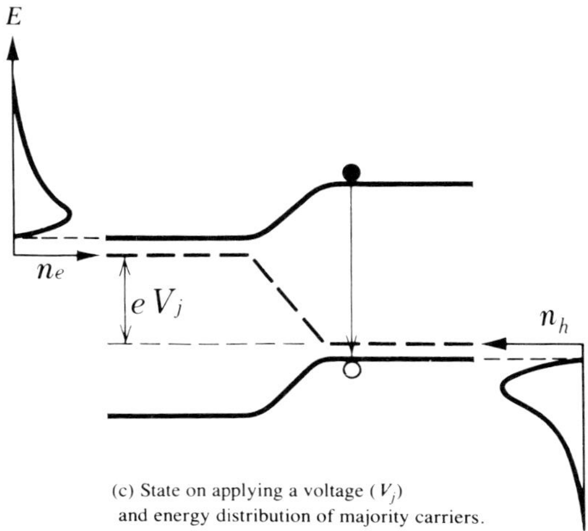

(c) State on applying a voltage (V_j) and energy distribution of majority carriers.

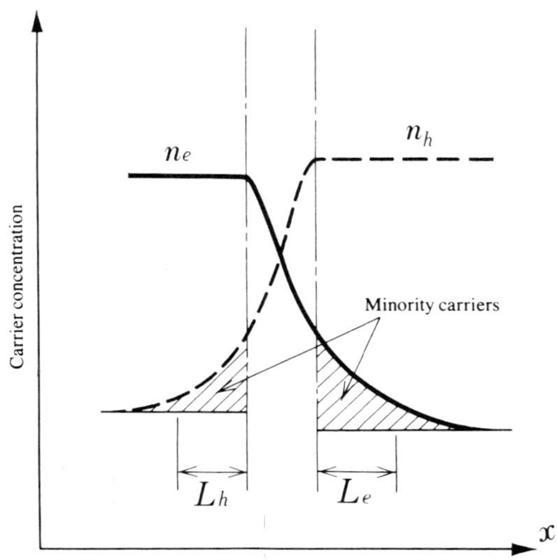

(d) Spatial distribution of majority carriers and minority carriers in (c)

L_e: Electron diffusion length

FL: Fermi level

E: Energy

V_j: Voltage applied to junction

***Figure** 7.5 (contd.)*

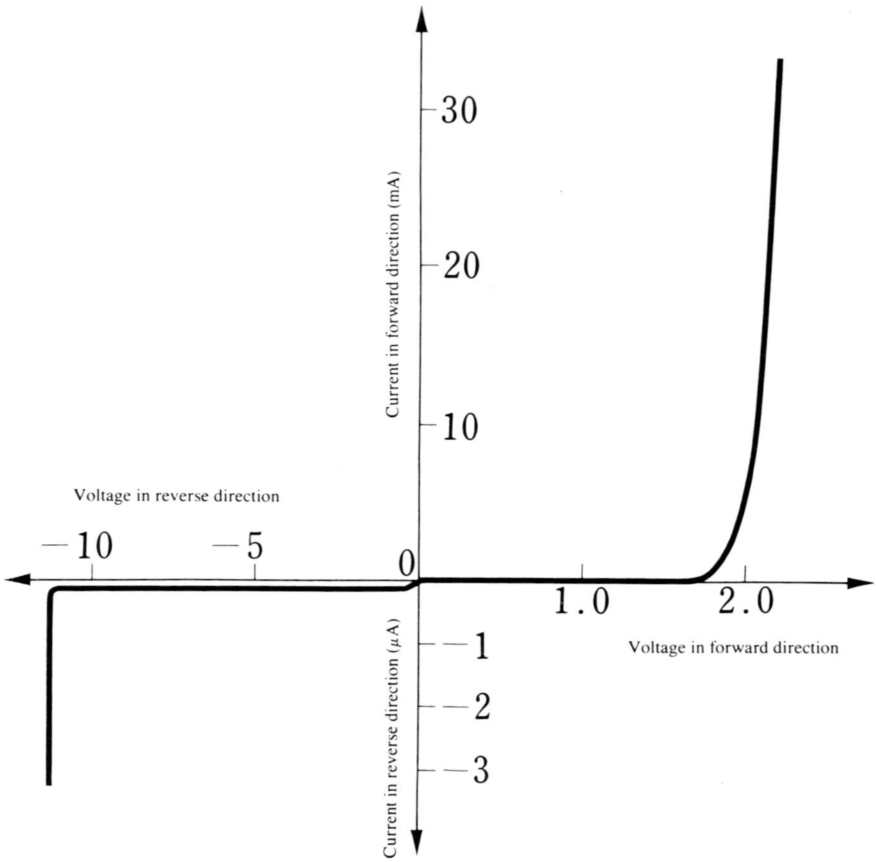

Figure 7.6 *Current–voltage characteristics of LED*

1 for the diffusion current component and 2 for the recombination current in the depletion region[6]. Therefore, for the region of low applied voltage, in which the depletion region recombination current is predominant, β approaches 2, whilst for the region of higher applied voltage, in which the diffusion current is predominant, β approaches 1. The recombination occurring in the depletion region proceeds via deep non-radiative recombination centres produced at the *pn* junction interface and does not in general contribute to light emission. Radiative recombination, therefore, has an intensity that is proportional to the diffusion current component. At low injection levels, the relationship between the intensity of the emitted light (L) and the current in the forward direction (I) is given by $L \propto I^2$, whilst at high injection levels this becomes $L \propto I$. Furthermore, at high injection levels, the influence of evolution of heat reduces the extent of light emission.

When using LEDs, it is necessary to consider first of all the current–voltage characteristic. Figure 7.6 shows a typical example. When a voltage is applied in the forward direction, it is found that when this applied voltage approaches the internal potential difference V_{bi} of the *pn* junction, the current in the forward direction increases rapidly. Accordingly, a constant input power is required to operate a LED. To achieve this, a resistance in series is normally added to a LED in order to operate the device by a constant drive.

7.2.3 Luminous efficiency

One of the quantities used to standardize the performance of an LED is its luminous efficiency. This is divided into two types of efficiency, quantum efficiency and power efficiency. The first named is expressed by the number of photons emitted in unit time divided by the current charge number (equation 7.5) and the last named is given by the power of the light output divided by the input power (equation 7.6):

$$\text{Quantum efficiency } \eta_q = \frac{\text{number of output photons (number } n/\text{s)}}{\text{current charge number } (A/e)}$$

$$= \int \frac{n(E)\mathrm{d}E}{I} \tag{7.5}$$

$$\text{Power efficiency } \eta_p = \frac{\text{light output power}}{\text{input power}}$$

$$= \int n(E) \frac{E\,\mathrm{d}E}{VI} \tag{7.6}$$

where $n(E)$ is the number of photons of energy E emitted in unit time. In the case of LEDs, the energy E of the emitted light is substantially equal to the applied voltage V and there is, therefore, no significant difference between quantum efficiency and power efficiency.

We shall now discuss the factors that determine the quantum efficiency of an LED. In general, the quantity referred to as the quantum efficiency of an LED is its external quantum efficiency. Even though light emission takes place with good efficiency in the interior of a semiconductor crystal, there remains the problem of a satisfactory emission of this light to the outside of the crystal. As seen from the following equation, the external photon efficiency η is expressed by the product of the internal photon efficiency η_{int} and the efficiency η of light extraction:

$$\eta_{ext} = \eta_x \, \eta_{int} \tag{7.7}$$

If we assume that the region of light emission of an LED is a diffusion current region of either *p* or *n*, η_{int} can be expressed as the product of injection efficiency γ of minority carriers into the light-emission region and the internal quantum efficiency η $(p \text{ or } n)_{int}$. The injection efficiency γ_p of positive holes into an *n* region and the injection efficiency γ_n of electrons into a *p* region are given respectively by the following equations[7,8]:

$$\gamma_p = \frac{\mu_e L_n N_A}{\mu_p L_n N_A + \mu_n L_p N_D} \tag{7.8}$$

$$\gamma_n = \frac{\mu L_p N_D}{\mu_p L_n N_A + \mu_n L_p N_D} \tag{7.9}$$

where μ_p and μ_n represent the mobilities of the positive holes and of the electrons respectively, L_n and L_p are the diffusion lengths of electrons in the p region and of positive holes in the n region and N_A and N_D are the acceptor and donor concentrations in both the p and n regions. Thus, the external quantum efficiency of an LED is governed by various factors. Requirements may be summarized as follows: (a) a structure permitting the injection of minority carriers into the light-emission region, (b) crystal properties that enable the injected carriers to be converted to light efficiently, and (c) the structure of the element must be such that the converted light can be efficiently extracted outside the crystal, this being important for increasing the external quantum efficiency.

(1) *Internal quantum efficiency*

As factors determining the internal quantum efficiency of a semiconductor crystal, we shall consider the band structure of the crystal, the radiative recombination centres, and the non-radiative recombination centres. The band structures of semiconductor crystals are essentially of two types, the direct transition type and the indirect transition type. A representative example of the first-named type is GaAs, whilst GaP is an example of the second type. Their band structures are illustrated in Fig. 7.7.

Thus, if we plot the energy levels of the conduction band and of the valence electron band in momentum space, we find that whilst in the case of GaAs crystals the momentum in the lowest part of the conduction band coincides with that in the uppermost part of the valence electron band, in the case of GaP there is some distance between these two. Therefore, in interband transitions in GaAs, momentum is preserved, so these are permitted transitions, whilst in GaP they are forbidden transitions. Thus, in the case of recombination in GaP, a recombination process must be adopted that relies on the assistance of lattice vibrations once the impurity level has been passed. For this reason, transitions in the former case are referred to as direct transitions and in the latter as indirect transitions. Indirect transitions are secondary recombination processes so the probability for the occurrence of such transitions is generally small. Therefore, use of direct transition type crystals for producing LEDs is to be preferred. In practice, it is found that the internal quantum efficiency of direct transition type crystals is close to 100 percent [8].

However, bright LEDs are also produced from GaP indirect transition type crystals. This is the result of utilizing suitable luminescence centres. Nitrogen atoms added to a GaP crystal replace phosphorus atoms at their lattice sites. The

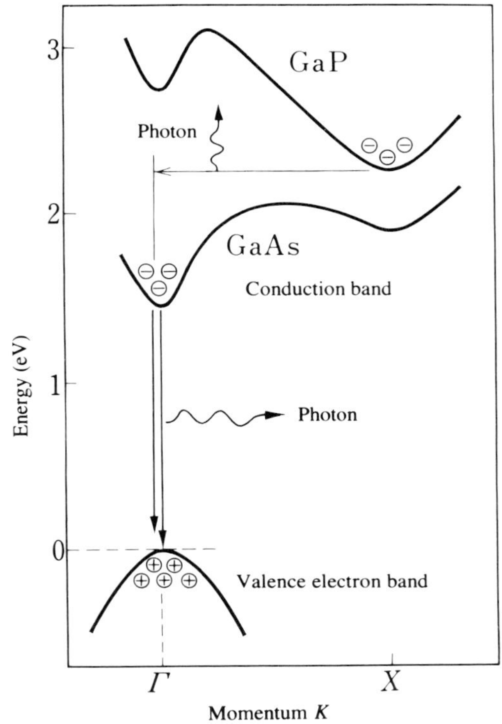

Figure 7.7 Band structure of GaAs and GaP crystals

nitrogen atoms which have taken the place of phosphorus atoms are electrically neutral, but since their electronegativity is greater than that of the phosphorus atoms, this substitution gives rise to a well type potential, referred to as an isoelectronic trap. The wave function of the electrons caught in this 'trap' is extended up to the τ point in momentum space (momentum position at the top of the valence electron band). The nitrogen atom trapping an electron becomes negatively charged, as a result of which a positive hole is also trapped. This electron–positive hole pair constitutes an exciton trapped by the nitrogen atom (a bound exciton).

Since the transition process involving recombination of this exciton has a direct transition component, the probability for its occurrence is high compared with that of recombinations proceeding via ordinary impurity centres [9]. The energy level created by the nitrogen atom is very slight (about 10 meV) for both electrons and positive holes [10], so its light emission reflects the forbidden band width of the GaP crystal (2.26 eV) substantially as such, resulting in emission of green light (5650 Å). This is the mechanism involved in the light emission of green LEDs consisting of GaP:N. Utilization of this type of excellent luminescence centre has reached the stage of practical application.

Zn–O complexes, formed from the closest bonded pairs of zinc and oxygen atoms when these two elements are added to a GaP crystal, also form isoelectronic traps of the same kind as those produced by nitrogen atoms. In this case, however, the energy level in respect of electrons is deeper and the light emitted as a result of recombination is red (7000 Å) For such GaP:ZnO red LEDs, the internal quantum efficiency is estimated at 35 percent [11].

An important factor in maintaining internal quantum efficiency, which cannot be disregarded, is the presence of non-radiative recombination centres. Recombination of minority carriers may be considered to comprise both radiative and non-radiative recombination processes. Denoting the corresponding recombination time constants by τ_r and τ_n, we obtain the following expressions for the lifetime τ_{mc} of the minority carriers and the quantum efficiency η:

$$\frac{1}{\tau_{mc}} = \frac{1}{\tau_r} + \frac{1}{\tau_n} \tag{7.10}$$

$$\eta = \frac{\tau_r^{-1}}{\tau_{mc}} = \frac{\tau_n}{\tau_r + \tau_n} \tag{7.11}$$

It is thus evident that in order to achieve a high internal quantum efficiency, it is especially important to reduce the number of non-radiative recombination processes and so increase the value of τ_n. Various types of crystal defects acting as non-radiative recombination centres include point defects such as vacant lattice sites [12] or impurities [13], linear defects such as dislocations [14], and planar defects such as dislocation loops [15], and so on.

(2) Efficiency of light extraction

Semiconductor crystals generally have a high refractive index. For instance, in the case of GaAs and FaP, its value is about 3.5. The critical angle within which light can be emitted from the interior of the crystal to outside is thus only 16°, which is very small. Furthermore, semiconductor crystals also have a certain absorption coefficient for the light which they have themselves emitted. As a result, practically all of the light generated in the interior of the crystal is attenuated by multiple reflections, without being emitted from the crystal. Direct transition type crystals have a particularly high absorption coefficient for the emitted light, since such absorption takes place by an interband transition. The light extraction efficiency η_x is therefore small, having a value of only about 4 percent. From this point of view, indirect transition type crystals give better results, compensating for their low internal quantum efficiency mentioned above. In the case of GaP:N, which is a typical example of an indirect transition type crystal, the light extraction efficiency is estimated at about 25 percent [16].

Raising the light extraction efficiency by some means is important for obtaining bright LEDs. Various means have, therefore, been adopted, e.g.

moulding LED crystals with a transparent resin or glass having a high refractive index, in order to ensure that the angle of launching of the light from inside the LED crystal to the outside lies within the critical angle or using dome-shaped semiconductor crystals [17], and so on.

7.2.4 *Light output and brightness*

The light-emission spectra of LEDs produced from different types of materials are shown in Fig. 7.8. The light-emission spectrum of GaAs ($E_g = 1.43$ eV) LEDs that have a small forbidden bandwidth has an emission peak in the vicinity of 900 nm. This is infrared light which cannot be seen by the human eye. Combining this type of IR LED with a photodetector yields devices used in optical communication and optical data processing. On the other hand, materials with a broad forbidden bandwidth, e.g. GaP ($E_g = 2.26$ eV) LEDs, give green or red LED materials.

Depending on whether a LED yields visible or infrared light, different photometric units are generally used. However, one unit used for both infrared and visible light is the watt (W). The radiant flux per unit solid angle is expressed by the radiation intensity (W/sr) and the amount of radiant flux per unit area is expressed by the radiant luminance ($\mathrm{W/st \cdot m^2}$) [8]. Characteristic of visible light photometry is the use of the human eye as a photodetector. The human eye has a spectral sensitivity characteristic, which is standardized by means of the standard spectral luminous efficiency curve (cf. Section 1.3 [3]).

The photometric unit used for LEDs will depend on their intended application. Depending on this, an optical system is installed at the front of the LED and in this case the spatial distribution of the radiant flux is not uniform. The photometric method and units used will vary, depending on whether it is intended to determine the energy radiated over the whole space or energy radiated into a particular region, or whether the human eye is used as a photodetector. In the case of visible LEDs used for display purposes, for pilot lamp type devices luminous intensity is generally used. For character displays in which the light-emitting part is a uniform diffusion surface, fL is used.

To quote a practical example, the power radiated to the whole of space by an LED capable of being approximated to by a line spectrum is expressed by the following equation:

$$W = \frac{I}{e} \times \eta \times E = \eta I E \qquad \mathrm{W} \tag{7.12}$$

where I is the current (A) passing through the LED, e is the unit charge, η the external quantum efficiency of the LED, and E the energy of the emitted light (eV). The conversion from watt to lumen can be made using the standard spectral luminous efficiency curve, so the luminous flux (lm) radiated from the LED into the whole of space can be found. The brightness in terms of candela (cd) units is found by taking into account the spatial intensity distribution. Thus the

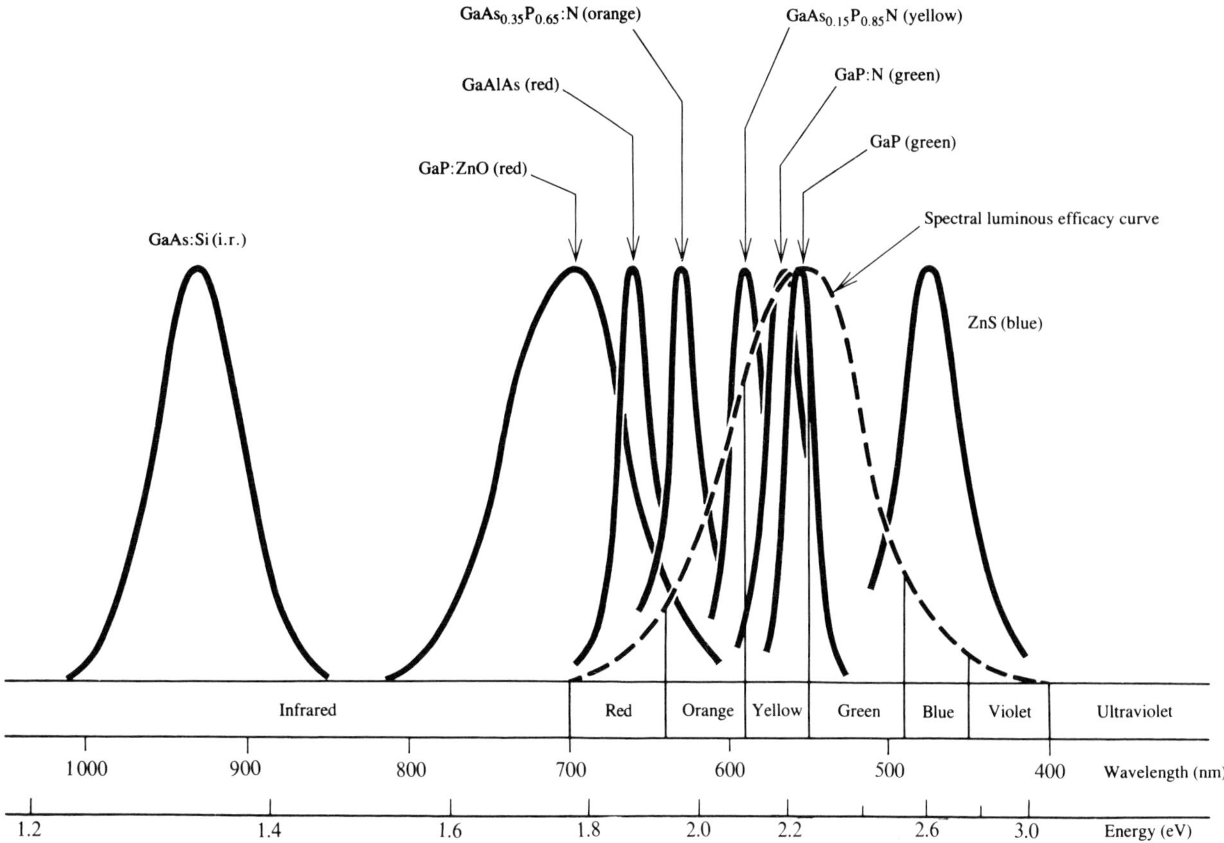

Figure 7.8 *Light-emission spectra of different types of LED*

brightness of an LED whose quantum efficiency has been measured is calculated in terms of candela units by measuring the energy of the emitted light, the current passed, and the spatial distribution of the emitted light.

Since an actual LED shows a certain spectral line width, strictly speaking it is necessary to carry out an integration at each stage of the above calculation. Basically, the light output and brightness of an LED are estimated by the above methods, in combination with different types of photometric instruments.

7.2.5 Modulation characteristic

Directly modulating the light-emission intensity by modulating the current passed through an LED is important in practical applications. The upper limit f_c of the modulation frequency of an LED is governed by the minority carrier lifetime (τ_{mc}). Their relationship is expressed by the following equation:

$$f_c = \frac{1}{2\pi\tau_{mc}} \tag{7.13}$$

For GaAs, which is a direct transition type crystal, τ_{mc} normally has a value of 1 to 10 ns. Therefore, when using GaAs, high-speed modulation within range of 16 to 160 MHz is possible. However, with many visible LEDs, consisting of indirect transition type crystals, τ_{mc} is made as large as possible in order to obtain good light-emission characteristics, so their modulation frequency is lowered to a range of several hundred kilohertz to several megahertz. Similar phenomena are also observed in practice in LEDs produced from direct transition type crystals, and in order to produce bright LEDs with rapid response, the structure of the element must be modified in various ways. An example of this is a double heterostructure similar to that used in semiconductor lasers (cf. Fig. 7.9). The lifetime of the minority carriers injected into the *pn* junction is given by the following equation[18]:

$$\tau_{mc} = \frac{1}{B(p_0 + \Delta n)} \tag{7.14}$$

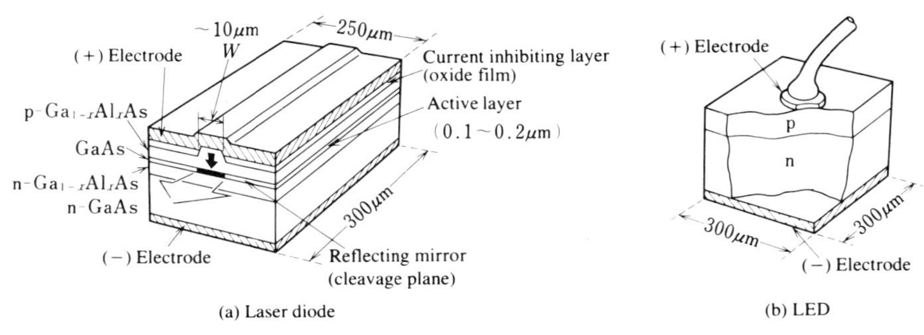

(a) Laser diode (b) LED

***Figure** 7.9 Comparison of the structures of laser diode and LED*

where B is the transition probability, p_0 the concentration of majority carriers in the thermal equilibrium state, and Δn the excess concentration of minority carriers produced by injection. The response can be made more rapid by increasing the concentration of majority carriers in the light-emission region or by increasing Δn.

In the double heterostructure the injected minority carriers are confined by the potential barrier into a region thinner than the diffusion length. This increases Δn and so improves the device response. Combining equation (7.14) with equation (7.10), we obtain the relationship

$$B = B_r + B_n$$

where B_r is the probability for a radiative transition and B_n the probability for non-radiative transition. From this it is evident that introduction of a non-radiative transition recombination process is effective in raising the frequency characteristic of an LED. However, in this case, it is necessary to select a region in which lowering of the luminous efficiency does not pose any problems.

7.2.6 *Light-emitting diodes and laser diodes*

Characteristic of the light emitted by LEDs is that its phase is random in time, i.e. spontaneously emitted light is utilized. On the other hand, resonance absorption and stimulated emission processes occur between the two energy levels E_1 and E_2 shown in Fig. 7.4. Denoting the numbers of electrons present in these respective energy states by n_1 and n_2, resonance absorption will be proportional to n_1I and stimulated emission to n_2I, where I is the intensity of the incident light. Under normal thermal equilibrium conditions $n_2 < n_1$ so incident light undergoes attenuation during its passage through the device. However, if it is possible to establish the condition $n_2 > n_1$ (population inversion), absorption will be exceeded by stimulated emission, giving rise to a light amplification effect. By setting up a stationary wave in a resonator made up of two mirrors and using this state of population inversion to obtain amplification of the light oscillations, we obtain a laser.

In a *pn* junction structure that is basically the same as that of an LED, oscillations can be obtained by expedients such as choice of a suitable type of semiconductor crystal and restricting the current injection region. Such a device is the laser diode. In practice, a laser diode is obtained by using a direct transition type material with a high transition probability, such as GaAs, adopting an element structure giving high current injection density and incorporating an optical waveguide. Light emitted from such an element has a higher directivity than in the case of LEDs and its time–space coherence is good. Light emitted by a laser diode can therefore in principle be condensed to the order of one wavelength. Thus, laser diodes for optical communication than can efficiently launch light into optical fibres of diameter under 50 μm and laser diodes for optical disk pick-ups used to pick up information using a beam of light as narrow as 1 μm or less are now in practical use.

Thus, compared with LEDs, laser diodes are eminently suitable for application in regions in which the nature of the light is in question. Such optical devices are attracting particular interest with the development of what is sometimes referred to as the 'information society' (i.e. a society in which exchange of information rather than of goods is of primary importance). Figure 7.9 shows schematic diagrams of the structure of LED and laser diode elements.

7.3 MATERIALS USED FOR LIGHT-EMITTING DIODES

Materials used for LEDs must satisfy the following conditions:
1 They must possess a forbidden band of adequate width to obtain the light emission required for the intended applications.
2. They must give good light emission.
3. Injection of minority carriers must proceed efficiently.
4. Light must be efficiently emitted to the outside from the interior of the crystal.

Materials satisfying these conditions, currently used in large amounts, are semiconductors consisting of compounds of elements of groups III–V of the periodic table, such as GaP and GaAs. However, in the case of some semiconductors of the above type, control of electrical conduction becomes difficult when the width of the forbidden band becomes broad, and there are also some materials where a *pn* junction cannot be produced in the same crystal. Mixed crystals of GaAs and GaP such as GaAsP also make good LED materials.

As discussed below, the manufacture of LEDs requires a crystal growth technique known as epitaxial growth. This technique involves growing on a single crystal a single-crystal thin film having the same crystal lattice regularity. A good expitaxial film is obtained when the crystal forming the base and the one grown on it have the same crystal structure and closely similar lattice constants. Semiconductor crystals consisting of compounds of elements of groups II–V with elements of groups II and VI have recently attracted attention as materials with broad forbidden bands for the production of blue LEDs.

Table 7.1 shows the crystal structures, lattice constants, forbidden bandwidths, transition type (direct or indirect), and the possibility of type control (whether *p*- or *n*-type crystals can be produced) for silicon and germanium, etc., element semiconductors and for binary compound semiconductors such as GaAs. Substances wherein type control is possible with the same crystal and which are of the direct transition type include AlSb, GaAs, GaSb, InP, InAs, InSb, and others. Among these compounds, GaAs is used to make IR LEDs emitting light at a wavelength of about 0.9 μm with good efficiency. InP has begun to play an important role as a substrate material for LED and laser diodes used for optical communication in the 1.3 to 1.6 μm band. InSb is used as a material for heat ray (about 4 μm) detectors.

In the case of indirect transition type materials, too, luminescence centres can conveniently be produced by inclusion of impurities in the crystal, and these substances can sometimes provide good LED materials. GaP is an important

Table 7.1 *Properties of representative element semiconductors and binary compound semiconductors*

SEMI-CONDUCTOR MATERIAL	CRYSTAL SYSTEM	LATTICE CONSTANTS (Å)	FORBIDDEN BANDWIDTH	TRANSITION TYPE	POSSIBILITY OF TYPE CONTROL
C	D	3.56679	5.47	Indirect	
Si	D	5.43086	1.12	Indirect	*
Ge	D	5.65748	0.803	Indirect	*
SiC	Z	4.358	2.2		
	W	$a = 3.0806,\ c = 2.52$	3	Indirect	*
AlN	W	$a = 3.111,\ c = 4.980$	6.2	Direct	
AlP	Z	5.451	2.52	Indirect	
AlAs	Z	5.6611	2.16	Indirect	*
AlSb	Z	6.1355	1.63	Direct	*
GaN	W	$a = 3.186, c = 5.176$	3.5	Direct	
GaP	Z	5.4505	2.24	Indirect	*
GaAs	Z	5.6534	1.43	Direct	*
GaSb	Z	6.0955	0.63	Direct	*
InN	W	$a = 3.54, c = 5.71$	1.9	Direct	
InP	Z	5.8688	1.29	Direct	*
InAs	Z	6.0585	0.33	Direct	*
InSb	Z	6.4788	0.16	Direct	*
	Z	5.42			
ZnS	W	$a = 3.82, c = 6.26$	3.6	Direct	
ZnSe	Z	5.671	2.67	Direct	
	Z	5.832			
CdS	W	$a = 4.16, c = 6.756$	2.42	Direct	
CdSe	Z	6.05	1.7	Direct	

D: diamond, Z: zincblende, W: wurzite. *Indicates type control possible.

LED material of this type. The forbidden bandwidth of GaP at room temperature is 2.24 eV. When this forbidden band is activated, green light is emitted. Green LEDs produced from GaP nowadays find widespread application and offer the shortest wavelengths that are presently available in practice from LEDs.

To obtain emission at even shorter wavelengths (blue LEDs), the forbidden bandwidth must be at least 2.6 eV. Materials with broad forbidden bands that have attracted attention for production of blue LEDs are GaN, ZnS, ZnSe, SiC, and the like.

The chief group of LED materials which have attained practical importance consists of mixed crystals of semiconductors consisting of compounds of elements of groups III–V. These include mixed crystals containing three or four elements such as mixed crystals of GaAs and GaP ($Ga_{1-x}As_xP$), mixed crystals of GaAs and AIAs ($Ga_{1-x}Al_xAs$), mixed crystals of GaAs and InAs ($Ga_{1-x}In_xAs$), and mixed crystals of GaAs and InP ($Ga_{1-x}In_xAs_{1-y}P_y$). Their crystal character-

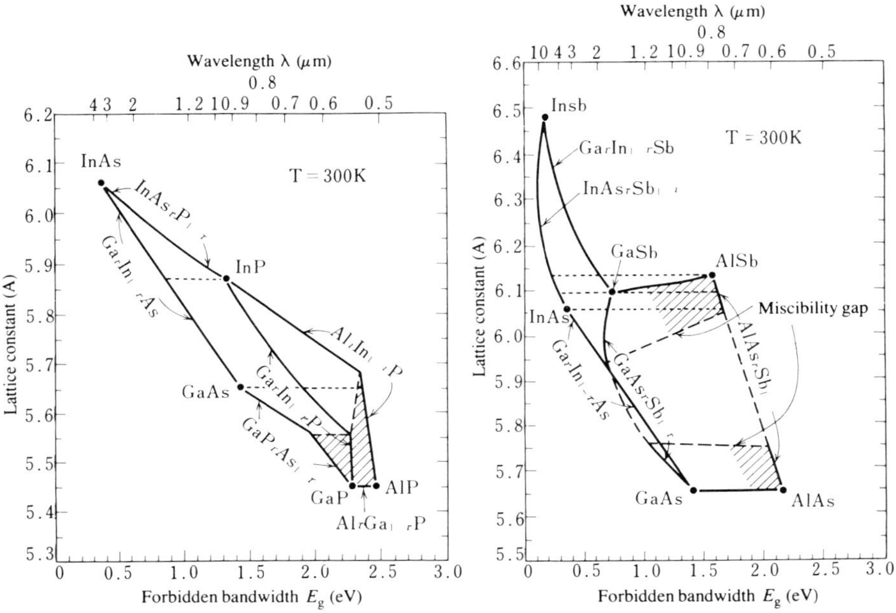

(a) InGaAsP and InGaAlP
(The shaded regions are indirect transition regions)

(b) InGaAsSb and GaAlAsSb
(The shaded areas indicate the indirect transition regions. The broken lines show regions where crystals of the corresponding compositions are not obtained, the so-called miscibility gap)

Figure 7.10 *Forbidden bandwidths and lattice constants of groups III–V mixed crystal compounds [19]*

istics lie between those of the constituent materials and correspond to the composition. GaAsP is used in visible LEDs emitting light in the red to yellow region, GaAlAs is used in LEDs in the red to IR region and in IR laser diodes (LD), InGaAs and InGaAsP are used as materials for LEDs and LDs emitting in the 1 to 1.6 μm band, all these becoming of increased importance. Figure 7.10 shows the relationship between the forbidden bandwidths of these mixed cyrstals and their lattice constants[19].

7.4 PROCESS TECHNIQUE FOR PRODUCING LIGHT-EMITTING DIODES

To produce bright LEDs, efficient minority carrier injection and efficient conversion of the injected minority carriers into light are necessary, as discussed above. For this purpose we require good-quality *pn* junctions and good-quality single crystals showing little non-radiative recombination. Furthermore, depending on requirements, it is important to have techniques available for type control and for the introduction of impurities into crystals for the purpose of forming luminescence centres.

Accordingly, techniques for manufacturing large single crystals assume particular importance. For this purpose, the epitaxial technique and diffusion technique are widely used. Furthermore, the production of LEDs requires process and assembling techniques for element making. Below, these techniques will be discussed one by one.

7.4.1 Techniques for producing single crystals

The following two types of methods have been put into practice for making semiconductor crystals of compounds of the elements of groups III–V. The first method, known as the Czochralski method, is the same as that used for silicon single crystals, and is a pulling method. The other one, referred to as the HB method, is the horizontal Bridgeman method. The pulling method involves bringing a fused liquid of a chemical compound semiconductor material, having for instance the composition GaP, into contact with a seed crystal, which is pulled up whilst being rotated (Fig. 7.11) [20]. In general, however, the dissociation pressure of the constituent elements of compound semiconductors is high in

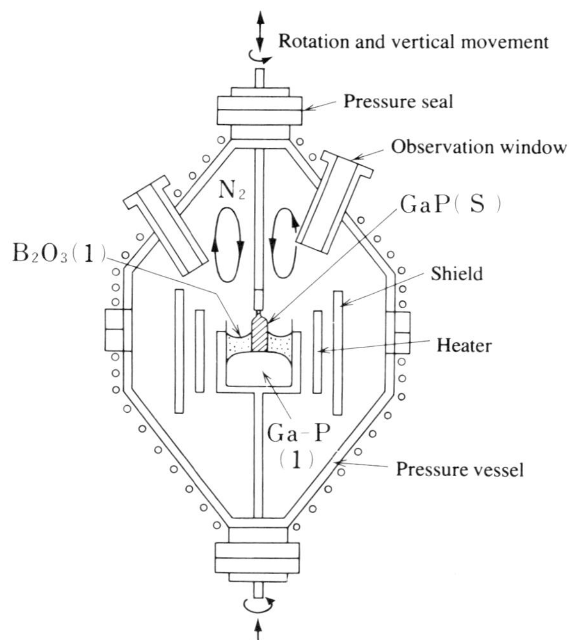

Figure 7.11 *Pulling of GaP single crystal (Czochralski method) [20]*

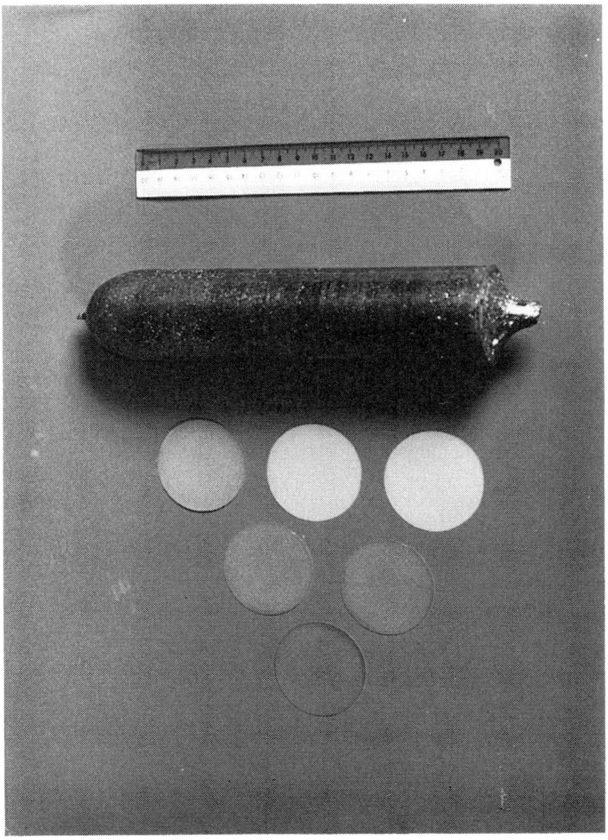

Figure 7.12 GaP LED crystal ingot and wafers

many cases,* so the pulling process is therefore carried out using a high-pressure vessel. To prevent dispersion of volatile material from the fused liquid, the surface of the liquid is covered with boron oxide (B_2O_3) which assumes a glass-like state and pulling is carried out from under this. This method is referred to as the LED (liquid encapsulated Czochralski) method since it involves pulling up material sealed in liquid.

The LED method was applied to growing GaP single crystals in 1968[20] and made possible a considerable advance in the practical application of GaP. Crystals produced by the LEC method include those of GaAs and InP, as well as GaP. Figure 7.12 shows an LEC single-crystal ingot of GaP and wafers of 300 μm thickness, cut from this ingot.

The method known as the horizontal Bridgeman method is used for materials

*For instance, the dissociation pressure of phosphorus in fused GaP attains values up to 35 atm.

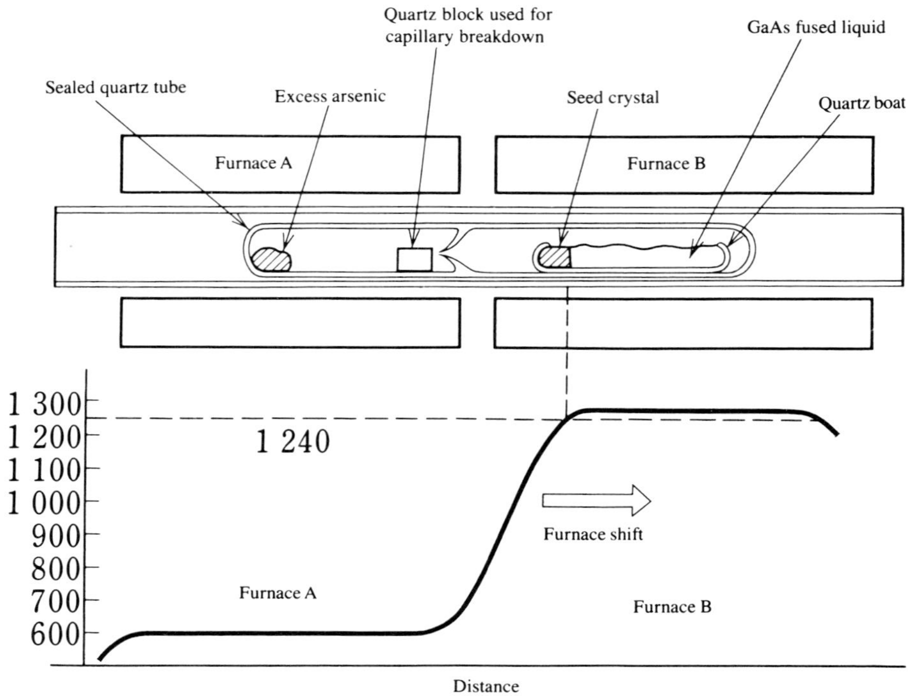

Figure 7.13 *Horizontal Bridgeman method [21]*

such as GaAs which have a comparatively low dissociation pressure. As shown in Fig. 7.13, single crystals are grown by bringing GaAs material, placed in an electric furnace with a temperature gradient and a long horizontal width, into contact with seed crystals on the low-temperature side. Since this involves comparatively mild single-crystal forming conditions, single crystals of GaAs used for GaAs substrates for laser diodes are mostly produced by this method rather than by the LEC method, as better crystal quality is obtained.

7.4.2 Epitaxial technique

Large single-crystal ingots are generally produced under severe conditions of high temperature and pressure. Therefore, it is found that, although single crystals are obtained, these contain numerous structural defects. Even if good-quality crystals are obtained, controlled addition of impurities to provide luminescence centres and for type control is difficult. The epitaxial growth technique, in which a single crystal wafer cut from a large single-crystal ingot is used as a base for the formation of a thin single-crystal film, is therefore important in both LED and LD fabrication.

Epitaxial growing techniques can be divided basically into those in which growth proceeds from the liquid phase and those in which it is from the vapour phase. Growth proceeding from the vapour phase may involve a vapour phase growth method utilizing a chemical reaction or the MBE (molecular beam epitaxy) method which makes use of physical adsorption processes. Of the first-named vapour phase growth methods, mention may be made of the MOCVD (metal organic chemical vapour deposition) method, which makes use of the thermal decomposition of organometallic compounds, and the conventional chemical vapour deposition (CVD) method.

(1) Liquid phase growth method

LPE (liquid phase epitaxy) is the oldest epitaxial growth method. It was first used by Nelson [22,23] (Fig. 7.14(a)) and was subsequently developed to a mass-production method. Considering GaP, it is known that the solubility of phosphorus in metallic gallium increases with the temperature. Therefore, when fused gallium which has been saturated with phosphorus at high temperature is brought into contact with a GaP single-crystal wafer and then gradually cooled, a single-crystal thin film grows on this single-crystal wafer. Since it is easy to introduce various impurities into this fused liquid, type control and control of impurities is readily achieved.

Since epitaxial growth conditions are comparatively moderate, compared with those of the LEC pulling method, the quality of the crystals is improved and they can be readily used as crystals for LEDs or LDs.

(2) Vapour phase growth method

Vapour phase growth methods make use of the vapour phase/solid phase precipitation reaction. GaAsP, important as a material for visible LEDs, is produced by a method known as the hydride method* (Fig. 7.15)[24]. The GaAs precipitation reaction proceeds in accordance with the following reaction scheme:

$$2\,AsH_3(g) \rightleftharpoons \tfrac{1}{2}As_4 + 3H_2(g) \tag{7.15}$$

$$2HCl(g) + 2Ga(l) \rightleftharpoons 2GaCl(g) + H_2(g) \tag{7.16}$$

$$2GaCl(g) + \tfrac{1}{2}As_4(g) \rightleftharpoons 2GaAs(s) + 2Hcl(g) \tag{7.17}$$

In the case of the precipitation of GaP, PH_3 is used instead of AsH_3 in equation (7.15). When preparing mixed crystals of GaAsP, both AsH_3 and PH_3 are used. Control of the composition of the mixed crystals is achieved by regulating the amounts of AsH_3 and PH_3 introduced. In this method, HCl gas is passed over the metallic gallium, resulting in the formation of GaCl. The hydrate of the group V

*One type of conventional CVD.

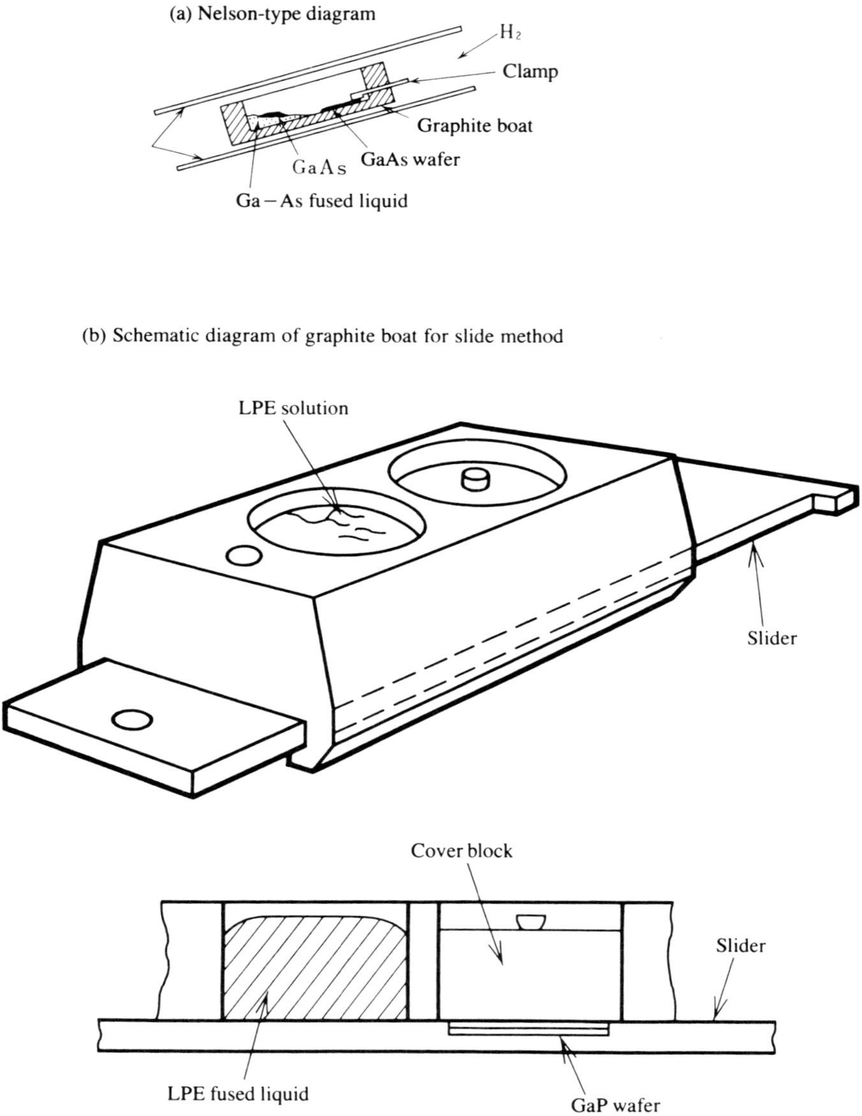

Figure 7.14 *(a) Nelson method [22, 23] and (b) slide method*

element is then introduced from a separate system, causing precipitation to take place.

In addition to this vapour phase growth method, there is the chloride method in which a chloride such as $AsCl_3$ is introduced to the metallic gallium. In recent years, the MOCVD method has attracted attention as a technique for manufacturing laser diodes, utilizing the thermal decomposition of organometallic

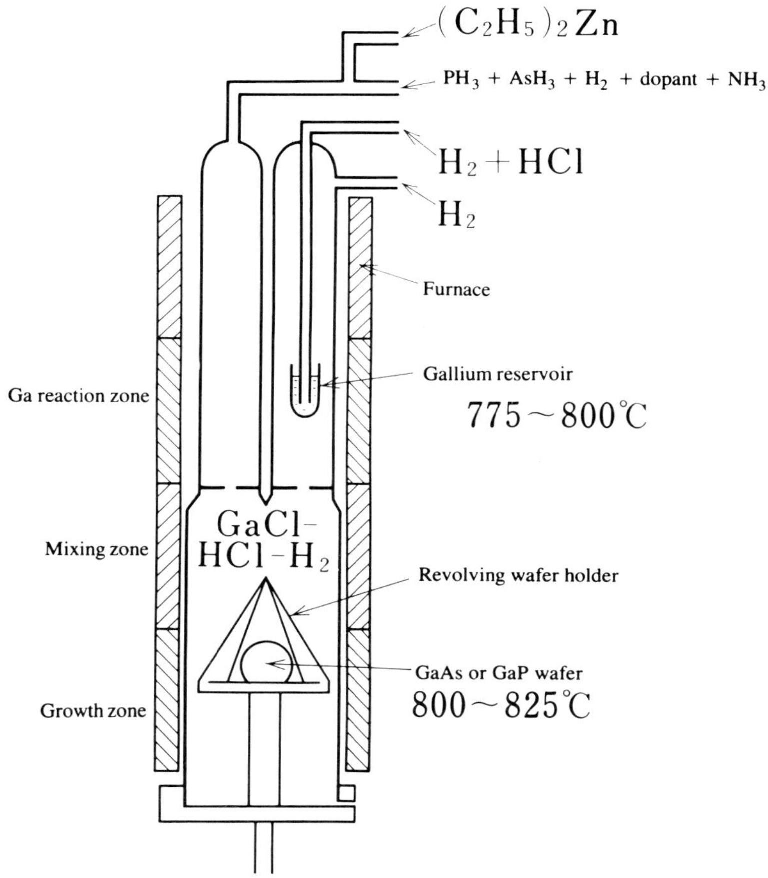

Figure 7.15 *Vapour phase growth–hydride method [24]*

compounds such as $Ga(CH_3)_3$ and AsH_3 to bring about vapour phase growth of GaAs (Fig. 7.16)[25].

The advantage of the epitaxial method lies in the fact that when a thin film consisting of a single crystal is grown on a substrate crystal, its quality is superior to that of the substrate crystal and in addition it is easy to introduce impurities. This method is, therefore, of importance both from the point of view of obtaining crystals of good quality and as a technique for producing *pn* junctions.

7.4.3 Diffusion technique

The diffusion method is of particular importance as a manufacturing process for silicon integrated circuits, but it is also important as a material for LEDs. The *pn* junctions of GaAsP LEDs are formed by the diffusion method.

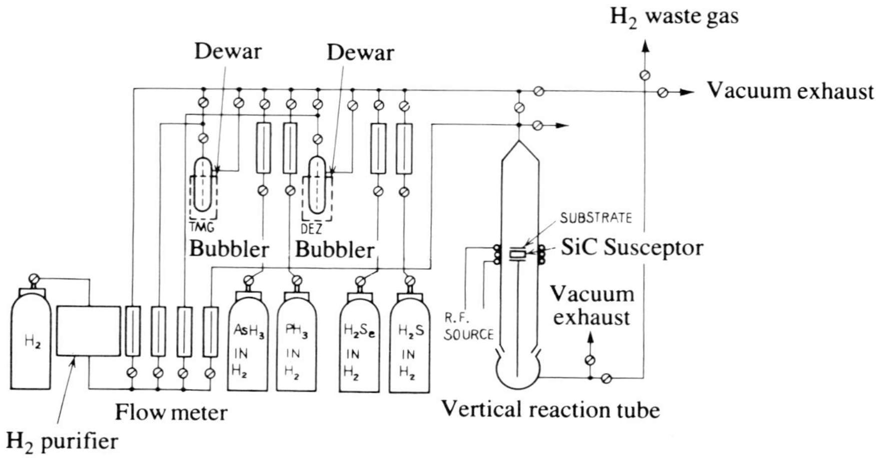

Figure 7.16 *Vapour phase growth–MOVCD method [25]*

The excellent characteristics of the diffusion method are due to the possibility of planar diffusion using a mask to prevent diffusion, selectively. A monolithic display can therefore be produced by forming multiple independent *pn* junctions on a single-crystal wafer. In the case of semiconductor crystals consisting of compounds of group III–V elements, diffusion of a group III element which acts as a acceptor is easier than that of a group VI element acting as donor. Therefore, diffusion of zinc as an acceptor impurity in *n*-type crystals is generally used.

7.4.4 Device-fabrication techniques

The functioning of LEDs requires efficient minority carrier injection at *pn* junctions when voltage is applied. For this it is necessary first of all to mount electrodes on *p*- and *n*-type crystals. The interface between these electrodes and the semiconductor crystals must not constitute a potential barrier but provide an ohmic contact. Alloys such as Au–Si and Au–Ge are generally used as ohmic electrodes for *n*-type crystals and Au–Be or Au–Zn for *p*-type crystals. Thin films of these alloys are evaporated onto group III–V compound semiconductor crystals and ohmic electrodes are then obtained by carrying out a heat treatment step.

These electrodes are patterned by the exposure technique, followed by dividing up into small LED pellets. The theoretical number of pellets of 0.3 mm square, cut from a wafer of 2 inch diameter, is about 35 000. When the diffusion method described above is used to obtain small pellets, 0.3 mm square, the dieing procedure is exactly the same. However, when a monolithic display is to be produced by planar diffusion, the size of the semiconductor pellets becomes close to that of the device that is to effect the display.

7.4.5 *Assembly techniques*

Diced LED crystals are installed on a lead frame or printed ciruit board depending on the intended application. Conducting epoxy resins and wire bonding are generally used to fix the dies. The LED pellets themselves are small semiconductor crystals of about 0.3 mm square, but when they are used in displays of various types, optical reflector plates are provided around the LED dies, as illustrated in Figs. 7.1 and 7.3, or an optical system is installed in front of them. How to fabricate the optical structure is an important assembly technique.

7.5 *VARIOUS TYPES OF LIGHT-EMITTING DIODES AND THEIR PERFORMANCE*

Table 7.2 shows the manufacturing methods, colour of the light emitted, peak wavelengths, luminous efficacy, external quantum efficiency and light-emission efficiency of various types of commercially available visible LED. Figure 7.17 illustrates the dependence of the total optical flux on the drive current. The light-emission characteristics and special features of the manufacturing methods for representative LEDs are described below.

7.5.1 *GaP:ZnO red LEDs*

GaP red LEDs to which Zn and O have been added are brightest in the current range below 10 mA (Fig. 7.17). They are used for d.c. driven displays.

Since the trap level with respect to the electrons of a Zn–O pair is deep (0.3 eV), the light emission accompanying the recombination of this bound exciton has low energy compared with the width of the forbidden band so red light is emitted (peak wavelength $\lambda_m = 690$ nm, photon energy $\Delta E = 1.8$ eV).

Red LEDs containing Zn–O pairs as light-emitting centres have up to now been reported as having a maximum light-emission efficiency of 12.6 per cent [26], though their commercially available light-emission efficiency is generally 2 to 3 per cent. Since in this type of LED the light-emission centres lie at levels deeper than the conduction band, current injection practically saturates the light-emission centres with trapped electrons. The light-emission efficiency of GaP red LEDs therefore has a maximum in the current region below 10 mA, with a drop at higher currents. Thus GaP red LEDs show the characteristics of low current drive LEDs. Since they emit light at a low energy compared with the forbidden band, there is little reabsorption of emitted light by the crystal. Therefore, red light emitted in a region within a few micrometres from the *pn* junction causes the whole of the crystal to light up and is discharged outside the crystal. This is a fact to which attention must be paid when deciding the construction of the display element—for instance, how to arrange the light reflection plate.

GaP red LEDs are produced by the liquid phase epitaxial method on crystals

Table 7.2 *Properties and constituent materials of visible LEDs in practical use*

Colour of emitted light	Wavelength of emitted light (nm)	Composition of LED material	Manufacturing method	Luminous efficacy (lm/W)	External quantum efficiency (%)		Light-emission efficiency (lm/W)	
					Commercial product	Maximum value	Commercial product	Maximum value
Red	700	GaP:ZnO/GaP	LPE*	20	2–4	15 (12.6)	0.4–0.8	3.0 (2.5)
Red	660	GaAlAs/GaAs	LPE	40	−3	−4	1.2	1.6
Red	650	GaAsP/GaAs	VPE† + diffusion	70	0.2	0.5	0.14	0.35
Orange	630	GaAsP:N/GaP	VPE + diffusion	190	0.2–0.3	0.65	0.4–0.6	1.2
Yellow	590	GaAsP:N/GaP	VPE + diffusion	450	0.15	0.3	0.66	1.35
Yellow	590	GaP:NN/GaP	VPE + diffusion	450		0.1		0.45
Green	565	GaP:N/GaP	LPE	610	0.3	0.7	1.8	4.9
Pale green	555	GaP/GaP	LPE	680	0.1		0.68	

*LPE: liquid phase epitaxial growth.
†VPE: vapour phase epitaxial growth.

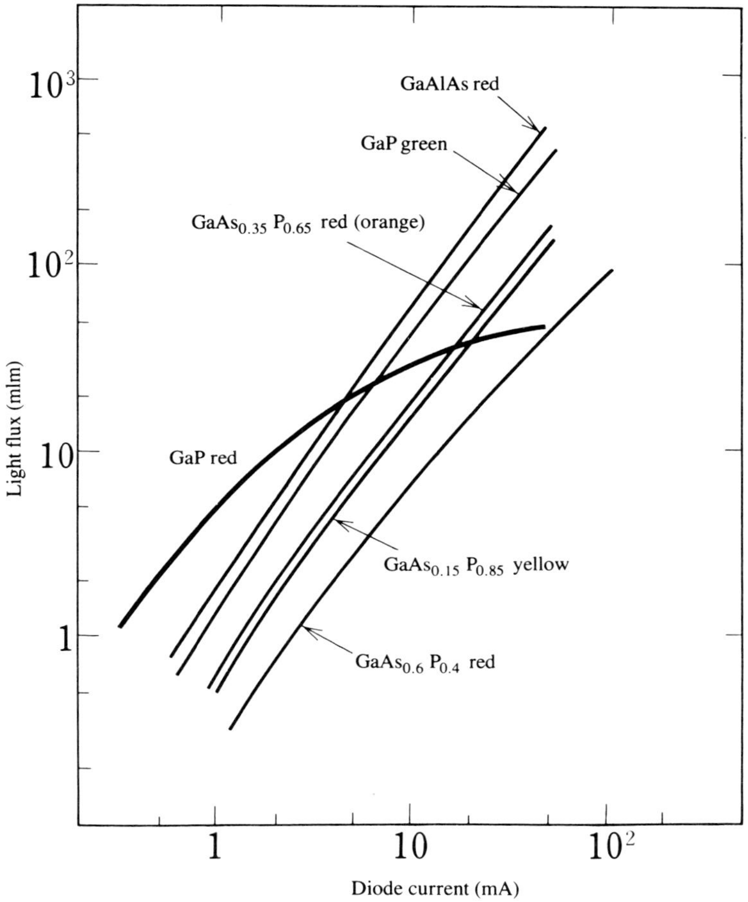

***Figure** 7.17 Comparison of light-emission characteristics of different types of LED*

obtained by the LEC method. Their structure is illustrated in Fig. 7.18. An *n*-LPE crystal is first formed on an *n*-LEC substrate, followed by formation of a *p*-LPE crystal on top of this. Zn is introduced into the *p*-LPE crystal as acceptor and oxygen is added at the same time. Zn–O pairs, acting as red light-emitting centres, are thus formed in a *p*-type region. Whether an efficient red LED is obtained is determined by whether electrons can be efficiently injected into the light-emission region.

7.5.2 GaP:N green LEDs

Thomas *et al.* [27] showed that nitrogen atoms can provide effective light-emission centres in GaP crystals, giving off an intense green light. This was based

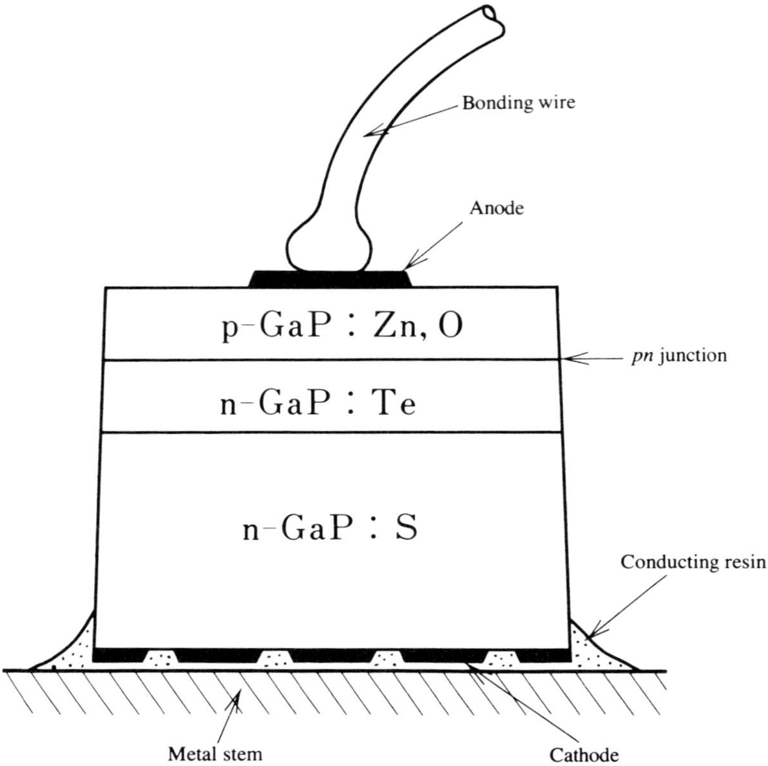

Figure 7.18 *Structure of GaP red LED*

on the fact that GaP crystals grown in a boron nitride (BN) crucible emit an intense green light. Subsequently, introduction of a trace of ammonia into the gas atmosphere present during the crystal-growing process was considered as an effective method of adding nitrogen, and this led to a major advance in the development of the production technique for green LEDs[28].

Nitrogen atoms added to the crystals form bound excitons with a very shallow depth of the bound level (approx. 20 meV)[29]. The emitted light is therefore green ($\lambda_m = 565$ nm, $\Delta E = 2.19$ eV). A state of thermal equilibrium exists between the conduction band, valence electron band, and the traps, and saturation phenomena of the type seen in the case of red LEDs are not observed. This is due to the fact that the light-emission intensity–current characteristic of green LEDs resembles that of other LEDs obtained from direct transition materials.

Green LEDs are produced on *n*-type LEC crystals by the LPE method. In this LPE method, nitrogen addition is carried out by introducing a trace amount of ammonia into the stream of hydrogen gas used as inert atmosphere. Nitrogen atoms are incorporated into the GaP crystal by the following chemical reaction:

$$\mathrm{Ga(l)} + \mathrm{NH_3} \rightleftharpoons \mathrm{GaN(s)} + \tfrac{3}{2}\mathrm{H_2} \qquad (7.18)$$

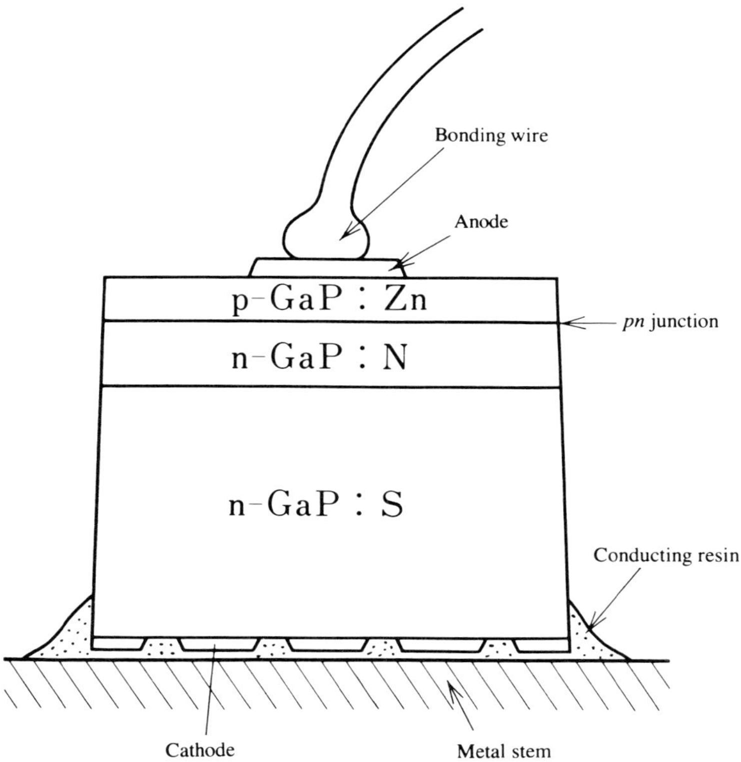

***Figure** 7.19 Structure of GaP green LED*

The concentration of nitrogen which can be introduced in the LPE method usually has a value of the order $10^{18}\,cm^{-3}$. The structure of a green LED is illustrated in Fig. 7.19.

The visual sensitivity of the human eye for green light is very good. However, the luminous efficacy distribution of a green LED is similar to the energy distribution of sunlight. It is therefore difficult to achieve contrast against a background of sunlight, which is a drawback. Endeavours are, therefore, continuing to develop bright green LEDs. One method for increasing brightness is to raise the concentration of light-emission centres. However, increasing the concentration of light-emission centres in a GaP:N LED results in a shift of the luminescence towards longer wavelengths, so that the light emitted comes closer to yellow and at the same time crystal defects also increase. As a result, light-emission efficiency does not exceed a maximum value of 0.7 per cent[30]. Furthermore, in the LPE method there is an upper limit to the concentration of nitrogen atoms that can be added. On the other hand, with the VPE method, of the order of $10^{20}\,cm^{-3}$ of nitrogen atoms can be added, enabling yellow LEDs to be produced.

7.5.3 GaAsP red LEDs

GaAs has a direct transition band structure and is an important substance as a material giving highly efficient light emission. However, its forbidden bandwidth is 1.43 eV so its light emission lies in the IR region. Broadening the forbidden bandwidth while retaining direct transition has been attempted by using GaAsP and GaAlAs.

As a result, red LEDs ($\lambda_p = 640$ nm) having the composition $GaAs_{0.6}P_{0.4}$ have been produced. An LED produced by diffusing zinc in a GaAsP crystal formed by the vapour phase growth method on a GaAs substrate crystal to create *pn* junctions was first brought on the market as a visible LED in 1970 by Monsanto Co. However, since the GaAs substrate crystal has a high absorption coefficient for the emitted light, this LED is less bright than other types of LED and at present is only used for small displays of special types, such as the date indication on cameras, and in some electronic calculators. On the other hand, the light-absorbing action of the GaAs substrate crystal means that these devices can be conveniently used to obtain planar displays.

7.5.4 GaAsP:N yellow–orange LEDs

Rapid development of GaAsP LEDs has been achieved by the use of nitrogen atoms as light-emission centres which proved to be as effective as in the case of green LEDs. Bright LEDs can be obtained by adding ammonia vapour to introduce nitrogen atoms in vapour phase growth of GaAsP (which has an indirect transition region similar to that of GaP). These have been practically applied as yellow ($\lambda_p = 590$ nm) and orange ($\lambda_p = 610$ to 630 nm) LEDs[32].

The structure of GaAsP visible LEDs is shown in Fig. 7.20[32]. Figure 7.20(a) shows a red LED produced on a GaAs substrate, which is suitable for a monolithic display, as the GaAs substrate acts as a light absorber. Figure 7.20(b) illustrates the structure of an orange–yellow LED produced on a GaP substrate. Since the substrate does not act as a light absorber in this case, a bright LED is obtained. Although GaP and GaAs have the same zincblende type crystal structure, their lattice constants differ by 3.7 per cent. Therefore, in order to grow epitaxial crystals of good quality, a transition zone with a gradually altering composition must first be established on the GaP substrate, followed by formation of a light-emission region of constant composition. In GaAsP visible LEDs produced on a GaP substrate, the reabsorption effect shown by the substrate is slight, so a bright LED is obtained. GaAsP/GaP LEDs can be used to give practically the same type of display as other GaP LEDs.

7.5.5 GaAlAs LEDs

The difference between the lattice constants of GaAs and AlAs is only 0.14 per cent, and this was noted from the initial stages of LED development onwards.

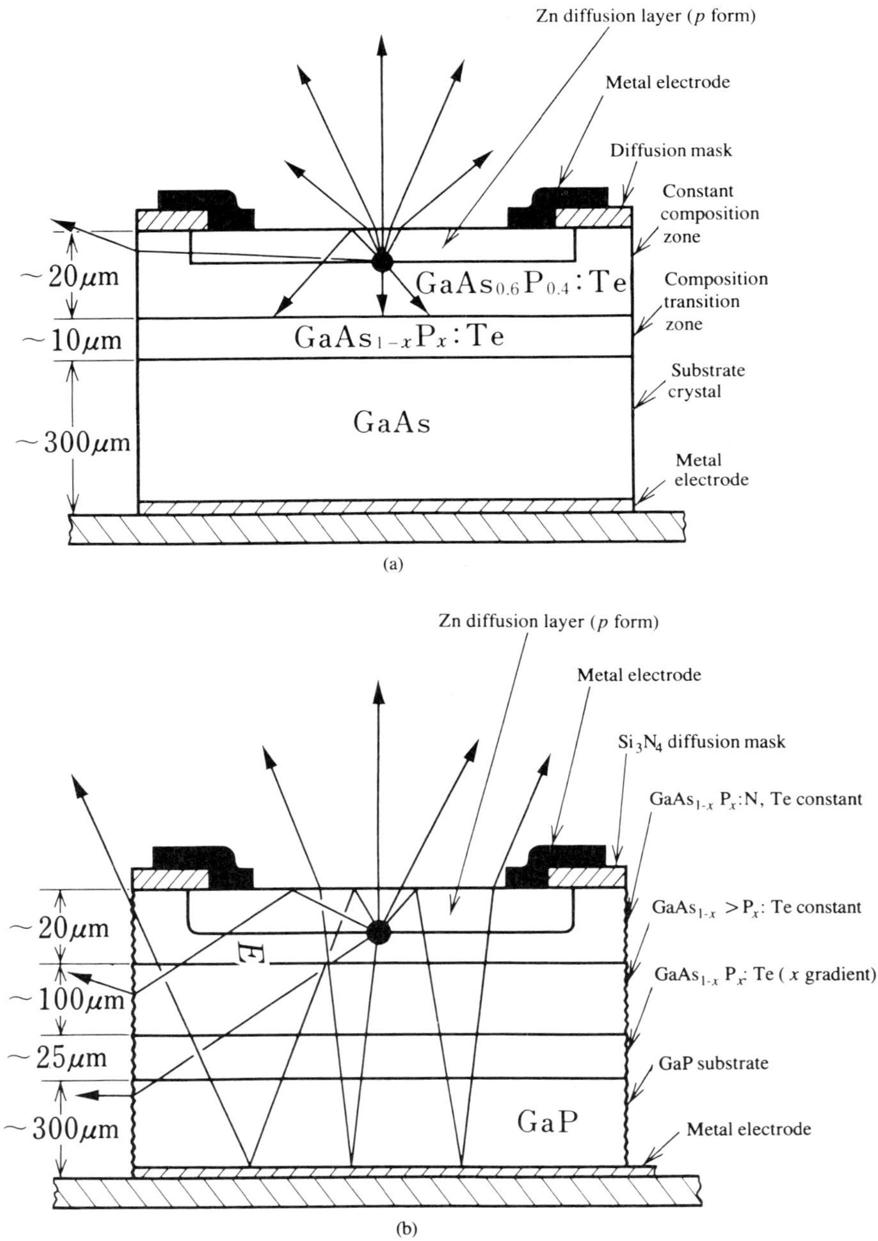

Figure 7.20 *Structure of GaAsP and GaAsP:N LED [32]*

However, in contrast to the situation regarding GaAsP, the Al present in the crystals is easily oxidized. This renders control of the composition during crystal growth difficult, and this must therefore be carried out by the LPE method. This has retarded the development of visible LEDs using this system. GaAlAs is, however, also excellent as a material for lasers emitting in the 0.8 μm band, and the development of an LPE process for lasers has reawakened interest in this material. The band structure of a material having an intermediate composition between GaAs and AlAs is shown in Fig. 7.21[33]. To obtain visible LEDs, we require an Al composition $x = 0.3$ giving a light-emission wavelength of 660 nm (1.88 eV).

The structure of the GaAlAs red LEDs that are currently most widely used is illustrated in Fig.7.22(a). They are produced by first growing p-type GaAlAs on a p-type GaAs substrate, followed by growing an n-type GaAlAs with a broad forbidden bandwidth. Injection of electrons into the p-type crystal is easy, and by using n-type crystal as a light-emitting window, a bright red LED is obtained. In contrast to GaP red LEDs, GaAlAs red LEDs do not display saturation of the light-emission intensity. They are, therefore, of importance as red LEDs of high luminance which can be operated by a high current drive, being much the same as GaAsP orange LEDs in this respect.

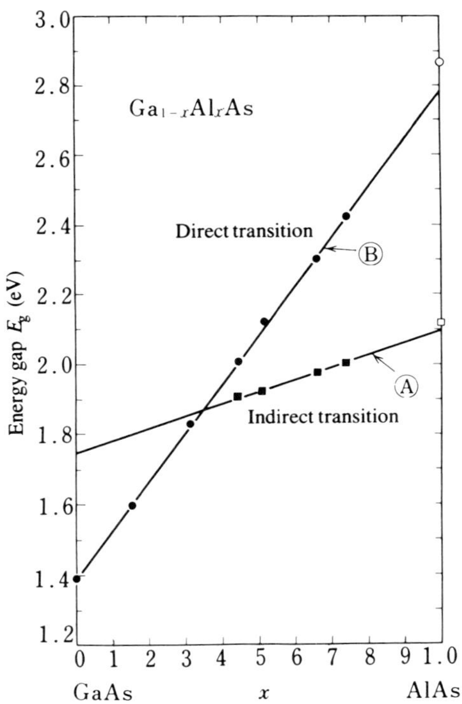

***Figure** 7.21 Band energy and mixed crystal ratio of GaAlAs mixed crystals [33]*

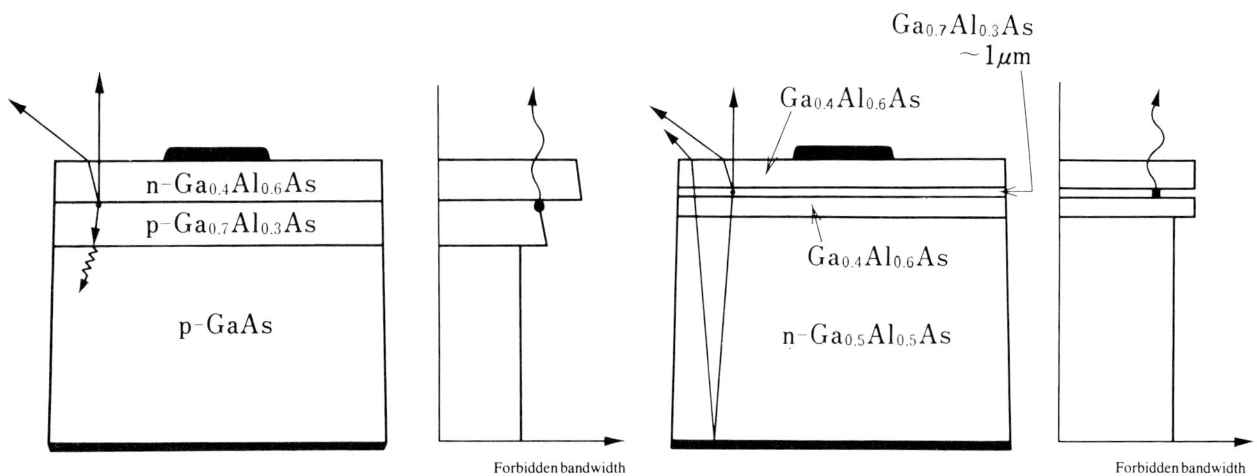

(a) Single heterostructure

(b) Double heterostructure

***Figure* 7.22** *Structure of GaAlAs red LED*

To reinforce these high luminance characteristics still further, a double heterostructure analogous to that used in semiconductor lasers may be employed in these GaAlAs visible LEDs. A cross-sectional diagram of this structure is shown in Fig. 7.22(b). In this structure, the light absorbing layer for the emitted light has been eliminated, and the carriers are confined into a double heterostructure. Light emission is therefore increased. At present, this type of LED offers the highest brightness.

7.5.6 IR LEDs

IR LEDs in practical use can basically be divided into three types. The first of these is GaAs:Si LEDs, the second GaAlAs LEDs, and the last one GaInAsP LEDs. Since the silicon that is added to GaAs is a group IV element, it can act as both donor and acceptor. Which of these roles is dominant will depend on the growth conditions such as the crystal growing temperature and so on. This property is made use of in the formation of *pn* junctions. When silicon is added to a Ga solution that is being used in LPE and the system is cooled gradually from about 900°C, *n*-type crystals separate out first of all, followed by separation of *p*-type crystals at a temperature below about 850°C. Formation of *pn* junctions is thus easy and the luminous efficiency of the resultant LED is also high (about 6 per cent). This device is important as a mass-produced IR LED, and is widely used as a light source in photocouplers, remote control, and so on, in combination with a photodetector.

However, the presence of silicon donors in high concentrations in GaAs causes a change in the configuration of the bottom of the conduction band (band 'tailing') and, in addition, silicon also becomes an acceptor, so that the emitted light has a longer wavelength than would be estimated from the forbidden bandwidth of pure GaAs (940 nm). Reabsorption of the emitted light is, therefore, considerable, and a photon recycling process takes place (light emission $\rightarrow$ absorption $\rightarrow$ light emission), and the attenuation life of the emitted light is long (about 1 μs). From the point of view of the transmission of information when the LED is combined with a photodetector, this imposes limitations on the application of this device, which make it unsuitable for use in optical communication. However, the efficiency of the effect has attracted interest and attempts have been made to produce blue–red LEDs by excitation of a phosphor using a GaAs IR LED, but this has not found practical application.

GaAlAs LEDs used for optical communication have practically the same double heterostructure as laser LDs. However, the active layer is not so thin as in the case of laser LDs.* Since they are frequently of the surface light-emitting type, their structure normally involves some means to assist light extraction (Fig. 7.23). In the optical communication use of LEDs, fine optical fibres having a diameter of 50 μm are required for optical coupling, so the light-emitting diameter must

*About 1 μm. For laser LDs it is about 0.1 μm.

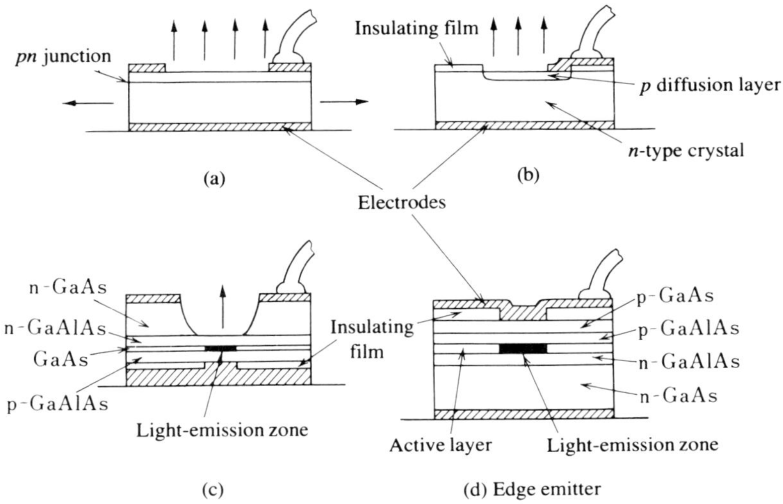

Figure 7.23 *Structure of GaAlAs system LED used for optical communication*

also be made as small as possible. The peak wavelength of the emitted light normally lies at 0.8 μm and the Al composition of the active layer corresponds to $x \cong 0.1$.

By modifying this double heterostructure LED, a high-output IR LED can be obtained. When the LED pellets are made dome shaped, the light output efficiency of the GaAs LED is improved, its quantum efficiency then exceeds 50 per cent, and it can be used as a spatial propagation type LED[17].

The improvement in the characteristics of quartz fibres used for optical communication has been accompanied by a shift in the minimum loss region from the 0.8 μm band to the longer wavelength region at 1.3 to 1.6 μm. For this reason, GaInAsP four-element mixed crystal LEDs emitting at 1.3 to 1.5 μm have become important together with the corresponding lasers.

7.5.7 ***Blue LEDs***

Red to green LEDs, for which semiconductors consisting of compounds of elements of group III–V are used, are currently being produced in large amounts, and their applications range from consumer applications to industrial applications. LEDs are important as completely solid state self-luminous devices. To complete their colour range we require blue LEDs. To obtain blue LEDs, semiconductors with a forbidden bandwidth in excess of 2.54 eV are needed. As blue LED materials, GaN as a group III–V compound semiconductor, ZnS and ZnSe as group II–VI element compounds, and Si–C as a semiconductor consisting of a compound of two group IV elements have attracted attention.

Table 7.3 *Blue LED and comparison of their constituent materials*

Material	Transition state	Forbidden bandwidth (eV)	Colour of emitted light and wavelength (nm)	Structure	Manufacturing method	Luminous efficacy (lm/W)	Quantum efficiency (%)	Light-emission efficiency (lm/W)
GaN	Direct	3	Blue–bluish green 490	MIS	VPE/sapphire	130	0.03	0.04
ZnS	Direct	3.7	Blue 465	MIS	Heat treatment/ Bridgeman	50	0.05	0.025
ZnSe	Direct	2.7	Blue 465	MIS	Implant/Bridgeman	50	0.001	0.001
SiC	Indirect	3	Blue 480	*pn*	LPE/Raleigh method, Achison method	80	0.013	0.01

The characteristics of elements developed with the use of these materials and the properties of the materials themselves are shown in Table 7.3.

Among these materials, Si–C is the only one for which *pn* junctions are produced from the same material. Whilst it is noteworthy in this respect, there is the problem that the crystal growth temperature of this material is high (1700 to 2000°C) so large single-crystal substrates cannot be obtained. Furthermore, since the band structure is of the indirect transition type, bright LEDs are difficult to obtain.

GaN, ZnS, and ZnSe all have a direct transition type structure which is a point in their favour, but there is the problem that *p*-type crystals are not formed as desired. Therefore, positive hole injection into the semiconductor crystal is performed from an electrode, using an MIS type LED structure, in which a thin layer of insulator is sandwiched between an *n*-type compound semiconductor crystal and a metallic electrode. In the case of GaN, large substrate crystals cannot be obtained, so the element is instead produced from a crystal grown from the vapour phase on a sapphire substrate[34]. However, the difference between the lattice constants of the substrate crystal and the GaN crystal is large (13 per cent) so raising the manufacturing yield of these elements poses problems.

Comparatively large single crystals of ZnS and ZnSe can be obtained by the Bridgeman method, but their quality is not good. It has recently been demonstrated that good-quality ZnSe single crystals can be obtained on GaAs single crystals by vapour phase growth using an organometallic compound. ZnS can also be grown on GaP crystals. The possibility of making a blue LED by growing a group II–VI compound single crystal on a group III–V compound single crystal has also been considered[35].

7.5.8 Reliability of LEDs

The stimulus for the development of LEDs was provided by the possibility of obtaining display elements having the high reliability possessed by solid state devices. Indeed, LEDs do have a much longer working life than tungsten lamps etc. However, they differ from other semiconductor elements in that their function depends on the light emission and recombination that occur as a result of injection of minority carriers. Accordingly, an increase in non-radiative recombination processes or an increase in the number of crystal defects will result in a deterioration of the functioning of the element, in particular its light-emission characteristics. When a current passes through the device, crystal defects increase and the light-emission characteristics show a deterioration. This is known from practice. When a stress acts on the crystals, this deterioration is accelerated[36,37]. It can be stopped by eliminating the stress. This type of deterioration, is known as the optical embrittlement effect and is caused by production of dislocations within the crystal by excitation by minority carriers.

Apart from this type of deterioration in properties, LED dies may possibly undergo mechanical damage caused by the assembling technique used. This can, however, be eliminated by improving the manufacturing process and by introducing reliability tests at the production stage.

7.6 APPLICATIONS OF LIGHT-EMITTING DIODE DISPLAYS

7.6.1 Pilot lamps

The simplest application of LEDs as display devices is in pilot lamps. These lamps may be of different types, as illustrated in Fig. 7.1, but the LED die always has a size of about 0.3 mm square. In some cases, the colour of the light emitted by a single lamp can be changed from red to yellow and green. This may be achieved either by placing pellets of two different types (red and green) in the same positions, or using pellets capable of emitting light of two different colours, of the type illustrated in Fig. 7.24[38].

The spatial distribution of the intensity of the light emitted to outside the crystal by an LED pellet closely approaches a Lambert distribution, if we consider the surface of the LED die as an assembly of uniform point light sources. In this case, the spatial distribution, expressed in polar coordinates, assumes a spherical shape (circular cross-section), as illustrated in Fig. 7.25, and the half-power angle is 120°C. Accordingly, in pilot lamps too, the spatial distribution (directionality characteristic) of the radiated light is controlled by mounting reflector plates around the LED pellet and fitting various types of optical systems at its front face. The minimum half-power angle used in visible LEDs employed for display purposes is about 20°. The solid angle ratio between half-power angles of 120° and 20° or 15° is 33 or 58. Assuming virtually all the radiated light flux lies within this half-power angle, making this angle small by installing a suitable optical system will be a very effective means of increasing brightness. In the case of a pilot lamp, however, problems occur when the directionality is raised excessively and a half-power angle of 40° is therefore frequently adopted. LED dies are in most cases installed in metal holders, which are in the forms of lead frames, produced from sheet metal by a punching-out operation.

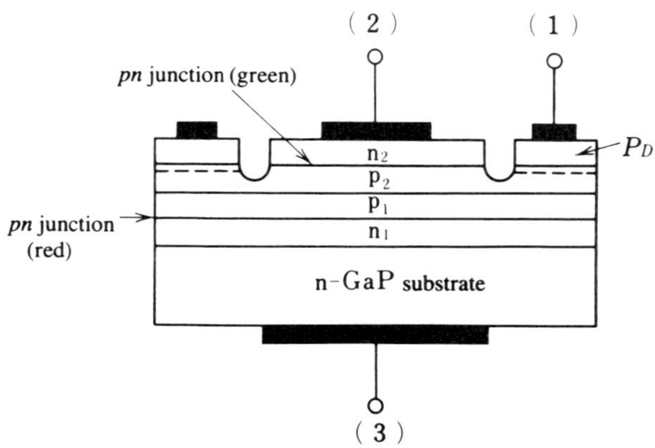

Figure 7.24 Two-colour LED [38]

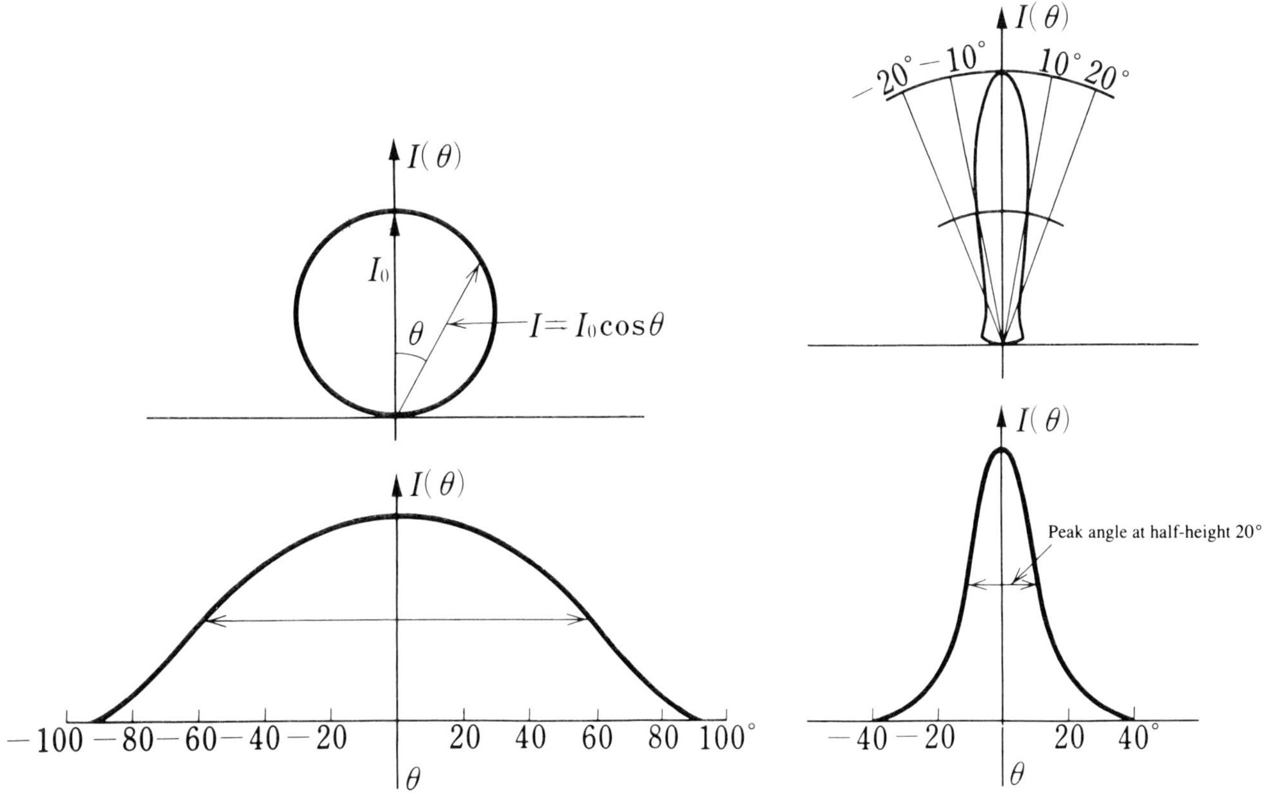

(a) (b)

Figure *7.25 Directional characteristics of LED. (a) Spatial distribution $I(\theta)$ of radiation intensity from LED pellet. (b) Directional characteristics of LED with installed optical system*

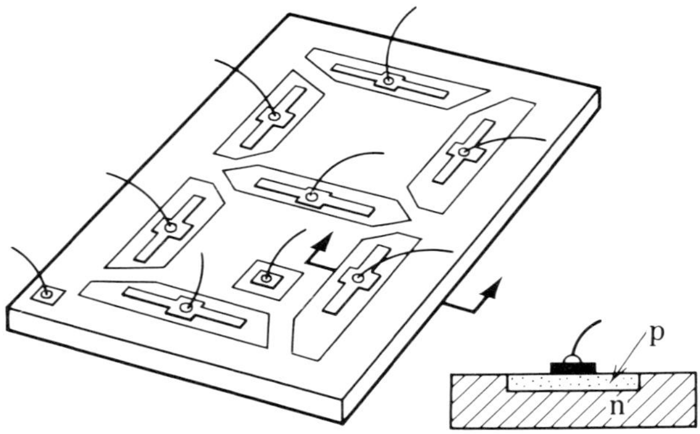

Figure 7.26 Typical monolithic display [39]

7.6.2 Monolithic displays

As discussed earlier on, *pn* junctions for GaAsP LEDs are produced by a diffusion method. With this method, selective diffusion can be achieved by application of a diffusion mask which partially prevents diffusion. LEDs produced by this method give monolithic displays. A typical example is shown in Fig. 7.26[39]. The advantage of a monolithic display is that an accurate light-emission pattern can be produced, since it is possible to use photographic techniques to define the light-emitting parts. On the other hand it is difficult to obtain large displays since a single-crystal wafer having the same size as the required display must be used. It is also difficult to produce displays of several colours. At present, practical application is restricted to date indications on cameras and to some extent in electronic calculators.

7.6.3 Hybrid displays

Most LED displays are of the hybrid type. Comparatively large displays are obtained by arranging 0.3 mm square small LED dies in mosaic form in various display patterns and using light reflector and diffusion plates (cf. Fig. 7.3). Alphanumeric display elements, level meters, or dot matrix type letter display elements can easily be produced as multicolour devices and of large size. Production of large displays is easily achieved by increasing the number of LED dies used at the centre of the light reflector plates.

7.6.4 Flat panel displays

Development of flat panel displays using LEDs is making steady progress. An example of an experimental monolithic type is shown in Fig. 7.27[40]. Figure

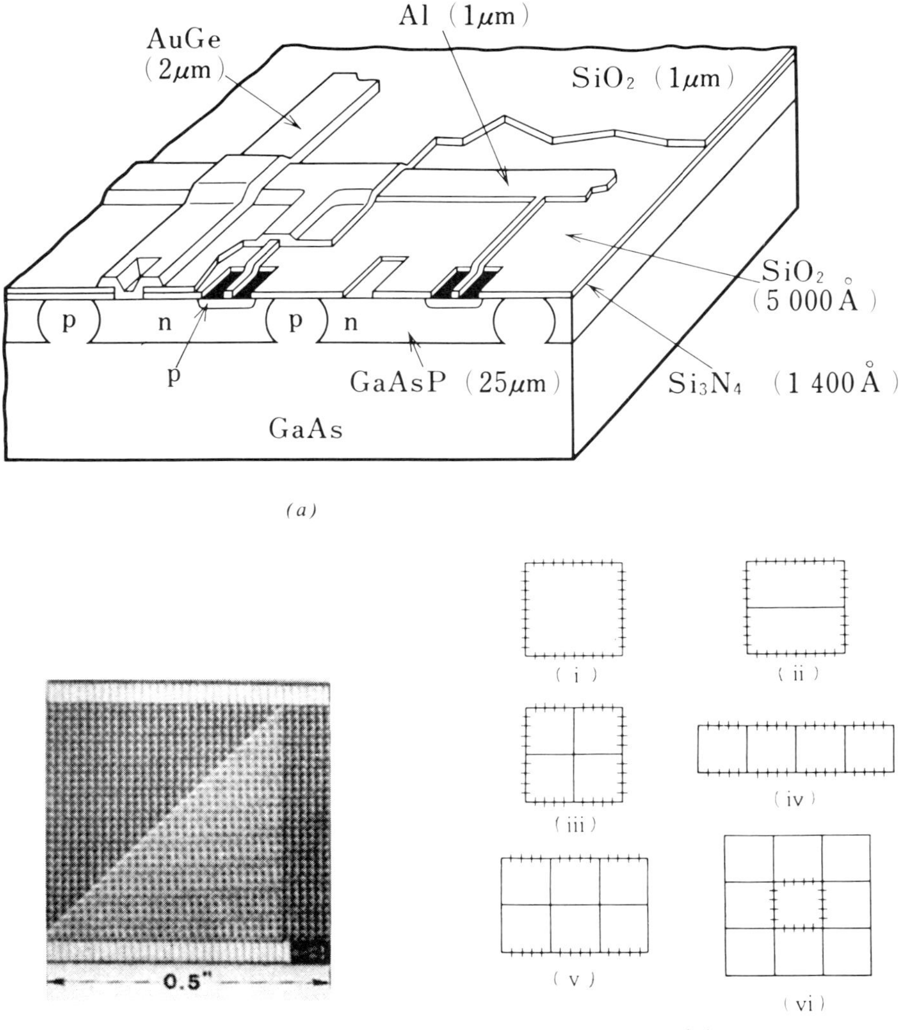

***Figure** 7.27 LED flat panel display with monolithic structure [40]*

7.27(a) shows the structure of the element and Fig. 7.27(b) an integrated circuit of size 0.5 × 0.5 inch, made up of 30 × 36 LEDs. Figure 7.27(c) is an explanatory diagram illustrating the combination of these monolithic displays to produce a large display. This was produced by application of a selective diffusion method to GaAsP. However, as discussed above, displays of large size are most easily produced using hybrid displays, and all recent experimental devices are of the hybrid type. Examples include a prototype TV display unit made up of

Figure 7.28 *TV display produced from GaP green LEDs with hybrid structure [41]*

320 × 240 = 38400 GaP green LEDs arranged with 0.5 mm pitch (Fig. 7.28)[41], a device made up of 64 × 64 GaP red LEDs and green LEDs, respectively arranged with 1.27 mm pitch[42,43] a device made up of 94 × 64 GaP LEDs capable of emitting both red and green light[44], and a 20 inch red-yellow-green multicolour display comprising 20 × 30 of the 20 mm square modules comprising 16 × 16 picture elements each (Fig. 7.29)[45].

Flat panel type LED displays are undergoing steady development, but lowering the cost of materials for the monolithic type and in particular the assembly costs of hybrid devices are likely to pose problems in practical application. Flat panel displays have possibilities for use in small multifunctional devices for OA and AV or for use in vehicles, and also in ultralarge high-performance displays.

7.6.5 Drive method for light-emitting diode displays

The current–voltage characteristics of an LED are shown in Fig. 7.6. It can be clearly seen that constant current drive is required. Figure 7.30 shows drive circuits for d.c. drive digital display elements. A constant current system is obtained by inserting a resistance in series with the LED. An important drive method for LED displays is the time-division lighting method.

The drive circuit for a clock display using this method is shown in Fig. 7.31.

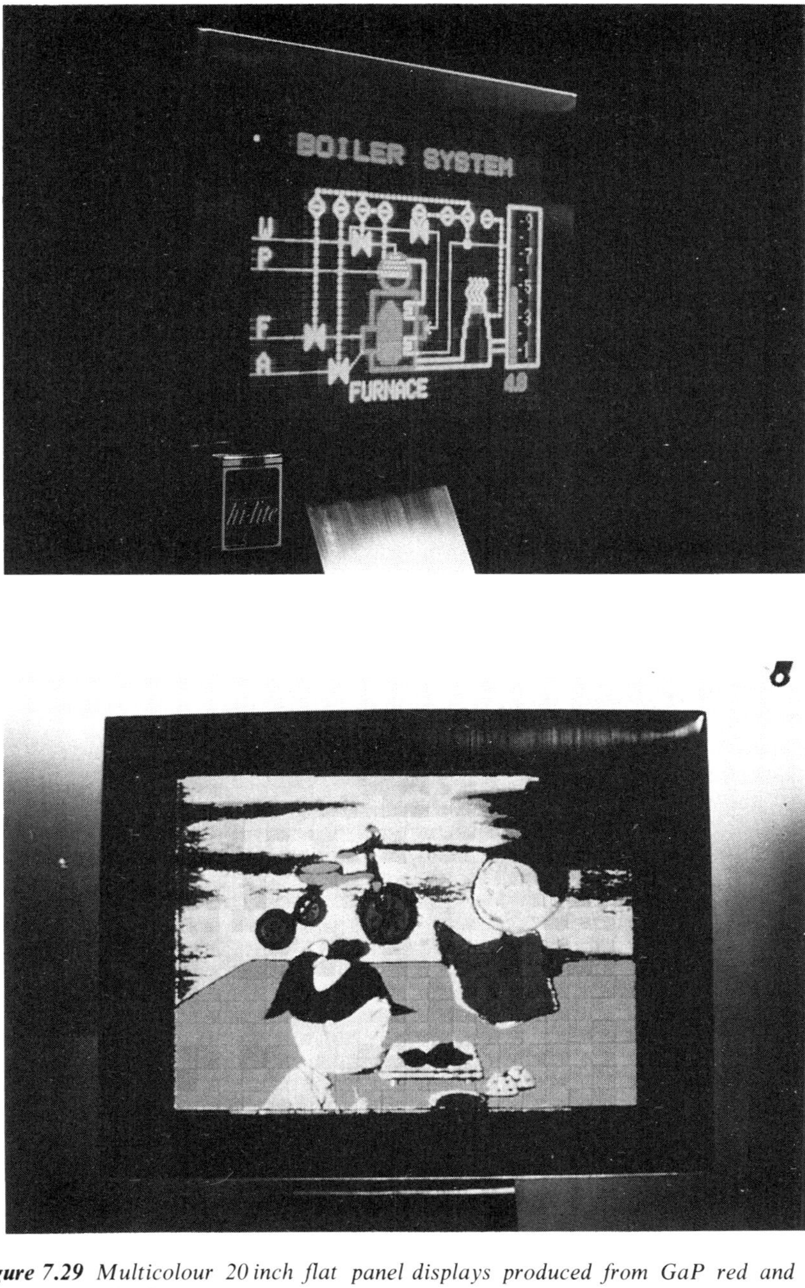

***Figure* 7.29** *Multicolour 20 inch flat panel displays produced from GaP red and green LEDs [45]*

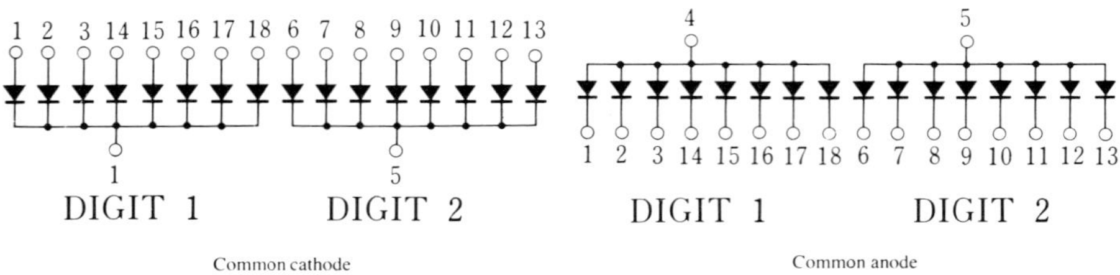

Figure 7.30 Drive connection scheme of digital display element with d.c. drive (connection for static drive)

Since the lighting probability is lower than with the d.c. drive method, the peak value of the drive current can be made correspondingly higher. This method is, therefore, useful for elements in which saturation of the light-emission intensity–current characteristic of the LED is not observed, or rather, for elements that have a superlinear charcteristic.

Drive systems for LED displays are basically of the above two types. The drive used for flat panel type displays with a large number of picture elements is basically the same, but depending on the size of the display, minor changes may be made to the actual drive circuit. A typical example is shown in Fig. 7.32.

7.7 APPLICATIONS OF LEDs OTHER THAN IN DISPLAYS

The most important application is in optical communications, especially GaAlAs ($\lambda_m = 0.8\,\mu m$) LEDs or GaInAsP ($\lambda_m = 1.3$ to $1.5\,\mu m$) LEDs coupled to optical fibres. In addition, GaAsSi ($\lambda_m = -0.94\,\mu m$) and GaAlAs high-output ($\lambda_m = 0.8\,\lambda m$) LEDs are used in optical data processing.

Specific applications include photocouplers, optical trigger light sources for electrical power control, spatial propagation type light sources, and others. LED printers used with a photosensitive drum and LED light sources for facsimile transmission terminal data reading are also made.

Various applications utilizing the special features of LEDs, other than display elements, are thus being developed.

7.8 OUTLOOK

The 1960s saw the development of the basic techniques. In the 1970s, mass-production methods were developed for visible LEDs produced from semi-conductors consisting of binary compounds of elements of group III and V. The 1980s witnessed a coming together of this development of LEDs with major research and development of semiconductor lasers for optical communication. These developments appear to be continuing. However, the most active period of

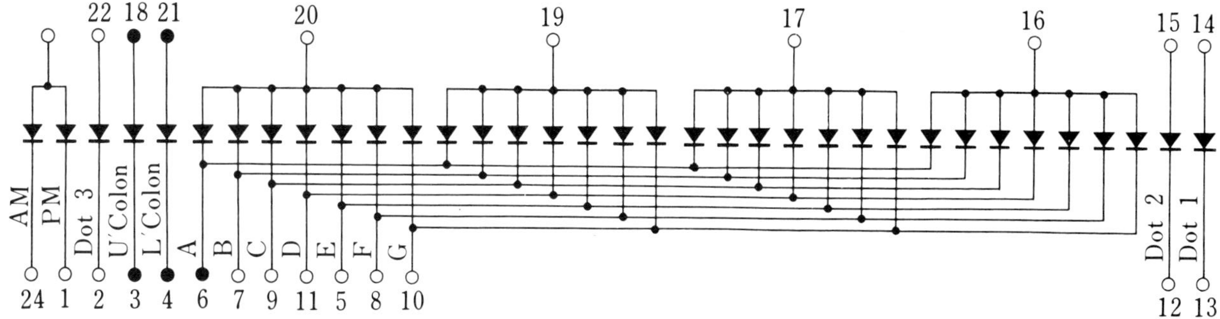

(a) Common anode dynamic drive connection

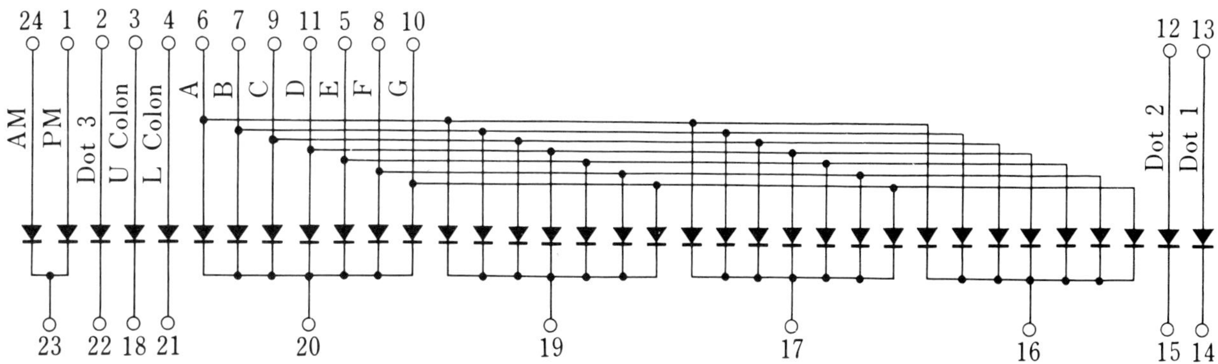

(b) Common cathode dynamic drive connection

***Figure* 7.31** *Connection scheme for clock display of time divison lighting type*

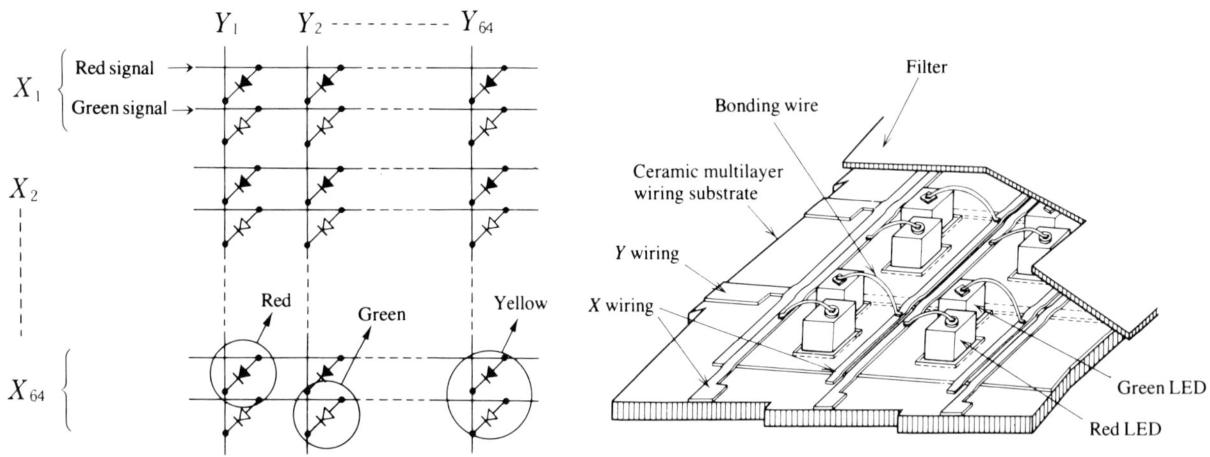

(a) (b)

Figure *7.32 (a) Matrix wiring and (b) layout of flat panel display (using red and green GaP LEDs)*

LED development as displays may have come to an end, though the development of a solid state equivalent to the CRT has not been achieved as yet.

One major problem remains with regard to the development of visible LEDs. This is the development of high-performance blue LEDs. The history of the development of blue LEDs began with the development of Si–C LEDs, but to date this development has not been brought to completion. The reasons for this are essentially bound up with the material. It seems likely that in order to break into this remaining unexplored region, further investigation of crystal physics will be needed.

Those who are currently engaged in the development of optical semiconductor devices are required not only to make steady progress in the development of light sources for optical data processing and optical communications (required as basic technical equipment for the construction of the information society) but also to have the originality to open up new fields. As one engaged in these matters, it is a pleasure to have the opportunity to increase their desire to conquer the mountains confronting them.

BIBLIOGRAPHY

Bergh, A. A., and P. J. Dean: *Light Emitting Diodes*, Clarendon Press, Oxford (1976).
Kressel, H., and J. K. Butler: *Semiconductor Lasers and Heterojunction LEDs*, Academic Press, New York (1977).
Sze, S. M.: *Physics of Semiconductor Devices*, Wiley–Interscience (1969).

REFERENCES

[1] O. W. Lossew: *Telegr. Telef.*, **18**, 61 (1923).
[2] R. J. Haynes: *Phys. Rev.*, **98**, 1866 (1955).
[3] H. Welker: *Z. Naturf.*, **7a**, 744 (1952).
[4] G. A. Wolff, P. H. Keck, and J. D. Broder: *Phys. Rev.*, **94**, 753 (1954).
[5] J. I. Pankove and M. Massooulie: *J. Electrochem. Soc.*, **11**, 71 (1962).
[6] W. Schockley: *Bell Syst. Tech. J.*, **28**, 435 (1949).
[7] S. M. Sze: *Physics of Semiconductor Devices*, Wiley–Interscience (1969).
[8] A. A. Bergh and P. J. Dean: *Light Emitting Diode*, Clarendon Press; Oxford (1976).
[9] R. A. Faulkner: *Phys. Rev.*, **175**, 991 (1968).
[10] D. G. Thomas and J. J. Hopfield: *Phys. Rev.*, **150**, 680 (1966).
[11] J. S. Jayson, R. Z. Bachrach, P. D. Dapkus, and N. E. Schumaker: *Phys. Rev.*, **136**, 2357 (1972).
[12] A. S. Jordan, A. R. Von Neida, R. Caruso, and M. Didomenico, Jr: *Appl. Phys. Lett.*, **19**, 394 (1971).
[13] P. D. Dapkus, W. H. Hackett, Jr, O. G. Lorimor, and R. Z. Bachrach: *J. Appl. Phys.*, **45**, 4920 (1974).
[14] T. Suzuki and Y. Matsumoto: *Appl. Phys. Lett.*, **26**, 431 (1975).
[15] A. J. R. Dekock, W. M. Van De Wijgert, J. H. T. Hengst, P. J. Roksnoer, and J. M. P. L. Huybregts: *J. Cryst. Growth*, **41**, 13 (1977).
[16] R. Z. Bachrach, W. B. Joyce, and R. W. Dixon: *J. Appl. Phys.*, **44**, 5458 (1973).
[17] K. Kurata *et.al.*: 4th Lecture Meeting of LED Experts, p. 21 (1980).
[18] K. Ikeda, S. Horiuchi T. Nakata, and W. Susaki: *IEEE Trans.*, **ED-24**, 1001 (1977).

[19] H. C. Casey, Jr. and B. B. Panish: *Heterostructure Lasers*, Academic Press (1978).
[20] J. S. Bass and P. E. Oliver: *J. Cryst. Growth*, **3** (4), 286 (1968).
[21] L. R. Weisberg, F. D. Rosi, and P. G. Herkart: Properties of elemental and compound semiconductors, in H. C. Gatos (ed.), Vol. 5, p. 25, Interscience, New York (1960).
[22] H. Nelson: *RCA Rev.*, **24**, 603 (1963).
[23] H. Nelson, U.S. Patent No. 3 565 702 (1971), received 14 Feb. 1969.
[24] J. W. Burd: *Trans. Metall. Soc. AIME*, **245**, 571 (1969).
[25] H. M. Manasevit and W. I. Simpson: *J. Electrochem. Soc.*, **116**, 1725 (1969).
[26] M. Naito and A. Kasami: IEEE Specialist Conference on the *Technology of Light Emitting Diodes*, Atlanta (1974).
[27] D. G. Thomas, J. J. Hopfield, and C. J. Frosch: *Phys. Rev. Lett.*, **15**, 857 (1965).
[28] R. A. Logan, H. G. White, and W. Wiegmann: *Solid State Electron.*, **14**, 55 (1971).
[29] D. G. Thomas and J. J. Hopfield: *Phys. Rev.*, **150**, 680 (1966).
[30] M. Iwamoto, M. Tashiro, T. Beppu, and A. Kascami: *Jpn. J. Appl. Phys.*, **19**, 2157 (1980).
[31] M. G. Craford, R. W. Shaw, W. O. Groves, and A. H. Harzog: *J. Appl. Phys.*, **43**, 4075 (1972).
[32] M. G. Craford: *IEEE Trans. Electron Dev.*, **ED-24**, 935 (1977).
[33] H. C. Casey, Jr. and M. B. Panish: *J. Appl. Phys.*, **40**, 4910 (1969).
[34] Y. Ohki, Y. Toyoda, H. Kobayashi, and I. Akasaki; GaAs and related compounds, pp. 479 (1981).
[35] K. Hirahara, A. Kamata, M. Kawachi, T. Sato, and T. Beppu: Extended Abstract of 15th Conf. on *Solid State Devices and Materials*, p. 349 (1983).
[36] W. A. Brantley and D. A. Harrison: Proc. IEEE Reliability Physics Symposium (1973), IEEE Cat. 73 CHO 755-9 PHY.
[37] M. Iwamoto and A. Kasami: *Appl. Phys. Lett.*, **28**, 591 (1976).
[38] T. Yamaguchi and T. Niina: Digest of the IEEE Specialist Conf. on LEDs, p. 158 (1980).
[39] David S. Cochran: Soc. Inform. Disp. 1973, *Int. Symp. Digest*, p. 82, Lewis Winner, New York (1973).
[40] B. L. Frescura; H. Luechinger, and C. A. Bittmann, *IEEE Trans. Electron Devices*, **ED-24**, 891 (1977).
[41] T. Niina, S. Kuroda, H. Yonei, and H. Takesada: *IEEE Trans. Electron Devices*, **ED-26**, 1182 (1979).
[42] Beppu and Ichikawa *et al.*: *Terebi Gaku Giho (TVScience Technical Reports)*, **3** (42), 37 (1980).
[43] O. Ichikawa, T. Sadamasa and T. Beppu: *Proc. 3rd Int. Dis. Res. Conf. 'Japan Display '83'*, p. 246 (1983).
[44] T. Niina *et. al.*: *SID 81 Digest*, p. 140 (1981).
[45] O. Ichikawa, T. Sadamasa, and T. Beppu: *Japan Display '83 Digest*, p. 246 (1983).

Chapter 8

Special displays

8.1 INTRODUCTION

This chapter is concerned with the principles, construction, and display characteristics, etc., of various types of electronic display, of the active type (light-emitting type) and passive type (non-light-emitting type), which were not covered in the previous chapters.

Various different principles are used in these displays. The display medium may be a solid, liquid, or a gas. Great efforts have been made to find ways of electrically controlling (in the case of the active type) the intensity of light emission or the emission spectrum, or (in the case of the passive type) the intensity or the spectrum of the reflected or transmitted light. Research along these lines is continuing, and further progress is expected.

Table 8.1 shows an overview of the progress of technical development of electronic displays and related technology. Table 8.2 shows the principles and construction of various types of electronic display technologies.

8.2 ELECTROPHORETIC IMAGE DISPLAY (EPID)

At the interface of two different phases, positive and negative electrical charges are normally generated on each side of the interface, and a layer is formed at which there is a potential difference. This layer is termed an 'electrical double layer'. In the case of the interface between a liquid and a solid, this electrical double layer gives rise to four different electrokinetic phenomena, as shown in Table 8.3 [1–3]. For a display, the electrophoretic or electroosmotic phenomena that occur on movement of one of the substances in response to an external voltage are utilized. First of all, we shall describe an electrophoretic image display (EPID) [4–10].

8.2.1 Operating principle of EPID

In a dispersed system in which colouring material, pigments, or the like are dispersed in a liquid in the form of fine particles, transfer of charge takes place

Table 8.1 *History of development of display technology*

Year	Active displays	Passive displays
1897	Invention of cathode ray tube (Braun)	
1926	Public testing of TV (UK)	
1933		Invention of SPD (Marks Corporation)
1936	Discovery of EL	
1937	Regular TV broadcasts started (UK)	Invention of Eidophore
1938	Invention of fluorescent light (GE Corporation)	
1947	Invention of transparent electrodes (Corning Corporation)	
1951	Colour TV perfected	
1952	Black and white TV transmission started (Japan)	
1956	Public testing of flat CRTs (Aiken)	Invention of DMD (RCA Corporation)
1957	Development of NIXI tube (Burroughs Corporation)	
1960	Ruby laser oscillation achieved (USA)	
	Colour TV broadcasts started (Japan)	
1962	GaAs semiconductor laser achieved (US)	
1964	Invention of AC-PDP (Illinois University)	
	Magnified projection of monochrome TV image using He–Ne laser (TI Corporation)	
1967	Announcement of flat CRT television (Sharp Corporation)	Development of Titus tube (Marie)
	Development of VFD (Ise Denshi Corporation)	

Table 8.1 (*contd.*)

YEAR	ACTIVE DISPLAYS	PASSIVE DISPLAYS
1968	Development of DC-PDP (Philips Corporation)	Announcement of DSM and GH liquid crystal display systems (RCA)
	Development of high-quality GaAs red LEDs	
1969		Announcement of WO_3 ECD
1970	Disclosure of projective type laser colour TV (Hitachi Corporation)	Development of transparent PLZT (Sandia Lab.)
1972		Announcement of TN and ECB liquid crystal display systems (Roche Corporation, Telefunken Corporation)
1973		Announcement of viologen ECD (Philips) Announcement of EPID (Matsushita Denki)
1974	Progress with thin-film AC-ELs (Sharp Corporation)	
1975		Announcement of MPD (Magnavox Corporation)
1976	Commercialization of dot matrix VFD (Ise Denshi Corporation)	Announcement of HAN type LCDs (Toshiba Corporation)
1977		Announcement of TBD (Xerox Corporation)
1978	Dot matrix TFEL module on sale (Sharp Corporation) Disclosure of colour TV using DC-PDPs (NHK)	
1980	Auroravision brought to practical form (Mitsubishi Electrical Corporation)	
1981		
1982	Flat CRT-TV on sale (Sony)	Watch liquid crystal TV on sale (Seiko Corporation)
1983		Liquid crystal black/white TV on sale (Casio Corporation)
1984		Liquid crystal colour TV on sale (Seiko Corporation)

Table 8.2 *Electronic display techniques*

	Name	Phenomenon	Construction
Light-emitting	CRT	Cathodoluminescence	Vacuum tube, phosphor layer
	LED	Injection luminescence	Semiconductor *pn* junction
	EL	Electric field luminescence	EL phosphor layer
	VFD	Slow cathodo-luminescence	Line hot cathode
			Slow excitation phosphor
	PDP	Gas discharge (glow)	Inert gas
	Nixie	Ditto	Ditto character type tube
	Tungsten lamp	Thermal radiation	Matrix of electric bulbs
Non-light emitting	LC	Birefringence scattering dichroism	Liquid crystal material
	ECD	Electrolytic coloration	ECD material
	EPID	Electrophoresis	Liquid dispersion
	SPD	Particle alighment	Ditto
	TBD	Particle rotation	Sheet with embedded multicoloured particles
	MPD	Magnetic particle	Sheet with embedded magnetic particles
		rotation	
	DFD	Foil electrostatic	Coloured liquid movable foil
		attraction	
	DMD	Ditto	Fine foil matrix
	EOD	Electroosmosis	Porous layer permeated by liquid
	EWD	Electrowetting	Ditto
	PLZT	Birefringence, scattering	Transparent strongly dielectric sheet
	MOD	Faraday rotation	Strongly magnetic thin film
	TCD	Thermal colour change	Thermal colour change material
	ED	Liquid evaporation	Liquid, heater
	Eidophore	Deformation of oil film	Oil film, electron beam, Schlieren optical system

Table 8.3 Four types of electrokinetic phenomena

	Movement of discontinuous phase (solid)	Movement of continuous phase (liquid)
Movement of substance by voltage	Electrophoresis	Electroosmosis
Voltage generation by movement of substance	Sedimentation potential	Flow potential

between the particles and the liquid, forming electrical double layers. As a result, the particles become positively or negatively charged. When a voltage is applied to such a dispersion, the Coulomb force causes the charged particles to migrate in the liquid. This phenomenon is called electrophoresis. This is used medically in the purification and separation of substances, or industrially in electrodeposition of coatings and in the wet developing processes in copying machines.

The construction of an EPID cell in which electrophoresis is used to provide a display panel is shown in Fig. 8.1. When voltage is applied across the dispersion between the electrodes, one of which is transparent, the particles migrate in a direction depending on the polarity of the voltage that is applied, producing a distribution as shown in the figure.

In Fig. 8.1, if we let the transmittance of the liquid be T, the reflectance of the white region is $I_0 R$ and the reflectance of the black region is $I_0\ R \cdot T^2$ so the contrast ratio of the display is $1/T^2$. For example, if the degree of coloration is $T = 18.2$ per cent, a contrast of 30 to 1 is obtained. Thus, with an EPID, high

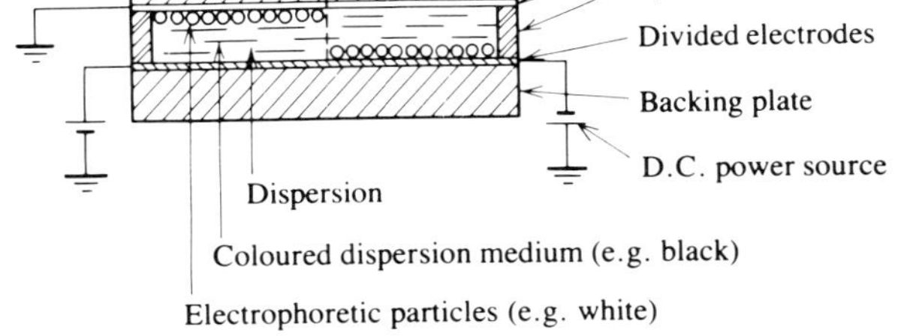

Figure 8.1 *Structure of EPID cell*

contrast is easily obtained. However, in some respects a compromise must be made between contrast and brightness of the white region [11,12]. An empirical relationship between the reflectance and contrast can be found [9] based on the analysis of Kubelka-Munk of the optical properties of printed layers.

8.2.2 Characteristics of EPID cells

The electrophoretic mobility μ of minute charged particles dispersed in a liquid is given by the following formula [2, 3].

$$\mu = \frac{\varepsilon\zeta}{6\pi\eta} \quad \text{(CGS system)} \tag{8.1}$$

where ε and η are the dielectric constant of the liquid and its viscosity, and ζ indicates the potential of the sliding surfaces of the particles, known as the 'zeta potential'.

The optical response speed t of the cell is determined by the time taken for the particles to cross between the two electrodes. From equation (8.1) above, the following equation is obtained:

$$t = \frac{6\pi\eta}{\varepsilon\zeta}\frac{d^2}{V} \tag{8.2}$$

where d and V indicate the interelectrode gap and the applied voltage.

Since the surface charge q of a single particle is proportional to the particle diameter a ($q = a\varepsilon_0\zeta$), and the frictional force f that acts during the migration is,

Table 8.4 *Characteristics of EPID cell*

Particles	Dispersion medium (dye)	Additive	Display colour	Cell thickness (μm)	Electrical strength (V/μm)
Composite particles (phenol + TiO_2)	Difluoil 0 m-xylene (Macrolex blue, oil red)	Polyethylene glycol; poly-ether amine salt	White/black	100	0.75
TiO_2	m-Xylene (Macrolex blue)	As above	White/blue	10	1
TiO_2	Trichloro-fluoroethane	Sodium organic sulphonate	Yellow/red	51	~ 1
Organic pigment	Perchloro-ethylene xylene (Sudan red 7B)	Imidopolymer	White/black	50	1
Composite particles (polyethylene + TiO_2)	Kerosene type		White/blue	50–100	1

from Stoke's law, also proportional to the particle diameter ($f = 6\pi a\eta$), particles of equal ζ potential show the same speed of migration irrespective of particle diameter, and since the particles were made small, the response cannot be rapid. Putting the matter the other way, although the size of the electrophoretic particles (about 1 μm) is enormous relative to the size of an ion, they show about the same mobility (10^{-4} to 10^{-6} cm^2/V·s) in the liquid as does an ion. From the above equation, as one example, if a dispersion with $\eta = 1$ cp, $\varepsilon = 2.5$, $\zeta = 50$ mV is inserted in a cell of $d = 10\ \mu$m, and a voltage $V = 10$ V is applied, $t \cong 14$ ms, which is a value that is practically equal to the experimental value. The characteristics of previously reported EPID cells are shown in Table 8.4 [7, 8, 10, 13, 14].

With an EPID, agglomeration of the particles results in unevenness of the display, and so must be avoided. The stability of colloid particles in respect of agglomeration is described by the DLVO theory [2, 3, 6, 15, 17] which was developed in the 1940s. This theory is based on calculating the combined potential energy, consisting of the Van der Waals attraction acting between the particles and the electrostatic repulsion due to the charge on the surface of the particles. If this energy barrier is sufficiently larger than the thermal energy obtained by Brownian motion, the dispersion is stable. Stabilization is also conferred by surfactants or polymeric materials adhering to the particles, preventing close contact between them. This is known as the steric hindrance effect. In the case of EPID, the dispersion is normally stabilized by a combination of both of these.

An example of EPID display is shown in Fig. 8.2. In an EPID cell, the layer of particles that accumulates on the electrode when voltage is applied exists stably over a long period even when the voltage is removed, giving the display a memory characteristic.

Table 8.4 *(contd.)*

Rise time (ms)	Fall time (ms)	Display contrast	Resistivity of dispersion (Ω·cm)	Power consumption Watts	Power consumption Joules	Life (number of times of switching)	Enterprise
300	200	25	5×10^{10}	11	3.3	$>2 \times 10^8$	Matsushita
30	20	5	6×10^9	1.6	0.05	$>10^7$	,,
9		10	7×10^9	71	0.64	$>10^7$	Xerox
30	20	11 (with filter)	10^{11}	5	0.15	$>2 \times 10^8$	Philips
100		20	2.5×10^9	200	20		Thompson CSF
		6	3×10^{10}			$>10^6$	Tektronix

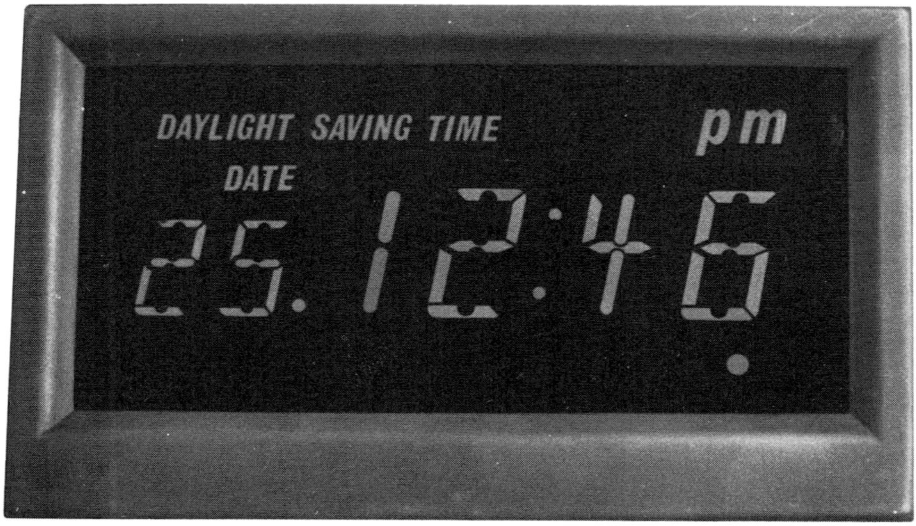

(*a*)

(*b*)

Figure 8.2 *Examples of EPID display panels. (a) Yellow–black display. (b) The back is black, the 'Technics' part changes from yellow to red, and the 'COMPONENT' part changes from white to blue. (c) White/black display*

8.2.3 Matrix display of EPID cells

The luminance–voltage characteristic of an EPID cell normally has no well-defined threshold value. Even low voltages, if applied for a long time, will produce particle migration. This makes it difficult to achieve a matrix display. EPID matrix displays have been realized by (a) giving the cell a threshold characteristic by adjusting the composition of the dispersion [6, 18–20], (b) giving the cell a threshold value by using a panel construction in which a grid electrode is inserted [8, 21], or (c) by preventing crosstalk by introducing non-linear elements such as varistors, MIMs, glow lamps, or TFTs [22].

In the case of method (a), there is the drawback of a slow response (several tens to several hundred of milliseconds per line). However, in the case of methods (b) and (c), the dispersion itself need not have a threshold characteristic, and addressing rates of several microseconds per line can be achieved. Xerox Corp. has realized a high-speed matrix display of 20 μs/line using an XY matrix EPID panel in which an array of non-linear elements is constituted on a varistor substrate [23].

8.2.4 Other applications of EPID cells

Prototype optical imaging [14, 24, 25] or X-ray image display memory plates [26–29] have been manufactured by laminating a photoconductor with an EPID dispersion. In one example of the former, an image of display contrast 6 to 1 is formed having intermediate tones, with a resolution of 5 lines/mm on a panel 28 × 38 cm. An X-ray image can be formed having a resolution of 4 lines/mm on 1 to 2.4 mR exposure. Using an electrostatic recording head placed above an EPID dispersion in sheet form, sandwiched between transparent insulating films, it is possible to form an image by applying an electrical field directly. Attempts have been made to apply this to create "soft copy" sheets that can be repeatedly rewritten and erased [30].

The internal phenomena that occur within an EPID during operation are extremely complicated. However, because of its attractive practical characteristics, such as the excellent visibility of the display, memory characteristic, low power, low cost, and the fact that large displays can easily be produced, development of improved cell structures and dispersion compositions, etc., is still continuing.

8.3 PARTICLE ROTATION DISPLAYS

8.3.1 Suspended particle displays (SPD)

A suspended particle display (SPD) cell* can be constructed, as shown in Fig. 8.3, by enclosing a suspension of colloidal needle- or plate-shaped light-absorbing particles in a transparent liquid between a pair of transparent electrodes. When no voltage is applied, due to the Brownian motion, the suspended particles are

*Marketed, for example, under the trade name 'Varad'.

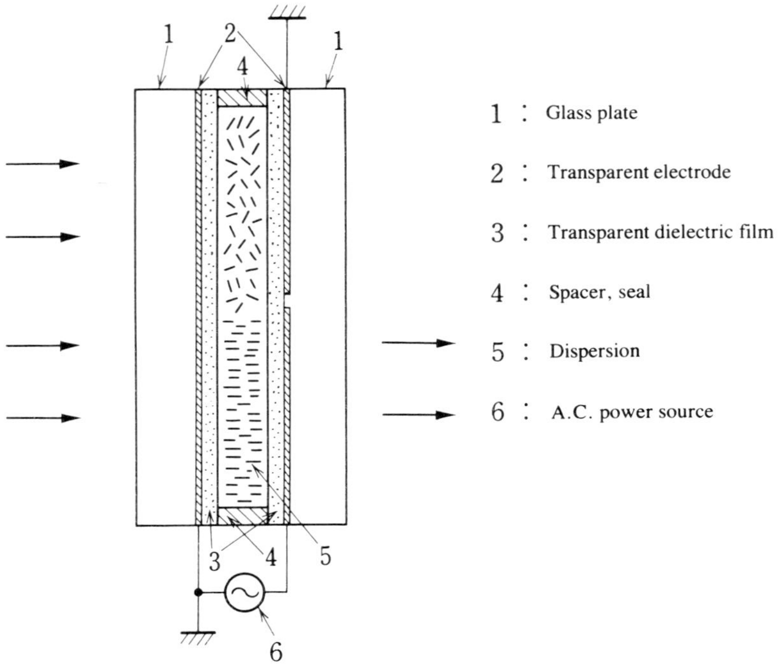

Figure 8.3 *Principle of SPD*

oriented in all directions, so incident light is absorbed or scattered by the particles, and the cell does not transmit light.

However, when a voltage (normally a.c.) is applied between the electrodes, the particles are polarized in the longitudinal direction and become aligned in the direction of the field. This results in a sharp decrease in the optical absorption cross-section, and light passes through the cell.

The interaction of light and particles can be analyzed by methods similar to the dipole antenna theory for microwaves [31]. Previously, the particles used were needle-shaped crystals of herapathite. Recently, iodine crystals (specific gravity about 1.54) in a base of hydrocinchonidine, in place of the previous quinine, have been used. Settling of the particles is prevented by matching the specific gravity by using a halogenated hydrocarbon solvent. It is reported that a dispersion of improved stability to heat and light can be obtained [32]. The characteristics of this cell are shown in Table 8.5.

The SPD cell is similar in operation to the GH type liquid crystal display described in Section 2.3.3, in that the absorption characteristic of the light is varied by dichroism of the particles, and consequently by their state of arrangement. However, unlike dichroic dye molecules, the size of the particles is enormous, so they have a large dipole moment. For this reason, there is no need to use a host liquid crystal, and ordinary liquids of low viscosity can be used.

Table 8.5 Characteristics of SPD

ITEM	CHARACTERISTICS
Interelectrode separation	50 μm
Voltage	5–10 V_{RMS}
Frequency	30–500 HZ
Display power consumption	About 1 $\mu W/cm^2$
Contrast	10–30
Rise time	100 ms
Fall time	300 ms
Viewing angle	Large
Panel size	1 cm square–50 cm square

(Research Frontiers Corporation) [32].

This method has the advantage that, if reliability can be sufficiently assured, a polarizer is unnecessary, so both transmission and reflection types are possible, and the contrast is higher than that of a GH type liquid crystal display. Cells produced by inserting SPD dispersions between mica plates and transparent electrode substrates are used as face plates of CRTs. A transmission type magnifying projection display has been used on a trial basis by RCA [33].

8.3.2 Rotation display using multicoloured particles (twisting ball display) (TBD)

This is the TBD (twisting ball display) reported in 1977 by Xerox Corp. and given the name Gyrikon [34]. As shown in Fig. 8.4, the construction is such that fine

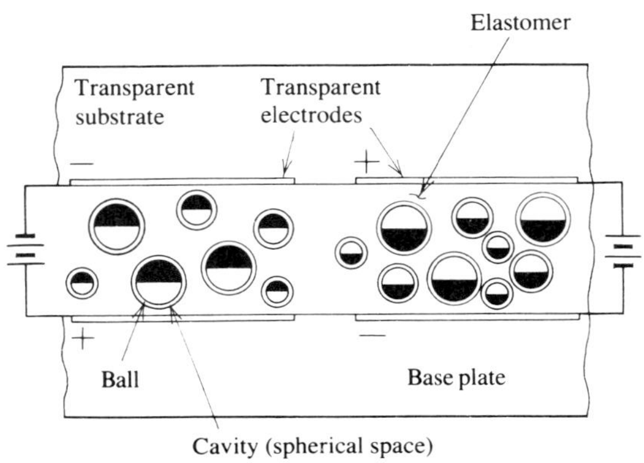

Figure 8.4 *Principle of TBD (Xerox Corporation) [34]*

spherical particles (diameter several 10 μm) whose hemispherical surfaces are of different colours can rotate freely in oil-filled spherical cavities formed in a layer of transparent elastomer (thickness 25 to 100 μm). As explained in Section 8.2 an electrical double layer is formed at the interface between the particles and the oil and, since the particles are constituted of different substances in each hemisphere so that they show different colours, the surface charges of the particles are also different in each hemisphere. Consequently, the particles have a dipole moment in the axial direction of the sphere. When an electrical field is applied, a torque acts such as to turn the axial direction in the direction of the electric field, so that all the hemispherical sides of the particles face the same way. If the direction of the electrical field is reversed, the inclination of the particles is also reversed, so the reflected colour of the display changes. The particles used in the TBD are comparatively large and normally they are stationary and in contact with the wall between the spherical cavities, so when the electrical field is removed, the particles do not rotate. Thus the display shows a memory characteristic.

Considerable progress has been made with TBDs, and Sony have recently developed a TBD with a new cell construction using an improved particle material and method of manufacture [35]. Table 8.6 shows the construction and characteristics of prototype TBD cells manufactured by Xerox Corp and Sony

Table 8.6 *Structure and characteristics of TBD cells*

	XEROX [34]	SONY [35]
Particles		
Materials	TiO_2 added glass	Li_2O–TiO_2–SiO_2 glass
Colouring material (evaporated)		Sb_2S_3, MgF_2
Particle diameter	25 μm	50 ± 5 μm
Elastomers		
Material	Rubber	PVA
Thickness	250 μm	150 μm
Particle layers	5 layers	2 layers
Liquid	Silicone	Toluene
Additives		Aerosol OT
Voltage	± 50 V	± 100 V
Display contrast	(3–5):1	(4–6):1
Response time	200 ms	20–30 ms
Resolution		0.3–0.5 mm
Conductance		$10^{-10}\,\Omega^{-1}/cm^2$
Capacitance		30 pF/cm^2
Power consumption		1 μW/cm^2
Number of times of switching	$> 10^7$	$> 10^8$
Viewing angle		$\pm 85°$
Panel size		10×14 cm

Figure 8.5 Examples TBD display (Sony) [35]

Corp. Figure 8.5 is an example of a prototype TBD numerical display made by Sony Corp. Figure 8.6 is a photomicrograph of the spherical particles (particle size about 50 μm) used in this display.

This type of display has the drawback that the particles and the method of cell

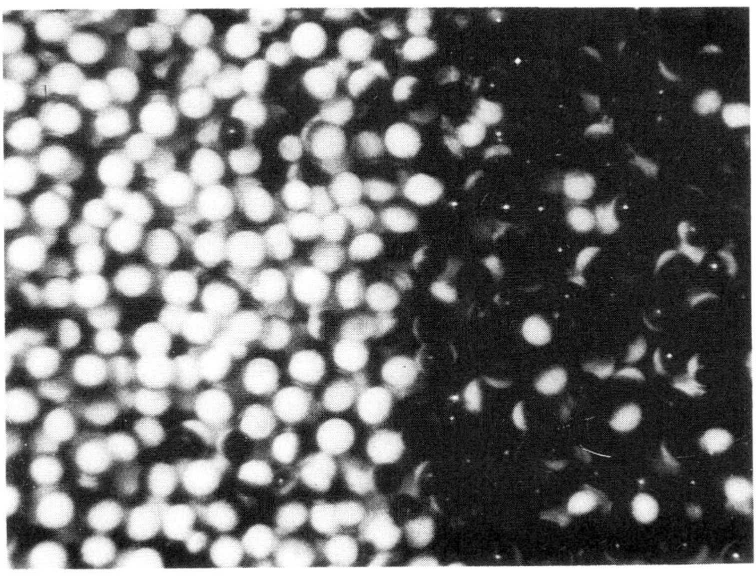

Figure 8.6 Electron photomicrograph of particles used in Figure 8.5 (Sony) [35]

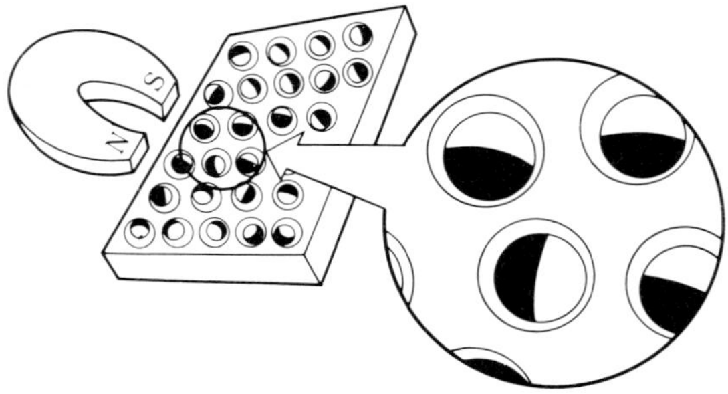

Figure 8.7 Principle of MPD (Magnavox Corporation) [36]

manufacture are somewhat complicated. However, it has the advantages that the display can easily be scaled up, is easy to view, and has a memory characteristic and low power consumption.

8.3.3 Magnetic particle display (MPD)

This is a method that was proposed by Magnavox Ltd. in 1975, before the TBD system described in the preceding section. Initially, it was investigated with TV display in view. The principles of an MPD (magnetic particle display) are shown in Fig. 8.7.

Magnetized spherical particles of which one hemisphere is white while the other hemisphere is coloured black are used as the display medium. Initially, the particles were used as aggregates, but, because, apart from the rotation of the particles, migration took place in the direction of the display faces, resulting in non-uniformity of particle concentration and non-uniformity of the display, the particles are now used in the form of individual egg-like microcapsules, which are spread out on a transparent sheet [39]. Addressing of the magnetic particles uses

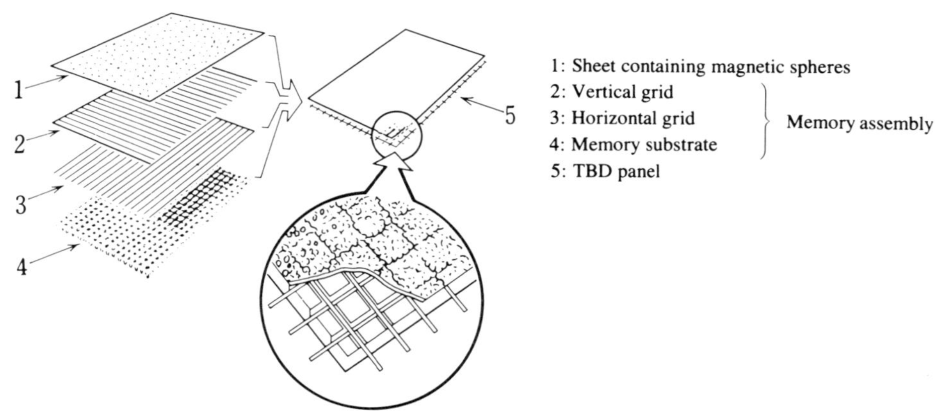

Figure 8.8 Construction of matrix type MPD panel (Magnavox Corporation) [37, 38]

Table 8.7 *Characteristics and outlook of MPD [38]*

CHARACTERISTICS	1977 LEVEL	IN PROSPECT
Particle diameter	30–400 μm	10 μm
Pixel size	0.15–1.7 mm	0.2 mm
Number of pixels	30 × 60	1000 × 1000
Display contrast	15	20–40
Crosstalk	None	—
Particle sensivity	7 G	3–5 G
Driver power	0.05 J/pixel (at 1.7 mm pixel)	Less than one-tenth of the 1977 figure
Temperature	−65–65 °C	Higher temperature than that achieved in 1977

the same technique as in the case of a magnetic core memory. As shown in Fig. 8.8, a memory assembly is constituted using vertical and horizontal grid conductors (2, 3) arranged in the form of an XY matrix, and a strongly magnetic memory substrate (4). At the points of intersection of these matrix conductors, corresponding to pixel regions, the direction of magnetization of the magnetic layer can be reversed by reversing the direction of the current flowing in these conductors. The layer of magnetic particles (1) can be rotated to present a white, a black, or a grey colour, by using the magnetic field of the strongly magnetic layer at the back to apply a rotary torque to the particles. This state is maintained. Assuming 500 × 500 pixels are point sequentially addressed by a TV signal, the address time of one pixel must be about 0.12 μs (= 63.5 μs/500). This is not long enough to rotate the particles, but a sufficient magnetization response of the ferromagnetic body ensures that, once the ferromagnetic body is magnetized, its magnetization continues to act on the particles, so it is sufficient if the particles can complete their rotation within the frame frequency (≅ 30 ms). Thus the memory assembly, besides giving a memory characteristic as already mentioned, plays) an important role in providing a threshold characteristic.

Techniques for manufacturing memory assemblies that have been investigated [39] include: forming XY matrix conductors on front and back surfaces by photoetching both faces of copper foil stuck on to polyimide and etching the polyimide into a concave shape in the pixel regions, these concavities then being filled with ferrite powder.

The stage of development that this technique had reached by 1977 and future prospects are shown in Table 8.7 [38]. Compared with a display method [40] in which magnetic particles of a magnetic powder dispersion known as a 'bitter solution' are oriented in a magnetic field, the method of the Magnavox Company, in which the particles have separate coatings, is characterized by improved viewing angle and contrast.

8.4 ELECTROSTATIC ATTRACTION DISPLAYS

One type of electrostatic attraction display is an Eidophore, which is used in magnified projection type TVs and employs deformation of an oil film corresponding to the magnitude of electrostatic charge deposited by an electron beam. However, we shall here discuss a display system wherein the optical reflection characteristic is altered by using electrostatic force to change or deform the position of attraction of a foil.

8.4.1 Foil attraction display (dye foil display) (DFD)

The DFD (dye foil display) was proposed by Philips Corp. [41] and Bendix Corp. [42] in 1980. As shown in Fig. 8.9, a metal foil (6) is immersed in a coloured liquid (5), being fixed at its right-hand end to a spacing element. The metal foil is electrically connected to an electrode arranged at, for example, its rear face. The metal foil and a front electrode form a capacitor, so, if the distance between the two electrodes is x, the foil is subject to an attractive force $f = (1/2)\varepsilon(V/d - X)^2$ and a restoring force Cx (where C is the modulus of elasticity). The effect is that the display becomes brighter because the foil approaches the surface electrode. When the voltage is turned off, the metal foil returns to its original position due to the strain-restoring force of the metal foil. The display will have the colour of the liquid. It is said that a well-defined threshold characteristic can be provided, by arranging for the foil to be attracted to the front electrode by a voltage at which the restoring force of the metal foil exceeds the limit of proportionality [41]. The characteristic of a prototype cell manufactured by Bendix is shown in Table 8.8 [42]. The metal foil required for the display is generally formed by photoetching.

1: Transparent base plate
2: Transparent electrode
3: Insulating spacer
4: Cell spacer and seal
5: Coloured liquid
6: Reflecting metal foil
7: Back electrode base plate
8: A.C. power source
9: Switch

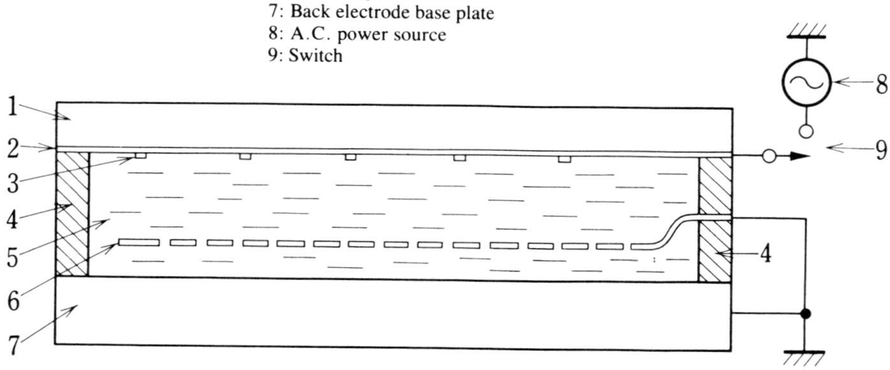

Figure 8.9 *Principle of DFD*

Table 8.8 Characteristics of DFD

Relative reflectivity in on state	80%
Contrast	> 13:1
Viewing angle	180°
Operating and storage temperature	−65–+100 °C
Element width	0.6 mm
Operating voltage	45 V_{AC}, 1 kHz
Power consumption	10 W/cm^2
Number of times of switching	> 10^7
Response (at 25 °C)	500 ms
Cell gap	100 μm

(Bendix Corporation) [42].

Another type of large display device that has been developed using the principle of electrostatic attraction is the MDD (moving dielectric display). As shown in Fig. 8.10, the display is effected by opening and closing a shutter of thin metal film by electrostatic force. Large displays can be fabricated with a simple construction. Seven-segment type numerical displays are commercially available [43, 44]. A movable electrode that performs the shutter action is produced by sputtering a 500 Å thickness of aluminium onto a 5 μm thick polyester film, and is arranged so as to surround a fixed electrode consisting of aluminium pipe coloured by paint. When the movable electrode and fixed electrode are at the

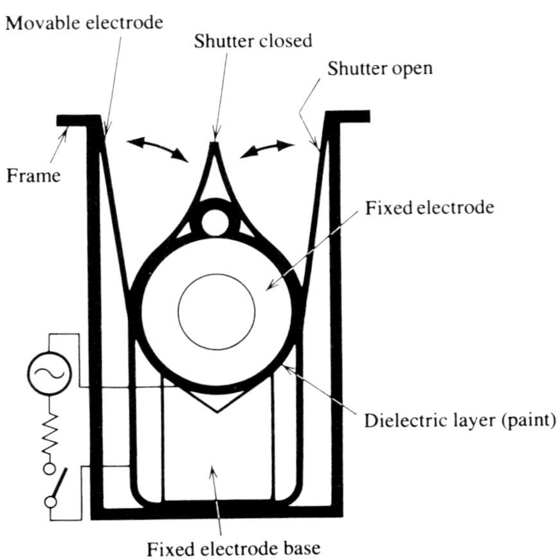

Figure 8.10 *Construction of MDD (Display Tech Corporation) [43]*

same potential, the shutter is opened by the elastic restoring force of the film. The shutter is only closed when a voltage (normally a.c.) is applied. In the prototype MDD, the height of the numbers ranged from several centimetres about 30 cm.

8.4.2 Foil deformation type display (deformable mirror display) (DMD)

The DMD (deformable mirror display) was invented by Auphan in 1956. Based on the DMD system, a prototype light valve display of the sealed-off type was manufactured in the 1970s by the RCA Corp. [45] and WH Corp. [46], etc. The object was to obtain a magnifying projection type TV in which the drawbacks of eidophores, due to denaturing of the oil film and contamination of the cathode, etc., could be avoided.

The WH Corp. system is shown in Fig. 8.11. Instead of the photoelectric surface of the ordinary 1.5 inch vidicon tube, minute mirror surfaces (corresponding to the pixels) are formed at 50 μm pitch on a matrix target formed by a sapphire substrate (Fig. 8.12). This target is scanned by an intensity-modulated electron beam. This deposits electrostatic charge onto this mirror matrix. As a result, each of four leaf-shaped mirrors receive a deformation of maximum angle 4° corresponding to the electrostatic charge, due to the Coulomb force between them and the substrate. A magnified image is projected using a reflective type

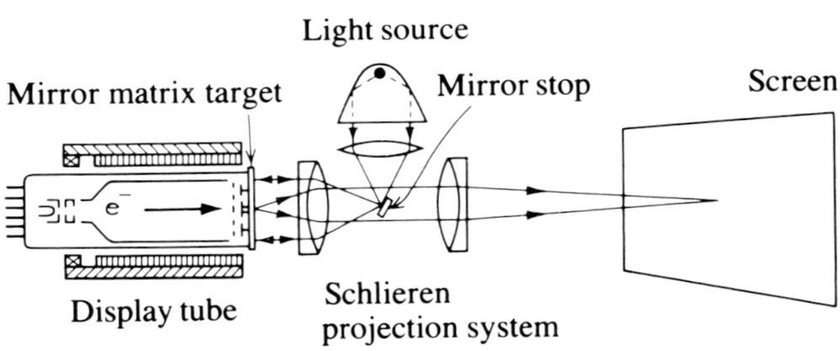

Figure 8.11 Projective display system using DMD (WH Corporation) [46]

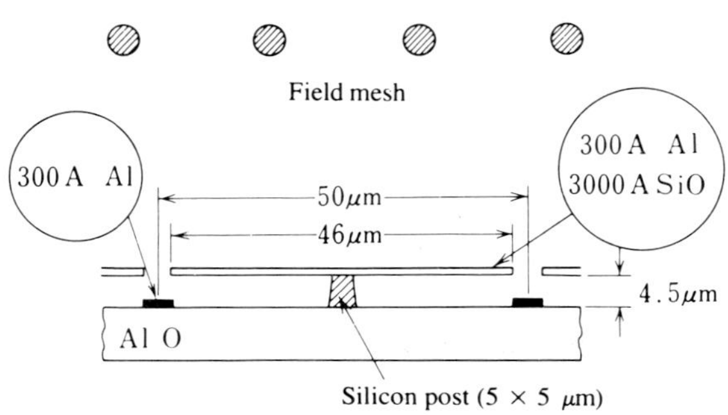

Figure 8.12 Detail view of one pixel of DMD target (WH Corporation) [46]

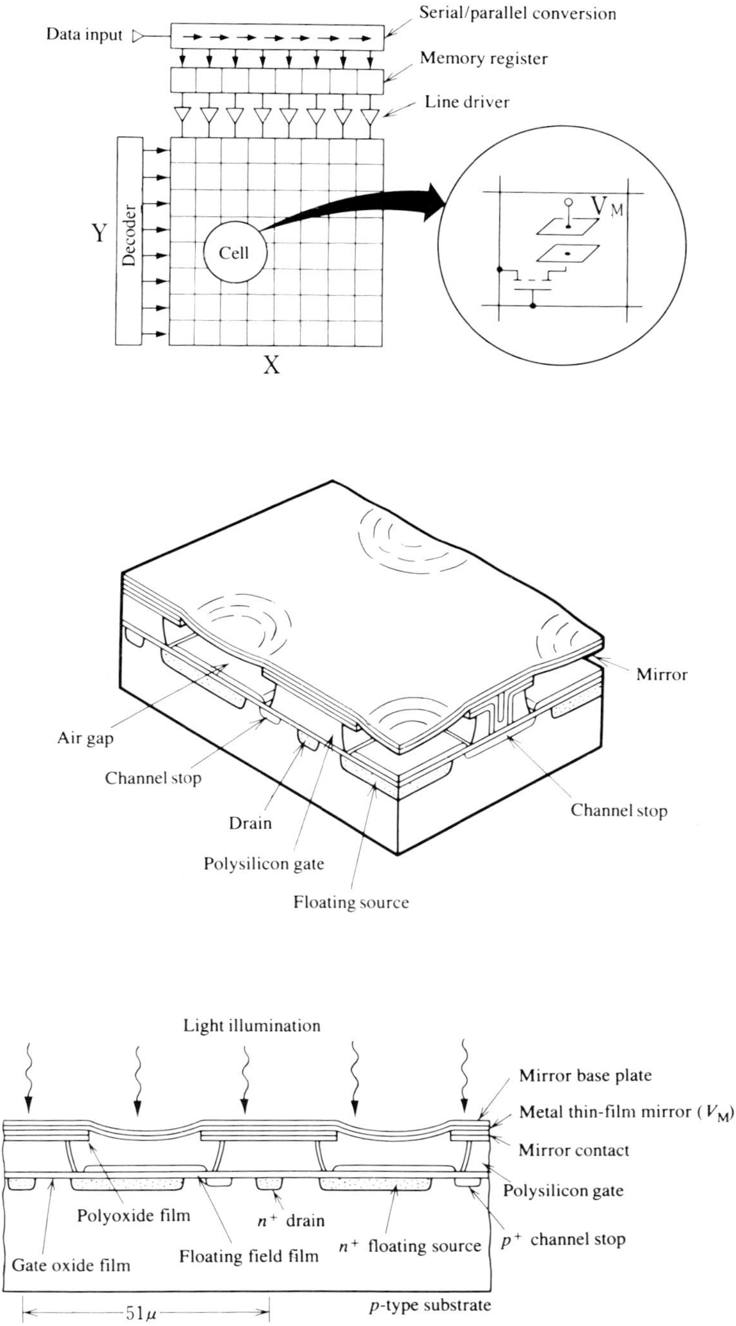

Figure 8.13 *Structure of solid state DMD (TI Corporation) [47, 48]*

Table 8.9 Characteristis of DMD projected image

Screen size	75 × 105 cm
Light source	150 W xenon (1000–1500 lm)
Effective optical flux to the target	135 lm
Effective optical flux to the screen	65 lm
Luminance	35 fL (screen gain 5)
Display contrast	15:1
Number of pixels	200 000
Resolution	400 TV tubes (vertical)
Grades	5–6

(WH Corporation) [46]

Schlieren optical system. The characteristics of this DMD projected image are shown in Table 8.9.

The targets of Fig. 8.12 are indeed manufactured using semiconductor process technology, but, insofar as deflection and modulation of the electron beam are employed to form an electrostatic charge, they may be considered as a kind of vacuum tube. A method of fabricating such devices entirely using solid state techniques has been tried by the TI Inc. [47, 48]. A solid state DMD construction is shown in Fig. 8.13.

An active matrix is constructed wherein, to prevent crosstalk, an FET is provided at each pixel. Analogue or digital signals are fed to floating sources in line sequential manner through drain lines. Specular deformable metal thin films, held at the same potential, are supported by side walls, which also serve as polysilicon gates. At each pixel, the potential difference between the metal film and the signal potential supplied to the source is therefore applied to an air gap capacitance of 0.62 μm. This produces a depression in the thin film, corresponding to the potential difference. Light directed onto the film will therefore be phase modulated. When an evaluation was carried out of the picture quality of the projected image produced in a Schlieren optical system by applying a TV speed signal to an array of 128 × 128 pixels with a pitch of 102 μm, a macro contrast of 10 to 1 with an aperture ratio of 32 per cent and mirror deformation response time of 25 μs were obtained [48]. This DMD device is intended for application to high-speed optical information processing or projection displays.

8.5 ELECTROOSMOTIC DISPLAY

As stated in Section 8.2, when an electrical double layer is formed at the interface between a liquid and a solid, the liquid itself becomes positively or negatively charged, and so moves when a voltage is applied to it. This electroosmotic phenomenon can be utilized in displays [49, 50]. For example, if a white porous substance is impregnated with a transparent liquid of equal refractive index, this porous substance will transmit light. However, if the liquid with which the porous substance is impregnated is expelled by electroosmosis, the light will be dispersed, because of the difference in refractive indices at the inside walls of the porous substance. Passage of the light is thereby cut off. A display based on this principle is called an EOD (electroosmotic display).

As an example, a numerical display has been achieved by the cell construction shown in Fig. 8.14 [51]. The cell constituent materials are shown in Table 8.10. The optical transmissivity of the porous layer at the front face when this is impregnated with liquid is practically 100 per cent. The transmissivity of the porous film on its own is 6 per cent so a maximum display contrast of 16.7 may be anticipated.

The response speed of optical control when a porous substance of mean capillary pore diameter $\phi = 0.5\,\mu m$ is used is about 20 to 30 ms both for rise (becoming non-transparent) and fall (becoming transparent) for a drive voltage of

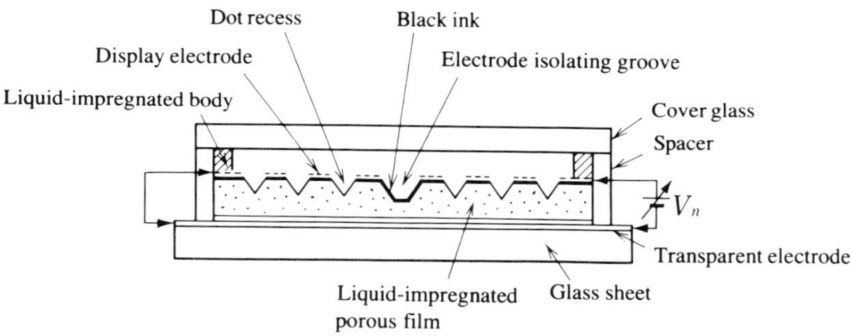

Figure 8.14 *Construction of EOD panel (Matsushita Denki) [51]*

Table 8.10 *Materials used in construction of EOD cells [51]*

Porous film	
Material	Cellulose acetate
Thickness	$120\,\mu m$
Refractive index	1.472
Capillary mean pore diameter	$0.2\,\mu m$
Void percentage	80%
Impregnated liquid	
Composition	Mixture of alkoxy silane ($n_d = 1.429$) and alkyl naphthalene ($n_d = 1.582$)
Refractive index	1.472
Viscosity	2 cs
Surface tension	27 dyne/cm

60 V. For the same drive voltage, porous bodies of smaller ϕ give a better display contrast, but their response speed is slower. Taking these opposing conditions into consideration, the optimum particle diameter is reported to be $\phi \cong 0.5\ \mu m$ [51].

Figure 8.15 shows an example of a display in which the EOD panel of Fig. 8.14 is projected using a slide projector. The effective panel area is $45 \times 35\ mm^2$, and the power consumption is about $100\ \mu W/cm^2$. In this EOD panel, the depressions at the dots are coloured by red, green, and blue coloured inks. The flat area outside the depressions is constituted as a black matrix in order to improve the display contrast.

A display that is very similar to the above EOD is the EWD electrowetting display (proposed by Bell Laboratories [52, 53]. 'Electrowetting' is the phenomenon whereby pressure at the interface between an electrode and an electrolyte is changed by application of a voltage, producing a change in the wetting. In order to apply this to a display, the cell structure of Fig. 8.16 has been proposed [52].

Liquid reservoir R is filled with electrolyte solution of refractive index equal to that of porous layer O. On the inside surface of the porous layer there are provided transparent electrodes (6) consisting of indium oxide or gold. By applying a voltage between these electrodes and the electrolyte solution in reservoir R, the interior of porous layer O is wetted with electrolyte, making it

Figure 8.15 *Example display produced by EOD (Matsushita Denki) [51]*

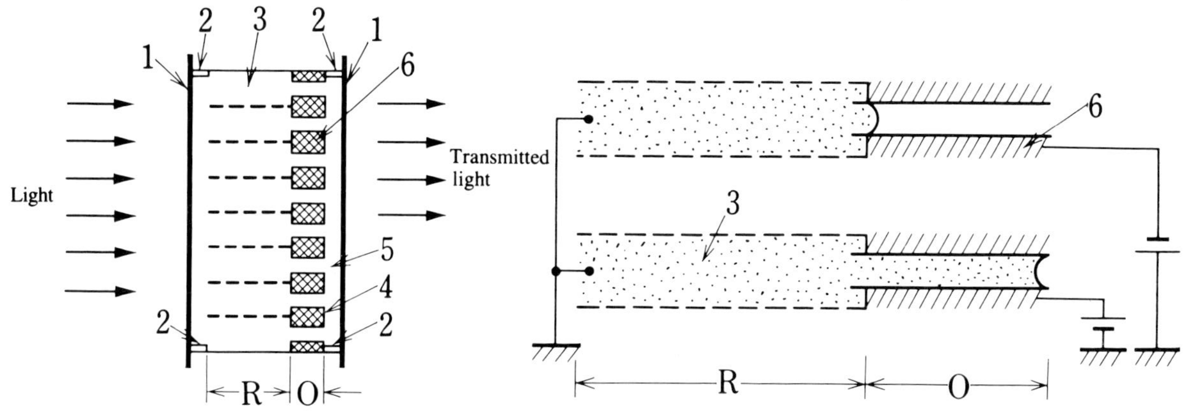

1: Transparent plate
2: Spacer
3: Electrolyte
4: White porous layer
5: Gas
6: Transparent electrode
O: Optical zone
R: Liquid storage zone

Figure 8.16 *Principle of EWD (Bell Research) [52]*

transparent. Applying a voltage that is not able to produce wetting produces a light-scattering condition. Thus the device operates as a light valve. According to calculation, the response speed and operating voltage should be respectively less than 1 ms and less than 1 V, and power consumption is estimated [52] as less than 1 mW/cm^2. However, this has not yet been practically verified.

8.6 *ELECTROMAGNETOOPTIC DISPLAYS*

Phenomena whereby the spectrum of light transmitted or reflected by a substance, its transmissivity, or polarization characteristic, etc., change in response to an applied electrical or magnetic field are called electromagnetic optical effects. Various types of displays have been proposed using crystals, ceramics, or thin metal films, etc.

8.6.1 Electrooptic display (ELOD)

The phenomenon of change of the refractive index of a substance in response to an electrical field is called the electrooptical effect. This change is divided into the linear electrooptical effect (Pockels effect), which is proportional to the first power of the electric field, and the quadratic electrooptic effect (Kerr effect), which is proportional to the second power. A display based on such electrooptic effects is called an ELOD (electrooptic display). Whereas the Kerr effect occurs in all substances, the Pockels effect only appears in 21 out of 32 types of crystal structure, and is associated with piezoelectricity. Typical piezoelectric crystals showing the Pockels effect are KDP, DKDP, ADP, CuCl, ZnSe, and GaAs.

(1) The principle of ELOD

A Pockels cell light valve, like the liquid crystal display discussed in Chapter 2, utilizes the change in birefringence produced by application of an electrical field. A transparent electrode is provided at both faces of a uniaxial crystalline plate wherein the optical axis is in the thickness direction of the plate (z axis) and a longitudinal* mode cell structure is produced, with crossed polarizer plates arranged on each side. When no voltage is applied, in the crystal plate, $n_x = n_y = n_o$, so there is no birefringence, and light is not transmitted through the crossed polarizers. On the other hand, when voltage is applied, n_x is not equal to n_y (biaxiality), and birefringence is produced. The light that passes through the crystal is elliptically polarized, and so is transmitted through the crossed polarizers. The birefringence Δn is given by the following equation [54]:

$$\Delta n = \gamma_{63} V \frac{n_0^3}{d} \tag{8.3}$$

*Mode in which the direction of the electric field is parallel to the direction of propagation of the light.

where γ_{63} is the electrooptic coefficient, V is the applied voltage, n_0 is the refractive index when there is no electrical field, and d is the thickness of the crystal.

Since the phase difference is $\Delta\phi = (2\pi/\lambda)\ \Delta n \cdot d$, if the azimuthal angle of the polarizers is set at $\pi/4$, the tramsmissivity T of the cell is,

$$T = \frac{\gamma^2}{2} \sin^2\left(\frac{\Delta\phi}{2}\right) = \frac{\gamma^2}{2} \sin^2\left(\frac{\pi}{\lambda}\Delta n \cdot d\right) = \frac{\gamma^2}{2} \sin^2\left(\frac{\pi}{\lambda}\gamma_{63} V n_0^3\right) \tag{8.4}$$

where γ indicates the maximum transmissivity of a linear polarizer for linearly polarized light.

When $\Delta\phi = \pi$, incoming linearly polarized light has its plane of rotation rotated through 90°, and maximum contrast is obtained. The voltage necessary for this 90° rotation is called the half-wavelength voltage $V_{\lambda/2}$, and is given by

$$V_{\lambda/2} = \frac{\lambda}{2} \frac{1}{\gamma_{63} n_0^3}$$

From equation (8.5), it can be seen that crystals of large γ_{63} and n_0 can be driven by low voltage.

(2) Applications of ELODs

Marie *et al.* constructed in 1967 a prototype reflection type projection display utilizing the Pockels effect of potassium dihydrogen phosphate (DKDP) crystals. They called this the Titus tube [55, 56]. Its construction is shown in Fig. 8.17.

A grid is arranged 50 μm from a target (crystal plate). An electrostatic charge image is formed on the target by using a video signal to vary the energy (secondary electron emission ratio) of an electron beam projected onto it. The DKDP crystal shows its highest anisotropy of refractive index at the Curie temperature (in the neighbourhood of −50 °C). Consequently, to obtain low operating voltage, the target ($30 \times 40 \times 0.25\ mm^2$) is cooled to −50 °C by a Peltier cell arranged in the tube. A reflecting arrangement is used, because it doubles the light path and enables a lower voltage to be used. In this way, the voltage required for 75 per cent optical modulation is lowered from 1.2 kV at room temperature to about 100 V at −50 °C. Using a 2.5 kW xenon lamp, a black

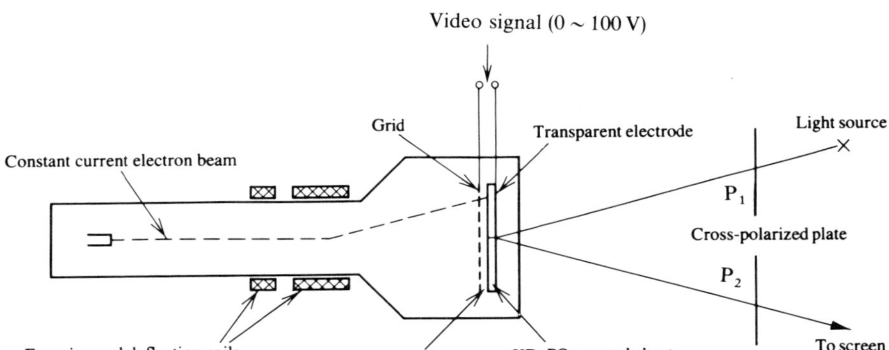

Figure 8.17 *Construction of Titus tube using Pockels effect*

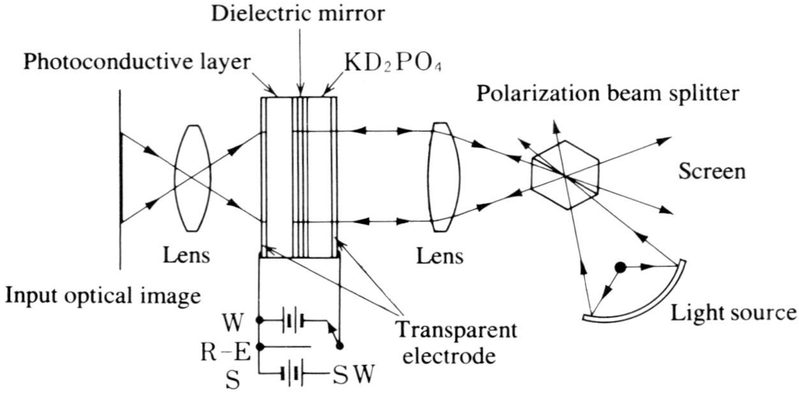

Writing: SW position = W (illumination with optical image)
Erasure: SW position = R-E (illumination by uniform light)
Projection: SW position = R-E
Superposition: SW position = S (illumination with second optical image)

Figure 8.18 *Construction of Phototitus system [55]*

and white TV image can be obtained with a light output of 2500 lm, display contrast 40:1, and resolution 700 points/line.

A prototype colour Titus system for projecting a colour TV image was manufactured in 1976 using two dichroic mirrors and three red, green, and blue light valves as shown in Fig. 8.17. Using a 4 kW xenon lamp, 2800 lm light output, and display contrast (30 to 60):1, a resolution of 750 dots/line was obtained [55, 56]. A trial optically addressable device (called a Phototitus) was also constructed by laminating a DKDP sheet with selenium (photoconductive film) (Fig. 8.18) and sandwiching this arrangement between transparent electrodes. Display contrast of better than 40:1 and a resolution of 1500 dots/line were achieved.

In about 1969, in research into composition and methods of synthesis, Sandia Lab. (USA) discovered that PLZT ceramic could be made transparent. This discovery was applied to optical shutters and displays etc.

An example of an optically addressable storage type imaging device using a PLZT sheet is shown in Fig. 8.19 [57]. A transparent conductive film (ITO) is provided on both faces of a PLZT sheet of thickness 200 μm laminated with polyvinyl carbazole photoconductive film. When an optical image is projected, using a slide projector or the like, onto the photoconductive film, whilst voltage is applied between the electrodes, a polarized image corresponding to the intensity of the light is formed and stored in the PLZT sheet. When such a cell is placed between crossed polarizers, or, as shown in Fig. 8.19, when dispersed light and non-dispersed light are separated using a Schlieren optical system employing an aperture, a projected image with excellent contrast is obtained. A clear image of resolution at least 20 lines/mm is obtained with a contrast ratio of 25 or more, using a 28 mm square PLZT plate. To remove the display image, a reverse bias

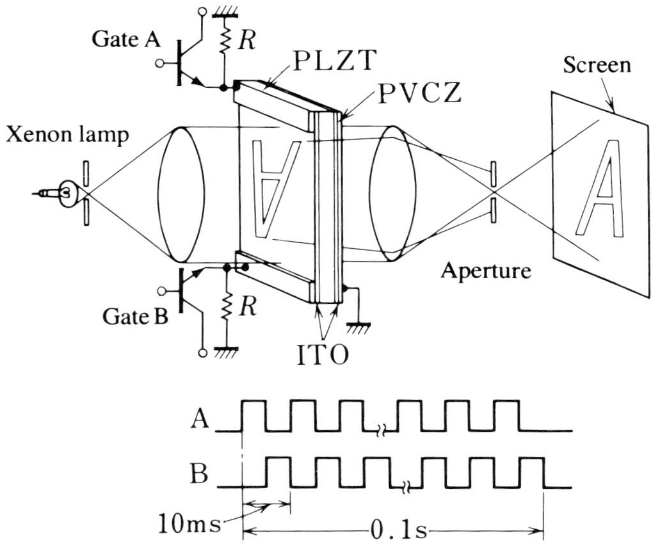

Figure 8.19 *PLZT image storage display device (Hitachi Central Research) [57]. (Signals A and B are applied to the ITO for image erasure, performed by heating and applying reverse bias)*

voltage may be applied while heating at the same time as current is passed to the ITO.

A TV display has been tried using PLZT elements in a matrix electrode arrangement [58, 59]. A real-time TV display has been achieved by line sequential driving a video signal with an XY matrix panel provided with 30 rows of ITO electrodes at the front and 30 columns of Al electrodes at the back of PLZT of thickness 200 μm and size 32×24 mm^2.

Apart from use in displays, the high-speed characteristics of PLZT cells are utilized, for example, in optical shutters, welding protection goggles, and three-dimensional TV glasses. In the case of the PLZT shutter [60] marketed by Motorola Inc., copper/nickel electrodes of 30 to 40 μm width are provided* in the form of a comb of 0.5 to 1 mm pitch on a PLZT wafer of thickness 0.3 mm with crossed polarizers arranged on either side. $V_{\lambda/2}$ is 280 V (in the case of 0.5 mm pitch), modulation frequency is maximum 10 kHz, and the optical transmissivities in the on and off states are 14 and 0.007 per cent respectively.

8.6.2 Magnetooptic display (EMOD)

Magnetooptical effects include the Faraday effect, magnetooptic Kerr effect, Cotton–Mouton effect, and oscillatory magnetooptical effect. Display systems

*Transverse mode.

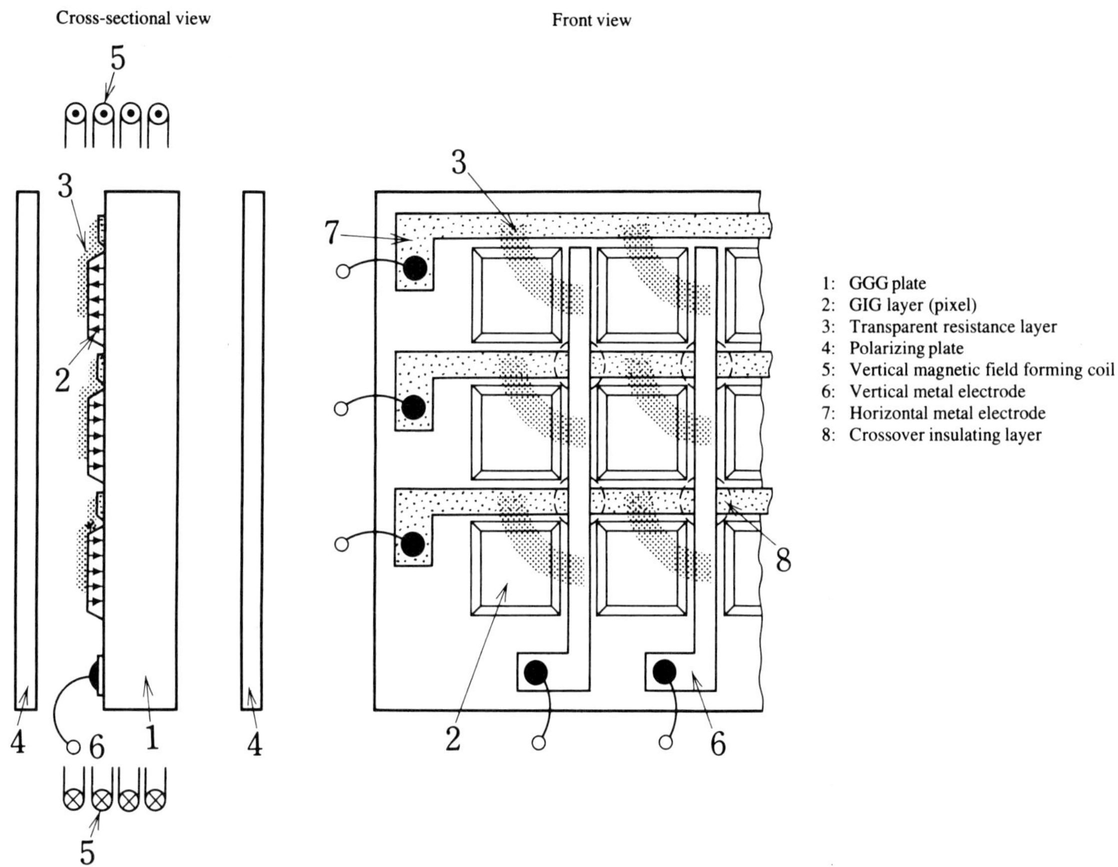

Figure 8.20 *Construction of EMOD (Philips) [61]*

using the Faraday effect have been proposed, and are called EMOD (electromagnetooptical displays). The Faraday effect is the effect whereby, when linearly polarized light passes through a magnetized substance, the plane of polarization is rotated to the left or to the right in accordance with the direction of magnetization.

The EMOD system will be described below with reference to Fig. 8.20, referring mainly to an example developed by Philips in Germany [61]. Like the magnetic bubble memory device, this uses an active substrate consisting of a bismuth-substituted iron–garnet thin film (3 to 5 μm) epitaxially grown on a single-crystal substrate (500 μm thickness) of gadolinium gallium garnet (GGG). The iron–garnet thin film is ferrimagnetic,* with its axis of easy magnetization perpendicular to the surface of the thin film. To stabilize the domain pattern, the thin film is not a continuous surface, but is patterned in the form of pixels. The XY matrix electrodes are connected to transparent electrodes for resistance heating on each pixel. The cell is designed so that magnetic field can be applied in the direction perpendicular to the substrate using a coil that is arranged around the periphery of the substrate.

Writing the signal to the pixels is achieved by reversing the direction of magnetization of specified pixels. This is done by applying voltage between the selected pair of XY matrix electrodes so that the selected pixel is heated by the Joule heat generated in the transparent electrode, whilst using the coil to apply a uniform magnetic field to the substrate. Crosstalk is prevented because of well-defined threshold characteristics for inversion of magnetization.

When 4 μm thick iron–garnet film is used, the overall percentage utilization of the light at 550 nm is 10 per cent and the Faraday rotation angle is 26°. A prototype 256 bit line shutter has been manufactured with a pitch of 60 μm and cell size 50 μm square. It is reported that, with a heater voltage of 10 V, pulses of 10 to 20 μs and power 15 to 20 mW, display contrast is 60:1, and a write speed of 20 Mbit/s is possible. A prototype 64 $\times$ 64 dot array for display use [62] has been manufactured.

A method of display in which domains are reversed simply by a magnetic field without using heat has been reported by Litton Inc. [63]. Pixel selection is performed by the magnetic field generated by the current flowing in XY matrix electrodes, in the same way as in a core memory. Since there is a well-defined threshold value for inversion of magnetization, the crosstalk magnetic field generated by electric current in either is insufficient to produce inversion. A dot matrix device of 48 $\times$ 48 pixels with pitch 125 μm and film thickness 5 to 10 μm has been manufactured on a trial basis. A display contrast of 20 to 1 is obtained with white light.

*When atoms or ions whose magnetic moments are in opposite directions and are of different magnitudes are contained in a crystal, the result is that it shows a ferromagnetic character. This is called ferrimagnetism. Most insulating ferromagnetic bodies are described as ferrimagnetic.

8.7 FLAT PANEL TYPE CRT

The CRT is most widely utilized in computer terminals, and is regarded as the king of electronic displays. Apart from its brightness, colour tone, display contrast, scanning, and ease of modulation, etc., at present it offers the optimum cost/performance ratio. However, its greatest drawback is its depth and bulkiness. For this reason, many attempts have been made for a long time to construct a flat CRT [64]. The arrangements that have been tried up to the present are shown in Table 8.11. Essentially, a CRT is a vacuum tube having an electron beam generating part, electron beam modulating part, electron beam deflecting part, and phosphor exciting part. The various systems shown in the table have different features due to the different ways in which the first three of these parts are constructed in each system.

8.7.1 Flat CRT

This was designed by W. R. Aiken, D. Gabor *et al.* as early as the 1950s, and was called, for example, the banana tube. Although it was subsequently researched in various places, it was not put into practical form until it was utilized by Sony Corp. in 1982 in a 2 inch monochrome TV.

Flat CRTs include the type (manufactured by Sinclair [65]) in which the electron source is transverse to the screen, the type (manufactured by Sony [67, 68] and Shin Nichiden [66]) in which it is below, the type in which the electron beam originates from the back face of the phosphor surface (near face type), and the type in which it originates from the observer side (front face type).

In all cases, since the electron source is positioned in the direction of the phosphor surface rather than behind it, there are large differences in the distances to be travelled by the electron beam depending on the position in the face. This produces deflection distortion and blurring of the focal point. To prevent this a horizontal/vertical compensation circuit is used.

Figure 8.21 shows the CRT of a portable 2 inch TV manufactured by Sony. The phosphor face is curved and recessed, so a display is obtained that is easy to see. Since the display is a front face type, rather than a rear face type, the luminance is about twice as great.

8.7.2 Digisplay

This was developed in 1971 by Northrop [69–71]. Subsequently, development aimed at producing a colour display [72] was continued by the TI Corp.

Essentially, this is a construction in which there is no need to deflect the electron beam (Fig. 8.22). Several hot cathodes are provided, forming a planar electron source, as in a VFD. The potentials of several X and Y grid electrode plates, called switch stacks, are controlled, to achieve a beam shutter and focusing

Table 8.11 *Various types of flat CRT*

System	Colour	Dots (vertical times horizontal)	Size	Power (panels only)	Brightness (nit)	Firm	Year	Stage of development
Flat CRT								
Rear face type	Monochrome					Aiken	1956	Exhibition demonstration
	Monochrome		8″			Sharp	1967	Exhibition demonstration
	Monochrome					Sony		Product
			3″			Shin Nichiden		Prototype
Front face type	Monochrome	180 × 271	2″		>86	Sinclair	1979	Product
	Monochrome	250 × 280	2″		100	Sony		Product
Display	Monochrome	17920	6.6″	(14 W)	822	Northrop	1973	Prototype
	Monochrome	160 × 256	5.4″	(18 W)	274	Northrop	1973	Prototype
	Colour	25600 triads	10″	(< 30 W)		TI	1978	Prototype
Beam guide	Monochrome	160 × 80	11″		350	RCA	1982	Prototype
	Colour	160 × 80 triads	11″		350	RCA	1983	Prototype
Plasma excitation	Monochrome	448 × 720	14″	(20 W)	200	Siemens	1982	Prototype
	Colour	288 × 240 triads	12″	(19 W)	100	Siemens	1983	Prototype
Channel multiplication	Monochrome		9″	(Several W)		Philips	1982	Prototype
HAVD	Monochrome	? × 96	2.3″	(<0.5 W)	34.3	Kanazawa Institute of Technology	1983	Prototype

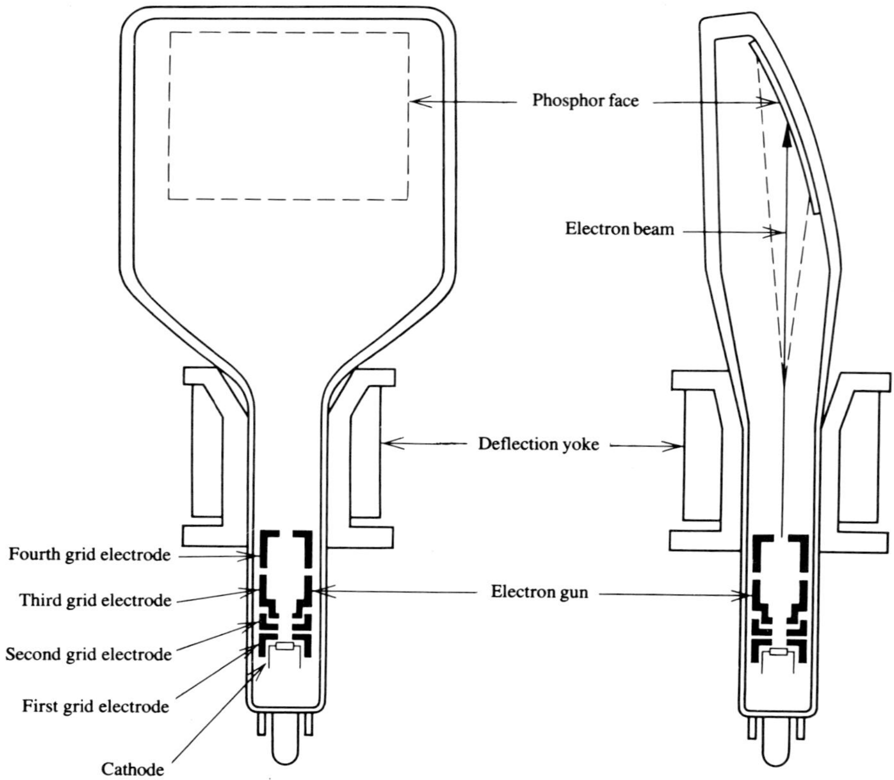

Figure 8.21 *Construction of flat CRT (Sony) [68]*

function. Analogue modulation and scanning of the electron beam is not performed as in the conventional CRT; digital addressing is used instead. The display is therefore called a 'digisplay'.

Figure 8.22 shows an example of a character display panel of the line sequential type. However, the number of line selection drivers can be considerably reduced (number of rows plus seven), thanks to the combination of row-selection electrodes with line-selection electrodes. Reduction in the number of signal drivers is also possible, albeit at the expense of luminance, if the signal side is also formed as a matrix.

8.7.3 Beam guide CRT

This is being developed [73–76] by RCA Corp., with the object of providing a 50 inch full-colour flat CRT for domestic use. The target values are shown in Table 8.12. The panel construction and principles of operation are shown in Figs. 8.23 to 8.26.

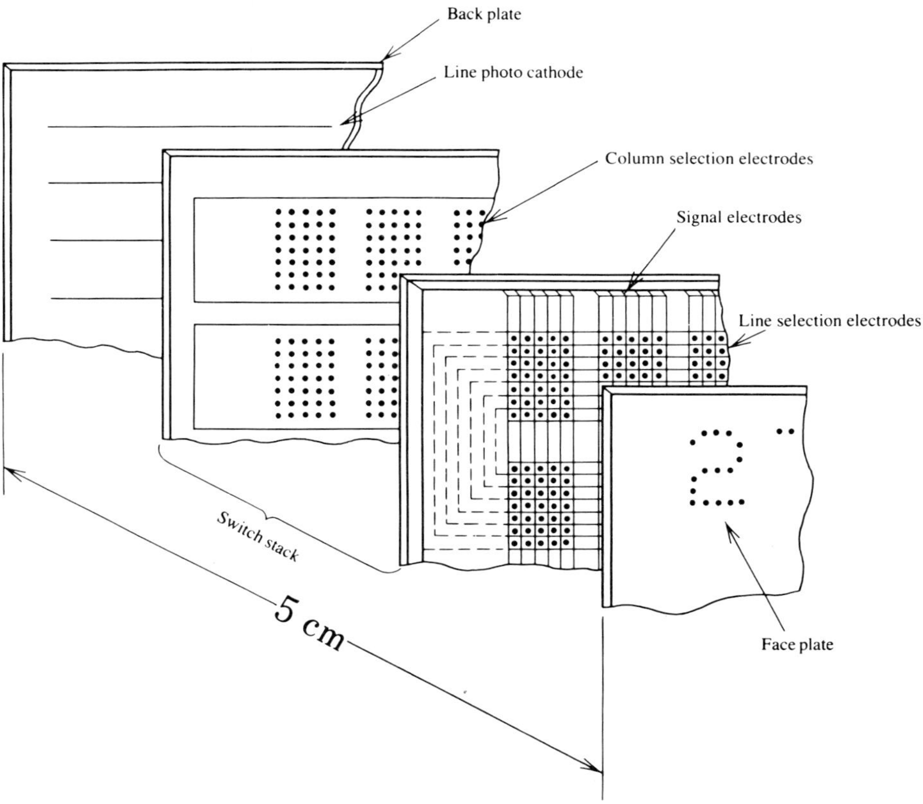

Figure 8.22 Construction of digisplay

Table 8.12 *Target for 50 inch full-colour flat CRT*

Screen size	$75 \times 100\,cm^2$ (corresponding to 50 inch)
Panel thickness	$\leqslant 10\,cm$
Pixel size	1.5 mm
Peak luminance	350 nit (100 fL)
Contrast	50:1
Resolution	400–500
Power consumption	< 300 W
Weight	< 50 kg

(RCA) [75]

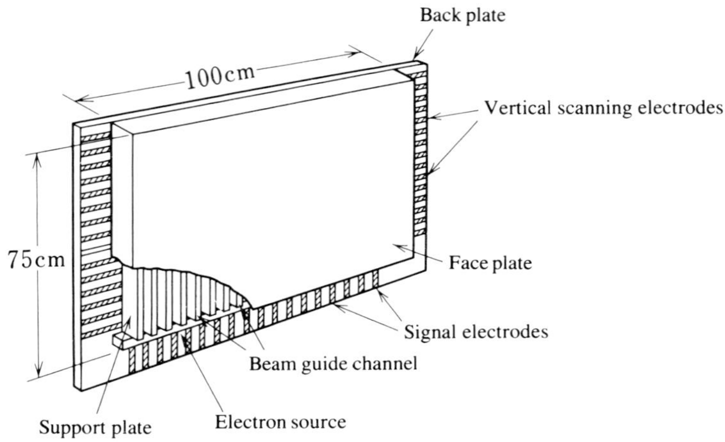

Figure 8.23 *Construction of beam guide type flat CRT (RCA) [76]*

The planar cathode, which is a point of difficulty in a digisplay, is replaced by a line cathode, and the heart of this system consists of pairs of grid plates constituting ladder beam guides forming periodic electrostatic fields, in which the electron beams are confined. The beam guides (Figs. 8.24 and 8.25) are used to transfer the electron beam with high efficiency over a long distance in a direction parallel to the display surface. Forty such channels are provided, each of width 2.5 cm, giving a transverse screen width of 1 metre. As shown in Fig. 8.26, three electron beams R, G, B are guided in each channel, and scanning of each channel is effected by a scanning electrode provided on a support plate.

This method of scanning is intermediate between the point sequential addressing and line sequential addressing types, and may therefore be called a

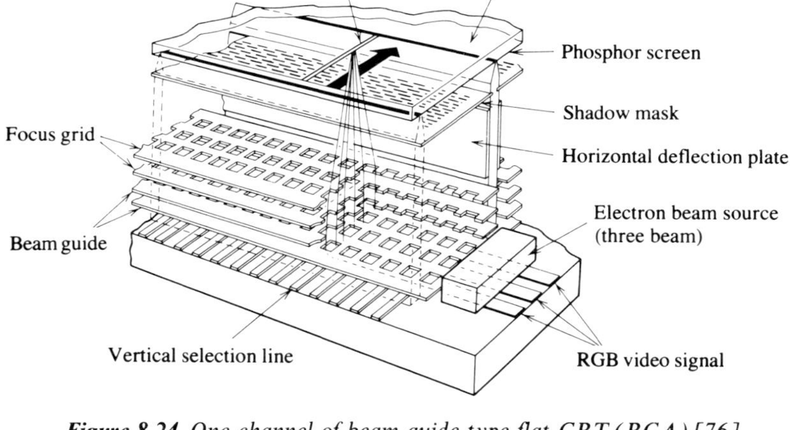

Figure 8.24 *One channel of beam guide type flat CRT (RCA) [76]*

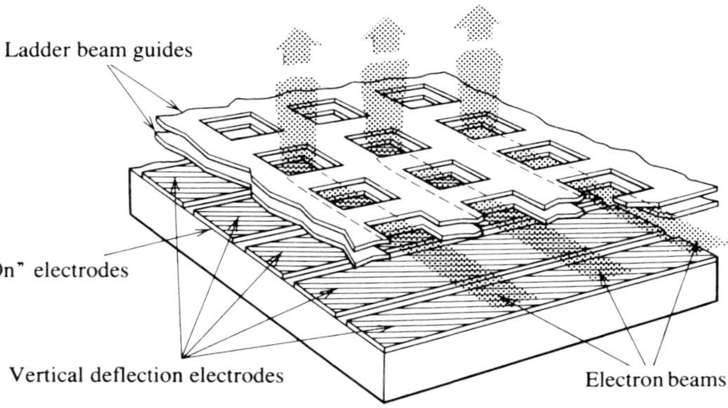

Figure 8.25 *Principle of beam vertical deflection (RCA) [76]*

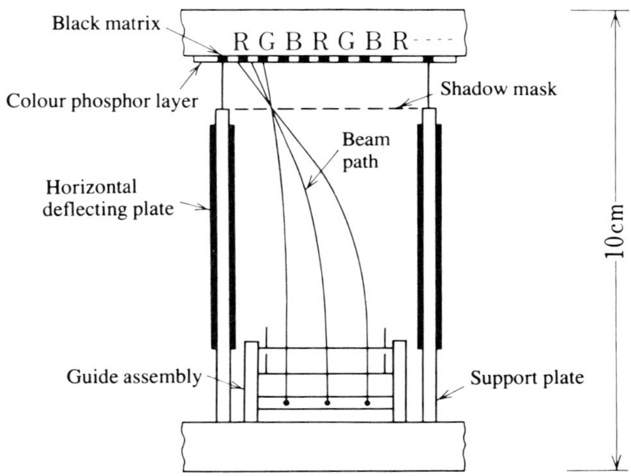

Figure 8.26 *Principle of beam horizontal deflection (RCA) 76]*

multichannel point sequential method. Since it is a multibeam method, the beam scanning speed required to obtain the same level of luminance as with a conventional CRT is M/N times (< 1) the speed needed in the conventional CRT (where M is the transverse size multiple of the screen and N is the number of channels). The load on the anode voltage or beam current can therefore be reduced. A test assembly has been constructed of height 25 cm and width 12.5 cm (four channels), giving a display area of about 4 per cent of the target value given in Table 8.12. This has been used to display a colour image (Fig. 8.27). A beam guide type flat CRT is an ingenious device, but the complexity of the cell construction is a problem. RCA has investigated a feedback type system in which

Figure 8.27 *Display of colour TV image by beam guide type flat CRT (prototype) RCA) [76]*

electron multiplication is performed by a dynode, and two other systems as well [77].

8.7.4 Plasma excitation CRT

As shown in Fig. 8.28, this system is the same as an ordinary CRT in that a phosphor is excited and made to emit light by an electron beam [78–80]. However, the electron source is not a hot cathode, but, instead, electrons are used that are generated by d.c. plasma emission of about 1 mb. The electrodes in the selected row are maintained at + 60 V, while the electrodes in the non-selected rows are held at + 10 V. Gas discharge occurs on application of 260 V between

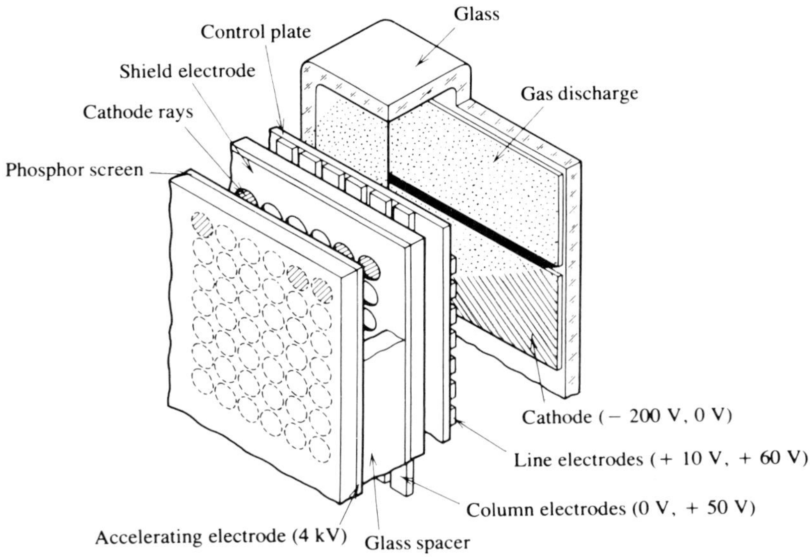

Figure 8.28 *Construction of plasma excitation flat CRT (Siemens) [80]*

the selected electrode and cathode. In response to the potential of the row electrode (on electrode: + 50 V, off electrode: 0 V), electrons generated by the plasma pass through holes in the electrode plate and are accelerated by an accelerating electrode held at 4 kV, causing excitation of the phosphor.

With a monochrome pane [79] of 720 × 448 dots, a display of 80 frames per second can be achieved, so the signal transfer rate is 80 × 720 × 448 = 25.8 MHz. The signal frequency can be reduced to one-half and the luminance raised by a factor of about two by dividing the screen into two (i.e. top and bottom portions), these top and bottom portions being accessed simultaneously by two sets of signal drivers. A 12 inch colour TV panel has been manufactured on a trial basis (Fig. 8.29), with separate bands of R, G, and B phosphor. In a graded display, 128 grades are generated by pulse width modulation of 0.5 to 64 μs with 7 bits. The characteristics of monochrome and colour panels manufactured on a trial basis are shown in Table 8.13.

This system makes effective use of Paschen's law. Specifically, with this panel, it is desired to generate discharge with as low a voltage as possible at the cathode. However, discharge must not occur in the cathode ray accelerating part. A gas pressure of 1 mb is therefore chosen, and the distances between the accelerating electrode and signal electrode and the signal electrode and cathode electrode are respectively chosen as 1 and 25 mm. Additionally, the discharge is isolated by means a of shield electrode.

Lucitron Inc. in the United States have attempted to eliminate the row electrode driver by introducing a plasma self-shift function on a similar principle. However, the panel construction is complicated. An attempt to utilize plasma

Figure 8.29 Examples of plasma excitation colour flat CRT display (Siemens) [80]

electrons for cathode luminescence was made in 1971 by Kazan and Pennebaker [81], but Siemens were the first to use high-speed cathode rays.

8.7.5 Channel multiplying CRT

In a photoelectron multiplier, the current of photoelectrons can be amplified by a factor of about 10^6, by using multiple dynode electrode stages provided with material (such as MgO) of high secondary electron emission ratio. A device that amplifies the electron current from the cathode by applying a d.c. voltage to both faces of a plate having multiple holes in the plate thickness direction (the inside surfaces of the holes being coated with secondary electron multiplying material) is called a channel plate. This is used for light amplification etc.

A flat CRT using an electron multiplying plate of this type has been developed by Philips [82]. Its construction is shown in Fig. 8.30. A channel multiplier plate is constructed by superimposing fourteen shadow masks with insulating material between them. A low-current electron beam (400 V, $< 1\ \mu A$), of low speed and with a horizontal position determined by a horizontal deflector plate, passes over the rear face of the screen and is changed in direction by 180° at its top edge. Vertical position specification of the electron beam is achieved by control of the potential of about ten vertical deflecting plates. The energy is amplified by a metal dynode channel amplifier plate, and then used to excite a phosphor. A

Table 8.13 Characteristics of plasma excitation flat CRT

	14 INCH MONOCHROME	12 INCH COLOUR
Number of vertical pixel	448	288
Number of horizontal pixels	720	720
Number of horizontal triples	—	240
Total number of pixels	322 560	69–120 triads
Raster	0.4 × 0.32 mm	0.6 × 0.32 mm
Effective screen	230 × 179 mm^2	231 × 173 cm^2
Effective diagonal	14 inch	12 inch
Panel dimensions	320 × 257 mm	320 × 257 mm
Panel thickness	6 cm	6 cm
Luminance	200 nit (green)	100 nit
Display contrast	20:1	100:1
Grades		128 (7 bits)
Efficiency (excluding driver)	6 lm/W	4 lm/W
Power consumption (panel only)	About 4 W	4 W
Power consumption (including drivers)	20 W	20 W
TV image		Non-interlaced
Bandwidth	Approx. 13 MHz	2.2 MHz

(Siemens Corporation) [78, 80]

monochrome TV image of 9 inch size can be displayed, using a channel amplifying plate of hole pitch 0.77 mm, with a power consumption of several watts.

8.7.6 HAVD type CRT (horizontal address vertical deflection flat CRT)

A prototype CRT of line sequential type, using a hot cathode in the form of a single line, has been made wherein vertical deflection is performed by modulating the electron currents for one line all at once by means of modulating electrodes. A 2.3 inch [83, 84] monochromatic TV picture was displayed. The construction is very simple, as shown in Fig. 8.31.

Barrier electrodes are provided, which are maintained at negative potential, to prevent the electron currents being affected by the neighbouring modulating electrodes. To raise horizontal resolution, the roles of these two types of electrodes are interchanged at each horizontal interval. A sawtooth voltage of +400 to −400 V is applied to the upper and lower pair of deflecting electrodes, so

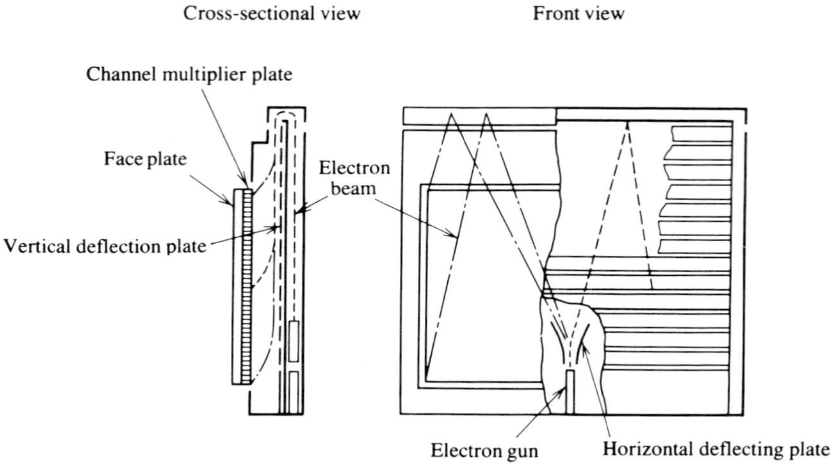

Figure 8.30 *Construction of channel multiplication type flat CRT (Philips) [82]*

that the beam is deflected in the vertical direction and the beam is focused by cooperating with the anode electrode. In a device of small depth (distance between anode modulating electrodes 19.5 mm), a luminance of 10 fL is obtained with anode voltage 1.8 kV for power of 0.5 W.

As explained above, various types of flat CRT have been tried, but the problem common to CRT displays is the conflict with atmospheric pressure. In order to achieve small depth and light weight, support pillars must be provided in order for the mechanical strength of the tube to be able to withstand atmospheric pressure. How to arrange the supporting pillars in order not to produce blurring of the display surface is a serious problem in aiming at increasing size and reducing weight.

8.8 LARGE-SCREEN DISPLAY

Large-screen displays are used to provide a large number of observers with guidance, announcements, messages, animation, or TV pictures, etc. They are arranged at such places as sports stadium scoreboards, airport and train terminals, outdoor advertising towers, wall surfaces of buildings, and underground streets. Various types of large display systems are shown in Table 8.14.

Projective displays have the advantage that a large screen can be conveniently obtained and are therefore widely used for video projectors. However, it is difficult to obtain sufficient luminance for a screen of several square metres or more. A further problem is the occupation of projection space. A flat plate type of non-projective display would be ideal, but the need to provide a large number of pixels pushes up the cost of the display body and drive circuitry. An image luminance of 15 to 60 nit is sufficient for a cinema or the like, but for outdoor use,

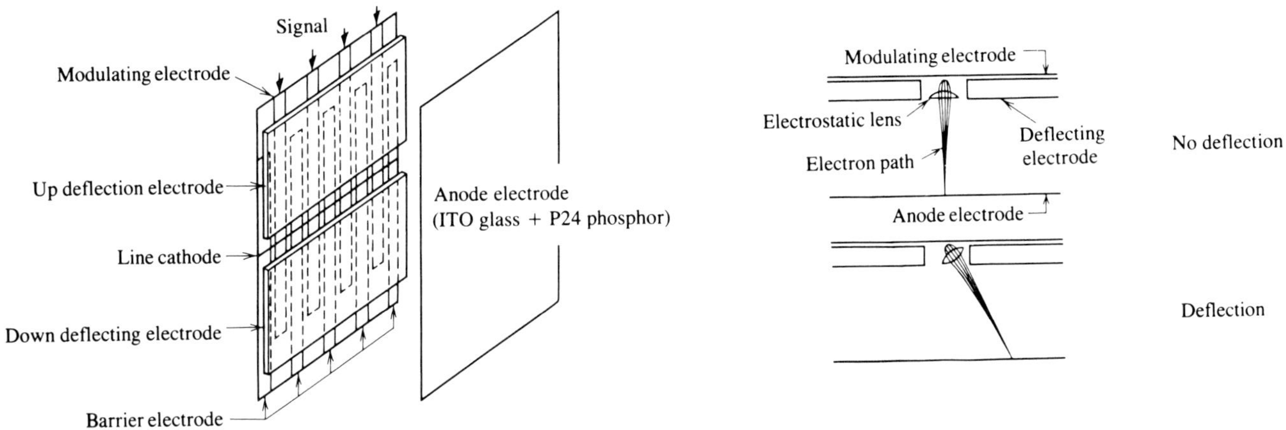

Figure 8.31 *Construction and principle of operation of horizontal address v ertical deflection flat CRT (Kanazawa Engineering University) [84]*

Table 8.14 *Various types of large-screen display system*

Type		Light Source	Construction		Full colour	Moving picture	Suitability for outdoor use	Trade name	Maker
Non-projective types	Active type	CRT	CRT matrix arrangement		○	○	×	Multiscreen	Various companies
Non-projective types	Active type	Flood-beam CRT	Matrix of small RGB CRTs (0.8–1.38)		○	○	○	Auroravision	Mitsubishi Electrical
Non-projective types	Active type	Incandescent bulb	Matrix of light bulbs fitted with RGB filters		○	○	○	Astrovision	Matsushita Communications Engineering
Non-projective types	Active type	High luminance fluorescent tube	Matrix arrangement of RGB fluorescent tubes		○	○	○	Super colour vision	Toshiba
Non-projective types	External light-controlled type	Neon tube	Electromagnetic opening and closing of shutters		×	×	△	Mosaic vision	Mediate
Non-projective types	External light-controlled type	Natural light	Magnetic inversion type	Four colours	×	×	○	NAP board	Nippon Advance Products
Non-projective types	External light-controlled type	Natural light	Magnetic inversion type	One colour?	×	×	○	Comment sign	Citizen
Non-projective types	External light-controlled type	Natural light	Magnetic inversion type	Two colours	×	×	○	Mag sign	Sumitomo Trading

		Fluorescent light	Magnetic inversion internal illumination type	×	×	○	Unex	Marubeni
		Natural light	Electrostatic attraction type: 2–4 colours	×	×	○	MDD	Fujitsu Kiden Display Tech
		Fluorescent light	TN liquid crystal cell and colour filter	○	○	×	Spectus	Mitsubishi Electric
			GH liquid crystal cell and colour filter	○	○	×	Liquid crystal Astro	Matsushita Communication Engineering
PROJECTIVE TYPES	Active type	Projective CRT	Refractive type or reflective type magnifying projection (45–400 inch)	○	○	×	Video projector	Various companies
		Laser	Polygonal mirror deflection scanning	○	○	×		Hitachi, NHK, etc. (prototypes)
	External light-controlled type	Xenon lamp	Oil film deformation, Schlieren optical system	○	○	×	Eidophore	Philips
			Pockels effect on DKDP crystal	○	○	×	Titus	LEP (France) (prototype)
			Mirror matrix tube	○	○	×		TI
			Oil film deformation, single gun colour	○	○	×	Tallaia	GE
			Writing by laser heating of liquid crystal	○	×	×		Various companies

Figure 8.32 Example of use of magnetic inversion display device (pixel dimensions: 16 mm square) (Nippon Advanced Products Corporation) [95]

a luminance of 1000 to 4000 nit is required.* This cannot be achieved by a projective display, so a non-projective display has to be used.

Mechanical types such as the magnetic inversion type do not have the speed to follow a moving picture display, so, for full-colour TV displays for outdoor use, at present only large-size systems consisting of a matrix arrangement of high luminance light sources such as small flat beam CRTs [85–87], incandescent electric light bulbs [88, 89], fluorescent lamps [90], etc., are used. Visual sensitivity is a maximum at 555 nm, and at this wavelength 1 W of optical energy corresponds to a light flux of 680 lm. However, in the case of equienergy white light, 1 W corresponds to only about 227 lm. Consequently, if 2000 nit white luminance is to be obtained with perfectly diffused light, a white light flux of 6289 lm/m^2 is necessary. † This corresponds to a white light energy of about 28 W. The electrooptic conversion characteristic of various types of light source is at present only about 70 lm/W, ‡ so for a light-emitting type device, a minimum power of 92 W/m^2 is necessary at the display surface.

*Normally, TV $\cong$ 200 nt, full moon $\cong$ 2500, fluorescent light tube surface = 10^4, 100 W bulb $\cong$ 5×10^4.

†For perfectly diffused light, 1 nt = 0.292 fL = 3.14 lm/m^2.

‡$\cong$ 30%. The light-emitting efficiency of present light sources when white light is to be obtained $\cong$ 30 lm/W in the case of cathode luminescence, $\cong$ 70 lm/W in the case of a fluorescent light, $\cong$ 15 lm/W in the case of a tungsten lamp, and $\cong$ 40 lm/W in the case of a xenon lamp.

In fact, when the power consumption of the drive circuitry is included, power consumption is more than 1 kW/m^2. Reducing power consumption is therefore an important task. Even in a back-lit passive display device, this power is necessary for the backlighting light source. Since the transmissivity of a light valve is normally somewhat below 100 per cent, obtaining the same degree of luminance as in the light-emitting type of display is difficult.

In contrast, in large display applications that do not necessarily require moving pictures, magnetic inversion display devices [93, 94] are widely used, for example in wall adverts or scoreboards. Thanks to the fact that the display is of a

Figure 8.33 *Examples of use of Auroravision (screen dimensions: 8 × 6 m) (Mitsubishi Electrical)*

Figure 8.34 *Large liquid crystal colour TV (1.6 × 2.1 m)(Matsushita Communications Industries, Matsushita Electronic Components)*

passive type, and since it has a memory and the power consumption is only that required to renew the display, power costs are less than 1/100 of those of a light-emitting type display. An example of the use of a four-colour magnetic inversion display device is shown in Fig. 8.32, an example of the use of a large colour TV [95] for a flat beam CRT is shown in Fig. 8.33, and an example of a liquid crystal full colour large TV [96] is shown in Fig. 8.34.

In the case of projective display systems using smectic liquid crystals which are thermally written by laser light, several seconds to several tens of seconds are required to write one screen, but since a large display* of high accuracy not attainable with the above-described video projector can be realized, development is continuing, aimed at applications such as remote conferencing, CAD, air traffic control, and newspaper compilation.

8.9 OTHER TYPES OF DISPLAY

Various types of display system were described in the previous sections, including some that have not yet been put into practice. However, even more types of display have been proposed, so it would be difficult to say that the previous sections cover this area fully. In this section, we shall therefore give a simple explanation of other displays, excluding, however, three-dimensional display techniques.

1. Liquid vapour display [97]. A film of volatile liquid is formed between an ITO electrode and a glass sheet. The ITO is heated by an electric current, vaporizing the film in the form of minute bubbles. The scattering of the light resulting from the formation of these bubbles is utilized.
2. Christiansen display [98]. This utilizes the Christiansen effect, namely that the light scattering characteristic changes when the temperature changes (because the temperature dependence of the refractive index of dispersed particles and of the liquid in which they are dispersed are different).
3. Latex display. Particles can be regularly arranged in a crystal-like fashion using the interparticle electrostatic repulsive force in a latex liquid dispersion [99] such as polystyrene. The change in the optical scattering characteristic when the interparticle distance is altered by an external voltage is utilized.
4. Elastomer display [100]. The change in reflection characteristic of light due to deformation of an elastomer layer by electrostatic force is utilized. A laminated construction of elastomer layer and photoconductive layer provided with a metal reflective film is used in a light valve. A prototype projective display system (Ruticon) based on this principle was manufactured by Xerox Corp.
5. Cathodochromic display [101]. A storage type display tube is utilized, using as a face plate sodalite glass, which changes its optical absorption characteristic when irradiated by cathode rays.

*For example, a display of (2000 to 8000) × (2000 to 8000) pixels.

6. Bimorph display [102]. This utilizes the warping produced by application of voltage to a bimorph formed by superimposing two sheets of piezoelectric PVF_2 film.
7. Thermal colour change display [103–105]. This utilizes the phenomenon of reversible change of colour by temperature.

8.10 OUTLOOK

Many types of display system have been devised and manufactured on a trial basis by the efforts and intelligence of a large number of people. The objectives of use and modes of utilization of display devices continue to become more diverse.* As well as improvements in conventional devices, the appearance of new display systems is expected.

The ideal for a display device is that it should be clearly visible both in the dark and in sunlight of about 60 000 lx. In addition, an ideal display device should have high picture quality, full colour, a thin shape, light weight, low power, and low cost. Of these, brightness and power consumption are the most important factors in both active and passive displays.

Usually when we view a photograph or printed matter under indoor lighting of, for example, 500 lx, taking the reflectivity of the object as, for example, 100 per cent (perfect diffusion surface), the luminance of the white areas is 500 lx $\cong$ 159 nit $\cong$ 46 fL. Thus, whether the display is of the active or passive type, the required brightness for indoor use is about 160 to 200 nit white luminance.

The relationship between screen size and power consumption for an active display giving 200 nit luminance is shown in Fig. 8.35, taking the efficiency of light emission as parameter. If 2 W is taken as the upper limit for battery drive,† the screen size in an active device is restricted to the region shown by a broken line in the figure. In contrast, to obtain the same luminance in a passive display under 1000 lx illumination, a reflectivity of about 63 per cent is required. In an active display, the efficiency of light emission is therefore the key factor, while in a passive display, it is reflectivity.

In the case of active displays, the current challenge is to develop a flat panel type display matching the overall efficiency of a colour CRT (about 3 lm/W, including heater and deflection system). In the case of a passive display, the challenge is to develop a flat panel type display that operates at low voltage and low power, that has a viewing angle and reflectivity superior to a liquid crystal display element, and that has memory capability and superior multiplexing performance. In particular, in the case of a reflective type colour display, the magnitude of the coefficient of reflectivity is decisively important. Progress with display systems such as EPID, ECD, TBD, MPD, SPD, and DFD that do not require a polarizer or development of systems superior to these is desired.

*Such as on the arm, pocket, bag, desk-top, wall, hall, vehicles, and outdoors, etc.
†About 2 hours for three or four batteries.

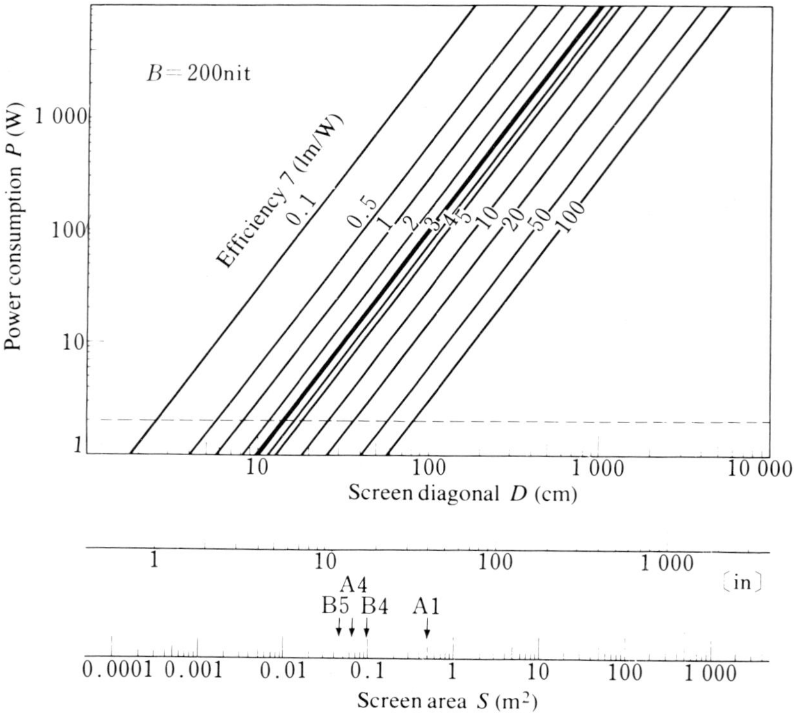

$$P[\mathrm{W}] = \frac{\pi B[\mathrm{nit}]}{\eta[\mathrm{lm/W}]} S[\mathrm{m}^2] = \frac{\pi \cdot B}{\eta} \cdot \frac{12}{25}\left(\frac{D[\mathrm{cm}]}{100}\right)^2$$

***Figure 8.35** Screen size and power consumption (vertical/horizontal ratio of screen dimensions is 3:4)*

Another important factor in display devices is multiplexing capability. At present most reflective displays are inferior in this respect. To improve this multiplexing performance, techniques using reflective or transparent display devices employing an active matrix of varistors, MIM, MOSFET, or TFT are therefore being developed. In these techniques, in order to achieve a high-density, high-capacity display, incorporation of drive circuitry is becoming increasingly important.

Display systems are currently being reassessed in the light of progress in the development of active matrix techniques, so those types which have highest compatibility with active matrix drives, i.e. low voltage, high resistance, preferably a memory characteristic, etc., as well as visibility (brightness, viewing angle, display contrast, etc.), will be sought after.

BIBLIOGRAPHY

Kazan, B.: *Advance in Image Pickup and Display*, Vol. **2**, Xerox Corporation (1975).
H. R. Luxenberg and R. L. Kuehn (compilers): *Display System Engineering*, McGraw-Hill (1968).
Pankove, J. I. (ed): *Display Devices*, Springer-Verlag (1980).
Sherr, S. *Electronic Displays*, John Wiley and Sons (1979).
Special Supplement, CRT versus flat panel displays, Nikkei Electronics, pp. 1–2 (1984).

REFERENCES

[1] J. H. Dessauer and H. E. Clark (compilers): *Xerography and Related Processes*, The Focal Press (1965).
[2] Y. Matsuura and T. Kondo: *Koroido Kagaku (Colloid Science)*, Hirokawa Shoten (1969).
[3] F. Kitahara and M. Watanabe (compilers): *Kaimen Denki Gensho* (*Interface Electrical Phenomena*), Kyoritsu Shuppan (1972).
[4] I. Ota, J. Ohnishi, and M. Yoshiyama: *Proc. IEEE*, **61**, 7 (1973).
[5] A. L. Dalisa: *IEEE Trans. Elect. Devices*, **24**, 827 (1977).
[6] J. C. Lewis, G. M. Garner, R. T. Blunt, and C. F. Carter: *Proc. SID*, **18**, 235 (1977).
[7] I. Ota, M. Tsukamoto, and T. Ohtsuka: *Proc. SID*, **18**, 235 (1977).
[8] A. L. Dalisa and B. Singer, *Proc. SID*, **18**, 255 (1977).
[9] D. W. Vance: *Proc. SID*, **18**, 267 (1977).
[10] A. Chiang: *Proc. SID*, **18**, 275 (1977).
[11] B. Fitzhenry-Ritz, 1980 Biennial Display Research Conference, p. 13 (1980); *IEEE Trans. Elect. Devices*, **ED-28**, 6 (1981).
[12] B. Fitzherry: *Appl. Optics*, **18**, 3332 (1979).
[13] J. L. Ploix, L. Colas, M. Moulin, and L'Onde: *Electrique*, **59**, 65 (1979).
[14] S. F. Blazo: *1982 SID Digest*, pp. 92–3 (1982).
[15] M. D. Croucher, J. Harbour, M. Hopper, and M. L. Hair: *Photogr. Sci. Engg.*, **25**, 80 (1981).
[16] A. L. Dalisa and R.A. Delano: 1974 SID Int. Symp., *Digest of Technical Papers* (1974).
[17] A. L. Dalisa: Electrophoretic displays, Ch. 6 in J. I. Pankove (compiler), *Display Devices*, Springer-Verlag (1980).
[18] P. Mürau, B. Singer: *J. Appl. Phys.*, **49**, 4820 (1978).
[19] A. Chiang: *Eurodisplay '81*, p. 107 (1981).
[20] I. Ota, T. Sato, S. Tanaka, T. Yamagami, and H. Takeda: Laser 75 Optoelectronics Conf. Proceedings (1975).
[21] R. Liebert, J. Lalak and K. Wittig: 1980 Biennial Display Research Conference, p. 10 (1980).,
[22] M. A. Hopper and V. Novotny: *IEEE Trans. Elect. Devices*, **ED-26**, 1148 (1979).
[23] A. Chiang: 1980 Biennial Display Research Conference, p. 10 (1980).
[24] K. Nakamura, N. Matzuura, and F. Tanaka: J. *Appl. Phys.* **51**, 5041 (1980.)
[25] N. K. Sheridon and M. A. Belkoviz: *1982 SID Digest*, 94–5 (1982).
[26] P. Mürau, R. Liebert, and B. Singer: *SID 79 Digest*, p. 46 (1979).
[27] P. C. Mürau, R. Liebert, and B. M. Singer: *IEEE Trans. Elect. Devices*, **ED-26**, 1153 (1979).
[28] R. K. Swank, A. R. Sears, and K. H. Yang: J. *Appl. Phys.*, **50**, 6534 (1979).
[29] P. Mürau: *1984 SID Digest*, p. 141 (1984).
[30] A. Chiang, D. Curry, and M. Zarzycki: *SID 79 Digest*, p. 45 (1979).

[31] Alvin M. Marks: *Appl. Optics*, **8**, 1397 (1969).
[32] J. S. Donal, Jr, and D. B. Langmuir: *Proc. IRE*, **31**, 208 (1943).
[33] R. L. Saxe, R. I. Thompson, and Matthew Forlini: 1982 International Display Research Conference, 175.
[34] N. K. Sheridon and M. A. Berkovitz: 1977 SID Digest, p. 114; *Proc. SID*, **18**, 289 (1987).
[35] M. Saitoh, T. Mori, R. Ishikawa, and H. Tamura: 1982 SID Digest, *Proc. SID*, **23** (4), 249 (1982).
[36] L. L. Lee: *IEEE Trans. Elect. Devices*, **ED-22**, 758 (1975).
[37] L. L. Lee: *1977 SID Digest*, (1977).
[38] L. L. Lee: *Proc. SID*, **18**, 283 (1977).
[39] L. L. Lee: *1980 SID Digest*, p. 128 (1980).
[40] R. E. Glusick and M. Fine: 1970 IEEE Conf. on Display Devices, p. 112.
[41] Ties S. te Velde: *1980 SID Digest*, p. 116 (1980).
[42] G. W. Goodrich and J. M. O'Connor: *1980 SID Digest*, p. 130 (1980).
[43] Nikkei Electronics, 12th December issue, p. 53 (1977).
[44] *Moving Dielectric Display*, Display Tech., Daiwa Vacuum Engineering Catalogue.
[45] J. A. Raalte: *Appl. Optics*, **9**, 2225 (1970).
[46] R. N. Thomas, J. Gulberg, H. C. Nathanson, and P. R. Malmberg: *IEEE Trans. Elect. Devices*, **ED-22** (1975).
[47] L. J. Hornbeck, J. T. Cario, M. W. Cowens, P. M. Osterberg. and M. M. De-Smith: *1980 SID Digest* (1980).
[48] Larry J. Hornbeck: 1982 International Display Research Conf. 76; *IEEE Trans. Elect. Devices*, **ED-30** (1983).
[49] T. Ohashi: 1981 Tsushin Gakkai Sozen Daikai 5 (General Meeting of the Communications Association, 5) (1981).
[50] T. Ohashi: 1982 Tsushin Gakkai Sozen Daikai 5 (General Meeting of the Communication Association, 5) (1982).
[51] T. Ohashi: IE 82-17 (1982).
[52] G. Beni and S. Hackwood: *Appl. Phys. Lett.*, **38**, 207 (1981).
[53] S. Hackwood, J. L. Jackel and G. Beni: 1981 Eurodisplay, p. 91.
[54] Luxenberg, Kuehn (ed.): *Display System Engineering* (1968).
[55] G. Marie and J. Donjon: *Proc. IEEE*, **61**, 942 (1973).
[56] G. Marie: *Ferroelectrics*, **10**, 9 (1976).
[57] A. Kumada: *Terebi Shi (Television Journal)*, **29**, 600 (1975).
[58] A. Kumada: *Denshi Zairyo (Electronics Materials)*, November issue, 60 (1977).
[59] A. Kumada, K. Suzuki, and G. Toda: *Ferroelectrics*, **10**, 25 (1976).
[60] *Eizo Joho (Image Data)*, June issue, 50 (1983).
[61] B. Hill and K. P. Schmidt: *Philips J. Res.*, **33** (1978).
[62] B. Hill and K. P. Schmidt: *1979 SID Digest*, p. 80 (1979).
[63] W. E. Ross, D. L. Cox, J. J. Fernandez, and G. R. Pullian: *1982 SID Digest*, p. 98 (1982).
[64] R. N. Jackson: *Electronics and Power*, p. 615 (1979).
[65] C. Sinclair: *1981 SID Digest*, 138 (1981).
[66] H. Nagai: Terebi Gakkai Gijutsu Hokoku (Technical Reports of the Television Association), ED 520 (1980).
[67] H. Sato and M. Hatanaka: *Denshi Gijutsu (Electronics Technology)*, **24**, 26 (1982).
[68] M. Maeda: *Japan Display '83*, PD2 (1983).
[69] W. F. Goede: L. A. Jeffries, and J. E. Gunther: *IEEE Trans. Elect. Devices*, **ED-18** (1971).
[70] W. F. Goede: *IEEE Trans. Elect. Devices*. **ED-20** (1973).
[71] L. A. Jeffries: *1973 SID Digest*, p. 96 (1973).
[72] W. C. Scott, W. C. Holton, W. G. Manns, D. F. Weirauch, M. R. Namordi, and F. Doerbeck: *1978 SID Digest*, p. 88 (1978).

[73] W. W. Siekanowicz, T. L. Credelle, F. E. Vaccaro, and C. H. Anderson: *1980 SID Digest*, p. 24 (1980).
[74] T. L. Credelle *et al.: 1980 SID Digest*, 26 (1980).
[75] T. L. Credelle: *Television*, May/June, p. 21 (1982).
[76] T. L. Credelle *et al.*: *Japan Display '83*, p. 26 (1983).
[77] J. G. Endriz *et al. IEEE Trans. Elect. Devices*, **ED-26**, 1320 (1979).
[78] A. Schauer: *IEDM '82*, p. 304 (1982).
[79] A. Schauer: *Electronics*, 15 Dec. p. 126 (1982).
[80] A. Schauer: *Japan Display '83*, p. 234 (1983).
[81] B. Kazan and W. B. Pennebaker: *Proc. IEEE*, July, p. 1130 (1971).
[82] A. Woodhead *et al.*: *1982 SID Digest*, p. 206 (1982).
[83] Y. Sakamoto and E. Miyazaki: T. V. Gakkai Zenkoku Daikai 109 National Congress of the T. V. Association) (1983).
[84] Y. Sakamoto and E. Miyazaki: *Japan Display '83, p. 30 (1983).*
[85] K. Kurahashi: Denki Gakkai Shi (Journal of the Electrical Association), 100 (1980).
[86] K. Kurahashi *et al.*: *1981 SID Digest*, p. 132 (1981).
[87] K. Kurahashi: *Japan Display '83*, p. 420 (1983).
[88] A. Takada and T. Ohashi: *JEE*, September (1982).
[89] 'Astrovision' pamphlet, Matsushita Communications Industries Ltd.
[90] T. Kaneko *et al.*: *Japan Display '83*, p. 428 (1983).
[91] O. Myodo *et al.*: *Japan Display '83*, p. 430 (1983).
[92] J. R. Burns and G. W. Taylor: *1983 SID Digest*, p. 40 (1983).
[93] Nikkei Sangyo Shinbun, 22 Apr. 1983.
[94] H. Peprnik: *1974 SID Digest*, p. 136 (1974).
[95] *Nap board*, Nippon Advanced Products KK Catalogue.
[96] Takada *et al.*: *Denshi Zairyo (Electronic Materials)*, January issue (1984).
[97] G. W. Taylor: *Proc. IEEE*, **61** (1973).
[98] R. K. Warning: *Proc. IEEE*, **65**, 1074 (1977).
[99] H. Fujita and A. Ametani: *Jpn. J. Appl Phys.*, **16**, 1907 (1977).
[100] D. W. Vance and P. B. Keenan: *1975 SID Digest*, p. 22 (1975).
[101] Nikkei Electronics, 4 Dec. 1972
[102] M. Toda and S. Osaka: *1978 SID Digest*, p. 18 (1978).
[103] S. Nakaya *et al.*: *Proc. SID*, **22** (1) (1981).
[104] S. Nakaya *et al.*: *1982 SID Digest*, p. 234 (1982).
[105] S. Nakaya *et al.*: *1983 SID Digest*, p. 58 (1983).

Index